Applied Physics and Technology

Applied Physics and Technology

Editor: Faith Sherman

NY RESEARCH PRESS

New York

Published by NY Research Press
118-35 Queens Blvd., Suite 400,
Forest Hills, NY 11375, USA
www.nyresearchpress.com

Applied Physics and Technology
Edited by Faith Sherman

International Standard Book Number: 978-1-63238-890-2 (Hardback)

Cataloging-in-Publication Data

Applied physics and technology / edited by Faith Sherman.
    p. cm.
Includes bibliographical references and index.
ISBN 978-1-63238-890-2
1. Physics. 2. Technology. I. Sherman, Faith.
QC23.2 .A67 2022
530--dc23

Contents

Preface

The branch of physics which is intended for a particular practical or technological use is termed as applied physics. It is rooted in the fundamental concepts of the physical sciences but focuses on the utilization of scientific principles in actual devices and systems. It functions as a bridge between physics and engineering. Applied physics contributes significantly towards technological advances such as the development of electronics and photonics. It is also involved in the improvement of practical investigations such as experimental particle physics and experimental nuclear physics. Some of the other major research areas where applied physics is used are microscopy, semiconductors, electromagnetic propulsion and lasers. This book discusses the fundamentals as well as modern approaches of applied physics and technology. It will also provide interesting topics for research which interested readers can take up. This book will help the readers in keeping pace with the rapid changes in this field.

After months of intensive research and writing, this book is the end result of all who devoted their time and efforts in the initiation and progress of this book. It will surely be a source of reference in enhancing the required knowledge of the new developments in the area. During the course of developing this book, certain measures such as accuracy, authenticity and research focused analytical studies were given preference in order to produce a comprehensive book in the area of study.

This book would not have been possible without the efforts of the authors and the publisher. I extend my sincere thanks to them. Secondly, I express my gratitude to my family and well-wishers. And most importantly, I thank my students for constantly expressing their willingness and curiosity in enhancing their knowledge in the field, which encourages me to take up further research projects for the advancement of the area.

Editor

SIR epidemics in monogamous populations with recombination

Damián H. Zanette[1]*

We study the propagation of an SIR (susceptible–infectious–recovered) disease over an agent population which, at any instant, is fully divided into couples of agents. Couples are occasionally allowed to exchange their members. This process of couple recombination can compensate the instantaneous disconnection of the interaction pattern and thus allow for the propagation of the infection. We study the incidence of the disease as a function of its infectivity and of the recombination rate of couples, thus characterizing the interplay between the epidemic dynamics and the evolution of the population's interaction pattern.

I. Introduction

Models of disease propagation are widely used to provide a stylized picture of the basic mechanisms at work during epidemic outbreaks and infection spreading [1]. Within interdisciplinary physics, they have the additional interest of being closely related to the mathematical representation of such diverse phenomena as fire propagation, signal transmission in neuronal axons, and oscillatory chemical reactions [2]. Because this kind of model describes the joint dynamics of large populations of interacting active elements or agents, its most interesting outcome is the emergence of self-organization. The appearance of endemic states, with a stable finite portion of the population actively transmitting an infection, is a typical form of self-organization in epidemiological models [3].

Occurrence of self-organized collective behavior has, however, the *sine qua non* condition that information about the individual state of agents must be exchanged between each other. In turn, this re-

*E-mail: zanette@cab.cnea.gov.ar

[1] Consejo Nacional de Investigaciones Científicas y Técnicas, Centro Atómico Bariloche e Instituto Balseiro, 8400 Bariloche, Río Negro, Argentina.

quires the interaction pattern between agents not to be disconnected. Fulfilment of such requirement is usually assumed to be granted. However, it is not difficult to think of simple scenarios where it is not guaranteed. In the specific context of epidemics, for instance, a sexually transmitted infection never propagates in a population where sexual partnership is confined within stable couples or small groups [4].

In this paper, we consider an SIR (susceptible–infectious–recovered) epidemiological model [3] in a monogamous population where, at any instant, each agent has exactly one partner or neighbor [4, 5]. The population is thus divided into couples, and is therefore highly disconnected. However, couples can occasionally break up and their members can then be exchanged with those of other broken couples. As was recently demonstrated for SIS models [6, 7], this process of couple recombination can compensate to a certain extent the instantaneous lack of connectivity of the population's interaction pattern, and possibly allow for the propagation of the otherwise confined disease. Our main aim here is to characterize this interplay between recombination and propagation for SIR epidemics.

In the next section, we review the SIR model and its mean field dynamics. Analytical results are then provided for recombining monogamous populations in the limits of zero and infinitely large recombination rate, while the case of intermediate rates is studied numerically. Attention is focused on the disease incidence –namely, the portion of the population that has been infectious sometime during the epidemic process– and its dependence on the disease infectivity and the recombination rates, as well as on the initial number of infectious agents. Our results are inscribed in the broader context of epidemics propagation on populations with evolving interaction patterns [4, 5, 8–11].

II. SIR dynamics and mean field description

In the SIR model, a disease propagates over a population each of whose members can be, at any given time, in one of three epidemiological states: susceptible (S), infectious (I), or recovered (R). Susceptible agents become infectious by contagion from infectious neighbors, with probability λ per neighbor per time unit. Infectious agents, in turn, become recovered spontaneously, with probability γ per time unit. The disease process S $\to$ I $\to$ R ends there, since recovered agents cannot be infected again [3].

With a given initial fraction of S and I–agents, the disease first propagates by contagion but later declines due to recovery. The population ends in an absorbing state where the infection has disappeared, and each agent is either recovered or still susceptible. In this respect, SIR epidemics differs from the SIS and SIRS models, where –due to the cyclic nature of the disease,– the infection can asymptotically reach an endemic state, with a constant fraction of infectious agents permanently present in the population.

Another distinctive equilibrium property of SIR epidemics is that the final state depends on the initial condition. In other words, the SIR model possesses infinitely many equilibria parameterized by the initial states.

In a mean field description, it is assumed that each agent is exposed to the average epidemiological state of the whole population. Calling x and y the respective fractions of S and I–agents, the mean field evolution of the disease is governed by the equations

$$
\begin{aligned}
\dot{x} &= -k\lambda xy, \\
\dot{y} &= k\lambda xy - y,
\end{aligned}
\tag{1}
$$

where k is the average number of neighbors per agent. Since the population is assumed to remain constant in size, the fraction of R–agents is $z = 1 - x - y$. In the second equation of Eqs. (1), we have assigned the recovery frequency the value $\gamma = 1$, thus fixing the time unit equal to γ^{-1}, the average duration of the infectious state. The contagion frequency λ is accordingly normalized: $\lambda/\gamma \to \lambda$. This choice for γ will be maintained throughout the remaining of the paper.

The solution to Eqs. (1) implies that, from an initial condition without R–agents, the final fraction of S–agents, x^*, is related to the initial fraction of I–agents, y_0, as [1]

$$
x^* = 1 - (k\lambda)^{-1} \log[(1 - y_0)/x^*].
\tag{2}
$$

Note that the final fraction of R–agents, $z^* = 1 - x^*$, gives the total fraction of agents who have been infectious sometime during the epidemic process. Thus, z^* directly measures the total incidence of the disease.

The incidence z^* as a function of the infectivity $k\lambda$, obtained from Eq. (2) through the standard Newton–Raphson method for several values y_0 of the initial fraction of I–agents, is shown in the upper panel of Fig. 1. As expected, the disease incidence grows both with the infectivity and with y_0. Note that, on the one hand, this growth is smooth for finite positive y_0. On the other hand, for $y_0 \to 0$ (but $y_0 \neq 0$) there is a transcritical bifurcation at $k\lambda = 1$. For lower infectivities, the disease is not able to propagate and, consequently, its incidence is identically equal to zero. For larger infectivities, even when the initial fraction of I–agents is vanishingly small, the disease propagates and the incidence turns out to be positive. Finally, for $y_0 = 0$ no agents are initially infectious, no infection spreads, and the incidence thus vanishes all over parameter space.

III. Monogamous populations with couple recombination

Suppose now that, at any given time, each agent in the population has exactly just one neighbor or,

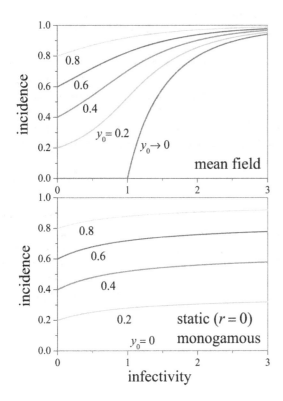

Figure 1: SIR epidemics incidence (measured by the final fraction of recovered agents z^*) as a function of the infectivity (measured by the product of the mean number of neighbors times the infection probability per time unit per infected neighbor, $k\lambda$), for different initial fractions of infectious agents, y_0. Upper panel: For the mean field equations (1). Lower panel: For a static (non-recombining) monogamous population, described by Eqs. (3) with $r = 0$.

in other words, that the whole population is always divided into couples. In reference to sexually transmitted diseases, this pattern of contacts between agents defines a *monogamous* population [5]. If each couple is everlasting, so that neighbors do not change with time, the disease incidence should be heavily limited by the impossibility of propagating too far from the initially infectious agents. At most, some of the initially susceptible agents with infectious neighbors will become themselves infectious, but spontaneous recovery will soon prevail and the disease will disappear.

If, on the other hand, the population remains monogamous but neighbors are occasionally allowed to change, any I–agent may transmit the disease several times before recovering. If such changes are frequent enough, the disease could perhaps reach an incidence similar to that predicted by the mean field description, Eq. (1) (for $k = 1$, i.e. with an average of one neighbor per agent).

We model neighbor changes by a process of couple recombination where, at each event, two couples (i,j) and (m,n) are chosen at random and their partners are exchanged [6,7]. The two possible outcomes of recombination, either (i,m) and (j,n) or (i,n) and (j,m), occur with equal probability. To quantify recombination, we define r as the probability per unit time that any given couple becomes involved in such an event.

A suitable description of SIR epidemics in monogamous populations with recombination is achieved in terms of the fractions of couples of different kinds, m_{SS}, m_{SI}, m_{II}, m_{IR}, m_{RR}, and $m_{SR} = 1 - m_{SI} - m_{II} - m_{IR} - m_{RR}$. Evolution equations for these fractions are obtained by considering the possible transitions between kinds of couples due to recombination and epidemic events [7]. For instance, partner exchange between two couples (S,S) and (I,R) which gives rise to (S,I) and (S,R), contributes positive terms to the time derivative of m_{SI} and m_{SR}, and negative terms to those of m_{SS} and m_{IR}, all of them proportional to the product $m_{SS}m_{IR}$. Meanwhile, for example, contagion can transform an (S,I)–couple into an (I,I)–couple, with negative and positive contributions to the variations of the respective fractions, both proportional to m_{SI}.

The equations resulting from these arguments read

$$\begin{aligned}
\dot{m}_{SS} &= rA_{SIR}, \\
\dot{m}_{SI} &= rB_{SIR} - (1+\lambda)m_{SI}, \\
\dot{m}_{II} &= rA_{IRS} + \lambda m_{SI} - 2m_{II}, \\
\dot{m}_{IR} &= rB_{IRS} + 2m_{II} - m_{IR}, \\
\dot{m}_{RR} &= rA_{RSI} + m_{IR}, \\
\dot{m}_{SR} &= rB_{RSI} + m_{SI}.
\end{aligned} \quad (3)$$

For brevity, we have here denoted the contribution of recombination by means of the symbols

$$A_{ijh} \equiv (m_{ij}+m_{ih})^2/4 - m_{ii}(m_{jj}+m_{jh}+m_{hh}), \quad (4)$$

and

$$B_{ijh} \equiv (2m_{ii} + m_{ih})(2m_{jj} + m_{jh})/2$$
$$-m_{ij}(m_{ij} + m_{ih} + m_{jh} + m_{hh})/2, \quad (5)$$

with i, j, $h \in \{S, I, R\}$. The remaining terms stand for the epidemic events. In terms of the couple fractions, the fractions of S, I and R–agents are expressed as

$$\begin{aligned} x &= m_{SS} + (m_{SI} + m_{SR})/2, \\ y &= m_{II} + (m_{SI} + m_{IR})/2, \quad (6) \\ z &= m_{RR} + (m_{SR} + m_{IR})/2. \end{aligned}$$

Assuming that the agents with different epidemiological states are initially distributed at random over the pattern of couples, the initial fraction of each kind of couple is $m_{SS}(0) = x_0^2$, $m_{SI}(0) = 2x_0 y_0$, $m_{II}(0) = y_0^2$, $m_{IR}(0) = 2y_0 z_0$, $m_{RR}(0) = z_0^2$, and $m_{SR}(0) = 2x_0 z_0$, where x_0, y_0 and z_0 are the initial fractions of each kind of agent.

It is important to realize that the mean field–like Eqs. (3) to (6) are *exact* for infinitely large populations. In fact, first, pairs of couples are selected at random for recombination. Second, any epidemic event that changes the state of an agent modifies the kind of the corresponding couple, but does not affect any other couple. Therefore, no correlations are created by either process.

In the limit without recombination, $r = 0$, the pattern of couples is static. Equations (3) become linear and can be analytically solved. For asymptotically long times, the solution provides –from the third of Eqs. (6)– the disease incidence as a function of the initial condition. If no R–agents are present in the initial state, the incidence is

$$z^* = (1 + \lambda)^{-1}[1 + \lambda(2 - y_0)]y_0. \quad (7)$$

This is plotted in the lower panel of Fig. 1 as a function of the infectivity $k\lambda \equiv \lambda$, for various values of the initial fraction of I–agents, y_0. When recombination is suppressed, as expected, the incidence is limited even for large infectivities, since disease propagation can only occur to susceptible agents initially connected to infectious neighbors. Comparison with the upper panel makes apparent substantial quantitative differences with the mean field description, especially for small initial fractions of I–agents.

Another situation that can be treated analytically is the limit of infinitely frequent recombination, $r \to \infty$. In this limit, over a sufficiently short time interval, the epidemiological state of all agents is virtually "frozen" while the pattern of couples tests all possible combinations of agent pairs. Consequently, at each moment, the fraction of couples of each kind is completely determined by the instantaneous fraction of each kind of agent, namely,

$$\begin{aligned} m_{SS} &= x^2, \quad m_{SI} = 2xy, \quad m_{II} = y^2, \\ m_{IR} &= 2yz, \quad m_{RR} = z^2, \quad m_{SR} = 2xz. \end{aligned} \quad (8)$$

These relations are, of course, the same as quoted above for uncorrelated initial conditions.

Replacing Eqs. (8) into (3) we verify, first, that the operators A_{ijh} and B_{ijh} vanish identically. The remaining of the equations, corresponding to the contribution of epidemic events, become equivalent to the mean field equations (1). Therefore, if the distributions of couples and epidemiological states are initially uncorrelated, the evolution of the fraction of couples of each kind is exactly determined by the mean field description for the fraction each kind of agent, through the relations given in Eqs. (8).

For intermediate values of the recombination rate, $0 < r < \infty$, we expect to obtain incidence levels that interpolate between the results presented in the two panels of Fig. 1. However, these cannot be obtained analytically. We thus resort to the numerical solution of Eqs. (3).

IV. Numerical results for recombining couples

We solve Eqs. (3) by means of a standard fourth-order Runge-Kutta algorithm. The initial conditions are as in the preceding section, representing no R–agents and a fraction y_0 of I–agents. The disease incidence z^* is estimated from the third equation of Eqs. (6), using the long-time numerical solutions for m_{RR}, m_{SR}, and m_{IR}. In the range of parameters considered here, numerical integration up to time $t = 1000$ was enough to get a satisfactory approach to asymptotic values.

Figure 2 shows the incidence as a function of infectivity for three values of the initial fraction of I–agents, $y_0 \to 0$, $y_0 = 0.2$ and 0.6, and several values of the recombination rate r. Numerically,

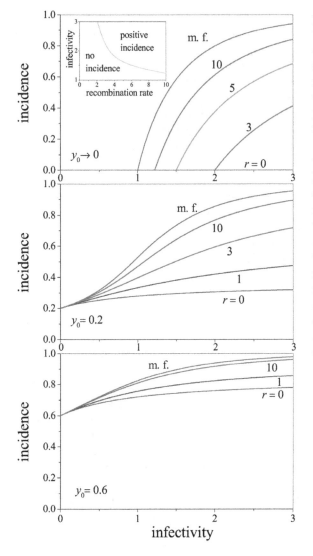

Figure 2: SIR epidemics incidence as a function of the infectivity for three initial fractions of infectious agents, y_0, and several recombination rates, r. Mean field (m. f.) results are also shown. The insert in the upper panel displays the boundary between the phases of no incidence and positive incidence for $y_0 \to 0$, in the parameter plane of infectivity vs. recombination rate.

the limit $y_0 \to 0$ has been represented by taking $y_0 = 10^{-9}$. Within the plot resolution, smaller values of y_0 give identical results. Mean field (m. f.) results are also shown. As expected from the analytical results presented in the preceding section,

positive values of r give rise to incidences between those obtained for a static couple pattern ($r = 0$) and for the mean field description. Note that substantial departure from the limit of static couples is only got for relatively large recombination rates, $r > 1$, when at least one recombination per couple occurs in the typical time of recovery from the infection.

Among these results, the most interesting situation is that of a vanishingly small initial fraction of I–agents, $y_0 \to 0$. Figure 3 shows, in this case, the epidemics incidence as a function of the recombination rate for several fixed infectivities. We recall that, for $y_0 \to 0$, the mean field description predicts a transcritical bifurcation between zero and positive incidence at a critical infectivity $\lambda = 1$, while in the absence of recombination the incidence is identically zero for all infectivities. Our numerical calculations show that, for sufficiently large values of r, the transition is still present, but the critical point depends on the recombination rate. As r grows to infinity, the critical infectivity decreases approaching unity.

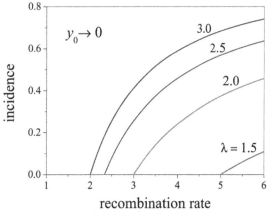

Figure 3: SIR epidemics incidence as a function of the recombination rate r for a vanishingly small fraction of infectious agents, $y_0 \to 0$, and several infectivities λ.

Straightforward linearization analysis of Eqs. (3) shows that the state of zero incidence becomes unstable above the critical infectivity

$$\lambda_c = \frac{r+1}{r-1}. \tag{9}$$

This value is in excellent agreement with the numerical determination of the transition point. Note also that Eq. (9) predicts a divergent critical infectivity for a recombination rate $r = 1$. This implies that, for $0 \leq r \leq 1$, the transition is absent and the disease has no incidence irrespectively of the infectivity level. For $y_0 \to 0$, thus, the recombination rate must overcome the critical value $r_c = 1$ to find positive incidence for sufficiently large infectivity. The critical line between zero and positive incidence in the parameter plane of infectivity vs. recombination rate, given by Eq. (9), is plotted in the insert of the upper panel of Fig. 2.

V. Conclusions

We have studied the dynamics of SIR epidemics in a population where, at any time, each agent forms a couple with exactly one neighbor, but neighbors are randomly exchanged at a fixed rate. As it had already been shown for the SIS epidemiological model [6,7], this recombination of couples can, to some degree, compensate the high disconnection of the instantaneous interaction pattern, and thus allow for the propagation of the disease over a finite portion of the population. The interest of a separate study of SIR epidemics is based on its peculiar dynamical features: in contrast with SIS epidemics, it admits infinitely many absorbing equilibrium states. As a consequence, the disease incidence depends not only on the infectivity and the recombination rate, but also on the initial fraction of infectious agents in the population.

Due to the random nature of recombination, mean field–like arguments provide exact equations for the evolution of couples formed by agents in every possible epidemiological state. These equations can be analytically studied in the limits of zero and infinitely large recombination rates. The latter case, in particular, coincides with the standard mean field description of SIR epidemics.

Numerical solutions for intermediate recombination rates smoothly interpolate between the two limits, except when the initial fraction of infectious agents is vanishingly small. For this special situation, if the recombination rate is below one recombination event per couple per time unit (which equals the mean recovery time), the disease does not propagate and its incidence is thus equal to zero. Above that critical value, a transition appears as the disease infectivity changes: for small infectivities the incidence is still zero, while it becomes positive for large infectivities. The critical transition point shifts to lower infectivities as the recombination rate grows.

It is worth mentioning that a similar transition between a state with no disease and an endemic state with a permanent infection level occurs in SIS epidemics with a vanishingly small fraction of infectious agents [6,7]. For this latter model, however, the transition is present for any positive recombination rate. For SIR epidemics, on the other hand, the recombination rate must overcome a critical value for the disease to spread, even at very large infectivities.

While both the (monogamous) structure and the (recombination) dynamics of the interaction pattern considered here are too artificial to play a role in the description of real systems, they correspond to significant limits of more realistic situations. First, the monogamous population represents the highest possible lack of connectivity in the interaction pattern (if isolated agents are excluded). Second, random couple recombination preserves the instantaneous structure of interactions and does not introduce correlations between the individual epidemiological state of agents. As was already demonstrated for SIS epidemics and chaotic synchronization [7], they have the additional advantage of being analytically tractable to a large extent. Therefore, this kind of assumption promises to become a useful tool in the study of dynamical processes on evolving networks.

Acknowledgements - Financial support from SECTyP–UNCuyo and ANPCyT, Argentina, is gratefully acknowledged.

---

[1] R M Anderson, R M May, *Infectious Diseases in Humans*, Oxford University Press, Oxford (1991).

[2] A S Mikhailov, *Foundations of Synergetics I. Distributed active systems*, Springer, Berlin (1990).

[3] J D Murray, *Mathematical Biology*, Springer, Berlin (2003).

[4] K T D Eames, M J Keeling, *Modeling dynamic and network heterogeneities in the spread of sexually transmitted diseases*, Proc. Nat. Acad. Sci. **99**, 13330 (2002).

[5] K T D Eames, M J Keeling, *Monogamous networks and the spread of sexually transmitted diseases*, Math. Biosc. **189**, 115 (2004).

[6] S Bouzat, D H Zanette, *Sexually transmitted infections and the marriage problem*, Eur. Phys. J B **70**, 557 (2009).

[7] F Vazquez, D H Zanette, *Epidemics and chaotic synchronization in recombining monogamous populations*, Physica D **239**, 1922 (2010).

[8] T Gross, C J Dommar D'Lima, B Blasius, *Epidemic dynamics in an adaptive network*, Phys. Rev. Lett. **96**, 208 (2006).

[9] T Gross, B Blasius, *Adaptive coevolutionary networks: a review*, J. R. Soc. Interface **5**, 259 (2008).

[10] D H Zanette, S Risau–Gusman, *Infection spreading in a population with evolving contacts*, J. Biol. Phys. **34**, 135 (2008).

[11] S Risau–Gusman, D H Zanette, *Contact switching as a control strategy for epidemic outbreaks*, J. Theor. Biol. **257**, 52 (2009).

2

Revisiting the two-mass model of the vocal folds

M. F. Assaneo,[1*] M. A. Trevisan[1†]

Realistic mathematical modeling of voice production has been recently boosted by applications to different fields like bioprosthetics, quality speech synthesis and pathological diagnosis. In this work, we revisit a two-mass model of the vocal folds that includes accurate fluid mechanics for the air passage through the folds and nonlinear properties of the tissue. We present the bifurcation diagram for such a system, focusing on the dynamical properties of two regimes of interest: the onset of oscillations and the normal phonation regime. We also show theoretical support to the nonlinear nature of the elastic properties of the folds tissue by comparing theoretical isofrequency curves with reported experimental data.

I. Introduction

In the last decades, a lot of effort was devoted to develop a mathematical model for voice production. The first steps were made by Ishizaka and Flanagan [1], approximating each vocal fold by two coupled oscillators, which provide the basis of the well known two-mass model. This simple model reproduces many essential features of the voice production, like the onset of self sustained oscillation of the folds and the shape of the glottal pulses.

Early analytical treatments were restricted to small amplitude oscillations, allowing a dimensional reduction of the problem. In particular, a two dimensional approximation known as the flapping model was widely adopted by the scientific community, based on the assumption of a transversal wave propagating along the vocal folds [2, 3]. Moreover, this model was also used to successfully

*E-mail: florencia@df.uba.ar
†E-mail: marcos@df.uba.ar

[1] Laboratorio de Sistemas Dinámicos, Depto. de Física, FCEN, Universidad de Buenos Aires. Pabellón I, Ciudad Universitaria, 1428EGA Buenos Aires, Argentina.

explain most of the features present in birdsong [4, 5].

Faithful modeling of the vocal folds has recently found new challenges: realistic articulatory speech synthesis [6–8], diagnosis of pathological behavior of the folds [9, 10] and bioprosthetic applications [11]. Within this framework, the 4-dimensional two-mass model was revisited and modified. Two main improvements are worth noting: a realistic description of the vocal fold collision [13,14] and an accurate fluid mechanical description of the glottal flow, allowing a proper treatment of the hydrodynamical force acting on the folds [8,15].

In this work, we revisit the two-mass model developed by Lucero and Koenig [7]. This choice represents a good compromise between mathematical simplicity and diversity of physical phenomena acting on the vocal folds, including the main mechanical and fluid effects that are partially found in other models [13, 15]. It was also successfully used to reproduce experimental temporal patterns of glottal airflow. Here, we extend the analytical study of this system: we present a bifurcation diagram, explore the dynamical aspects of the oscillations at the onset and normal phonation and study the isofrequency curves of the model.

This work is organized as follows: in the second section, we describe the model. In the third section, we present the bifurcation diagram, compare our solutions with those of the flapping model approximation and analyze the isofrecuency curves. In the fourth and last section, we discuss our results.

II. The model

Each vocal fold is modeled as two coupled damped oscillators, as sketched in Fig. 1.

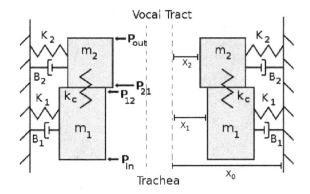

Figure 1: Sketch of the two-mass model of the vocal folds. Each fold is represented by masses m_1 and m_2 coupled to each other by a restitution force k_c and to the laryngeal walls by K_1 and K_2 (and dampings B_1 and B_2), respectively. The displacement of each mass from the resting position x_0 is represented by x_1 and x_2. The different aerodynamic pressures P acting on the folds are described in the text.

Assuming symmetry with respect to the saggital plane, the left and right mass systems are identical (Fig. 1) and the equation of motion for each mass reads

$$\dot{x}_i = y_i \qquad (1)$$
$$\dot{y}_i = \frac{1}{m_i}\left[f_i - K_i(x_i) - B_i(x_i, y_i) - k_c(x_i - x_j)\right],$$

for $i, j = 1$ or 2 for lower and upper masses, respectively. K and B represent the restitution and damping of the folds tissue, f the hydrodynamic

force, m is the mass and k_c the coupling stiffness. The horizontal displacement from the rest position x_0 is represented by x.

We use a cubic polynomial for the restitution term [Eq. (2)], adapted from [1, 7]. The term with a derivable *step-like* function Θ [Eq. (3)] accounts for the increase in the stiffness introduced by the collision of the folds. The restitution force reads

$$K_i(x_i) = k_i x_i(1 + 100x_i{}^2) \qquad (2)$$
$$+ \Theta\left(\frac{x_i + x_0}{x_0}\right) 3k_i(x_i + x_0)[1 + 500(x_i + x_0)^2],$$

with

$$\Theta(x) = \left\{ \begin{array}{ll} 0 & \text{if } x \leq 0 \\ \frac{x^2}{8\ 10^{-4} + x^2} & \text{if } x > 0 \end{array} \right. , \qquad (3)$$

where x_0 is the rest position of the folds.

For the damping force, we have adapted the expression proposed in [7], making it derivable, arriving at the following equation:

$$B_i(x_i) = \qquad (4)$$
$$\left[1 + \Theta\left(\frac{x_i + x_0}{x_0}\right)\frac{1}{\epsilon_i}\right] r_i(1 + 850x_i{}^2)y_i,$$

where $r_i = 2\epsilon_i\sqrt{k_i m_i}$, and ϵ_i is the damping ratio.

In order to describe the hydrodynamic force that the airflow exerts on the vocal folds, we have adopted the standard assumption of small inertia of the glottal air column and the model of the boundary layer developed in [7, 11, 15]. This model assumes a one-dimensional, quasi-steady incompressible airflow from the trachea to a *separation point*. At this point, the flow separates from the tissue surface to form a free jet where the turbulence dissipates the airflow energy. It has been experimentally shown that the position of this point depends on the glottal profile. As described in [15], the separation point located at the glottal exit shifts down to the boundary between masses m_1 and m_2 when the folds profile becomes more divergent than a threshold [Eq. (7)].

Viscous losses are modeled according to a bi-dimensional Poiseuille flow [Eqs. (6) and (7)]. The equations for the pressure inside the glottis are

$$P_{in} = P_s + \frac{\rho u_g^2}{2a_1^2}, \tag{5}$$

$$P_{12} = P_{in} - \frac{12\mu u_g d_1 l_g^2}{a_1^3}, \tag{6}$$

$$P_{21} = \begin{cases} \frac{12\mu u_g d_2 l_g^2}{a_2^3} + P_{out} & \text{if } a_2 > k_s a_1 \\ 0 & \text{if } a_2 \leq k_s a_1 \end{cases}, \tag{7}$$

$$P_{out} = 0. \tag{8}$$

As sketched in Fig. 1, the pressures exerted by the airflow are: P_{in} at the entrance of the glottis, P_{12} at the upper edge of m_1, P_{21} at the lower edge of m_2, P_{out} at the entrance of the vocal tract and P_s the subglottal pressure.

The width of the folds (in the plane normal to Fig. 1) is l_g; d_1 and d_2 are the lengths of the lower and upper masses, respectively. a_i are the cross-sections of the glottis, $a_i = 2l_g(x_i + x_0)$; μ and ρ are the viscosity and density coefficient of the air; u_g is the airflow inside the glottis, and $k_s = 1.2$ is an experimental coefficient. We also assume no losses at the glottal entrance [Eq. (5)], and zero pressure at the entrance of the vocal tract [Eq. (8)].

The hydrodynamic force acting on each mass reads:

$$f_1 = \begin{cases} d_1 l_g P_s & \text{if } x_1 \leq -x_0 \text{ or } x_2 \leq -x_0 \\ \frac{P_{in}+P_{12}}{2} & \text{in other case} \end{cases} \tag{9}$$

$$f_2 = \begin{cases} d_2 l_g P_s & \text{if } x_1 > -x_0 \text{ and } x_2 \leq -x_0 \\ 0 & \text{if } x_1 \leq -x_0 \\ \frac{P_{21}+P_{out}}{2} & \text{in other case} \end{cases} \tag{10}$$

Following [1, 7, 10], these functions represent opening, partial closure and total closure of the glottis. Throughout this work, piecewise functions P_{21}, f_1 and f_2 are modeled using the derivable step-like function Θ defined in Eq. (3).

III. Analysis of the model

i. Bifurcation diagram

The main anatomical parameters that can be actively controlled during the vocalizations are the subglottal pressure P_s and the folds tension controlled by the laryngeal muscles. In particular, the action of the thyroarytenoid and the cricothyroid muscles control the thickness and the stiffness of folds. Following [1], this effect is modeled by a parameter Q that scales the mechanic properties of the folds by a cord-tension parameter: $k_c = Qk_{c0}$, $k_i = Qk_{i0}$ and $m_i = \frac{m_{i0}}{Q}$. We therefore performed a bifurcation diagram using these two standard control parameters P_s and Q.

Five main regions of different dynamic solutions are shown in Fig. 2. At low pressure values (region I), the system presents a stable fixed point. Reaching region II, the fixed point becomes unstable and there appears an attracting limit cycle. At the interface between regions I and II, three bifurcations occur in a narrow range of subglottal pressure (Fig. 3, left panel), all along the Q axis. The right panel of Fig. 3 shows the oscillation amplitude of x_2. At point **A**, oscillations are born in a supercritical Hopf bifurcation. The amplitude grows continuously for increasing P_s until point **B**, where it jumps to the upper branch. If the pressure is then decreased, the oscillations persist even for lower pressure values than the onset in **A**. When point **C** is reached, the oscillations suddenly stop and the system returns to the rest position. This onset-offset oscillation hysteresis was already reported experimentally in [12].

The branch **AB** depends on the viscosity. Decreasing μ, points **A** and **B** approach to each other until they collide at $\mu = 0$, recovering the result reported in [3,10,14], where the oscillations occur as the combination of a subcritical Hopf bifurcation and a cyclic fold bifurcation.

On the other hand, the branch **BC** depends on the separation point of the jet formation. In particular, for increasing k_s, the folds become stiffer and the separation point moves upwards toward the output of the glottis. From a dynamical point of view, points **C** and **B** approach to each other until they collapse. In this case, the oscillations are born at a supercritical Hopf bifurcation and the system presents no hysteresis, as in the standard flapping model [17].

Regions II and III of Fig. 2 are separated by a saddle-repulsor bifurcation. Although this bifurcation does not represent a qualitative dynamical change for the oscillating folds, its effects are relevant when the complete mechanism of voiced sound

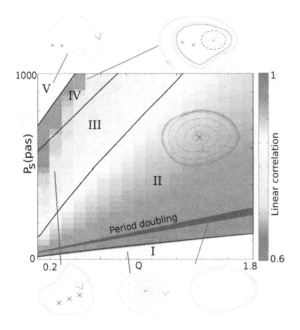

Figure 2: Bifurcation diagram in the plane of sub-glottal pressure and fold tension (Q, P_s). The insets are two-dimensional projections of the flow on the (v_1, x_1) plane, the red crosses represent unstable fixed points and the dotted lines unstable limit cycles. Normal voice occurs at $(Q, P_s) \sim (1, 800)$. The color code represents the linear correlation between $(x_1 - x_2)$ and $(y_1 + y_2)$: from dark red for $R = 1$ to dark blue for $R = 0.6$. This diagram was developed with the help of AUTO continuation software [20]. The rest of the parameters were fixed at $m_1 = 0.125$ g, $m_2 = 0.025$ g, $k_{10} = 80$ N/m, $k_{20} = 8$ N/m, $k_c = 25$ N/m, $\epsilon_1 = 0.1$, $\epsilon_2 = 0.6$, $l_g = 1.4$ cm, $d_1 = 0.25$ cm, $d_2 = 0.05$ cm and $x_0 = 0.02$ cm.

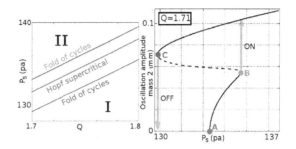

Figure 3: Hysteresis at the oscillation onset-offset. Left panel: zoom of the interface between regions I and II. The blue and green lines represent folds of cycles (saddle-node bifurcations in the map). The red line is a supercritical Hopf bifurcation. Right panel: the oscillation amplitude of x_2 as a function of the subglottal pressure P_s, at $Q = 1.71$. The continuation of periodic solutions was realized with the AUTO software package [20].

production is considered. Voiced sounds are generated as the airflow disturbance produced by the oscillation of the vocal folds is injected into the series of cavities extending from the laryngeal exit to the mouth, a non-uniform tube known as the vocal tract. The disturbance travels back and forth along the vocal tract, that acts as a filter for the original signal, enhancing the frequencies of the source that fall near the vocal tract resonances. Voiced sounds are in fact perceived and classified according to these resonances, as in the case of vowels [18]. Consequently, one central aspect in the generation of voiced sounds is the production of a spectrally rich signal at the sound source level.

Interestingly, normal phonation occurs in the region near the appearance of the saddle-repulsor bifurcation. Although this bifurcation does not alter the dynamical regime of the system or its time scales, we have observed that part of the limit cycle approaches the stable manifold of the new fixed point (as displayed in Fig. 4), therefore changing its shape. This deformation is not restricted to the appearance of the new fixed point but rather occurs in a coarse region around the boundary between II and III, as the flux changes smoothly in a vicinity of the bifurcation. In order to illustrate this effect, we use the spectral content index SCI [21], an indicator of the spectral richness of a signal: $SCI = \sum_k A_k f_k / (\sum_k A_k f_0)$, where A_k is the Fourier amplitude of the frequency f_k and f_0 is the fundamental frequency. As the pressure is increased, the SCI of $x_1(t)$ increases (upper right panel of Fig. 4), observing a boost in the vicinity of the saddle-repulsor bifurcation that stabilizes after the saddle point is generated.

Thus, the appearance of this bifurcation near the region of normal phonation could indicate a possible mechanism to further enhance the spectral richness of the sound source, on which the production of voiced sounds ultimately relies.

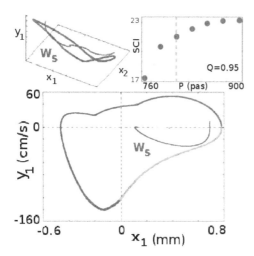

Figure 4: A projection of the limit cycle for x_1 and the stable manifold of the saddle point, for parameters consistent with normal phonatory conditions, $(Q, P_s) = (1, 850)$ (region III). Left inset: projection in the 3-dimensional space (y_1, x_1, x_2). Right inset: Spectral content index of $x_1(t)$ as a function of P_s for a fixed value of $Q = 0.95$. In green, the value at which the saddle-repulsor bifurcation takes place.

In the boundary between regions III and IV, one of the unstable points created in the saddle-repulsor bifurcation undergoes a subcritical Hopf bifurcation, changing stability as an unstable limit cycle is created [19]. Finally, entering region V, the stable and the unstable cycles collide and disappear in a fold of cycles where no oscillatory regimes exist.

In Fig. 2, we also display a color map that quantifies the difference between the solutions of the model and the flapping approximation. The flapping model is a two dimensional model that, instead of two masses per fold, assumes a wave propagating along a linear profile of the folds, i.e., the displacement of the upper edge of the folds is delayed 2τ with respect to the lower. The cross sectional areas at glottal entry and exit (a_1 and a_2) are approximated, in terms of the position of the midpoint of the folds, by

$$\begin{cases} a_1 = 2l_g(x_0 + x + \tau\dot{x}) \\ a_2 = 2l_g(x_0 + x - \tau\dot{x}) \end{cases}, \qquad (11)$$

where x is the midpoint displacement from equilibrium x_0, and τ is the time that the surface wave takes to travel half the way from bottom to top. Equation (11) can be rewritten as $(x_1 - x_2) = \tau(y_1 + y_2)$. We use this expression to quantify the difference between the oscillations obtained with the two-mass model solutions and the ones generated with the flapping approximation, computing the linear correlation coefficient between $(x_1 - x_2)$ and $(y_1 + y_2)$. As expected, the correlation coefficient R decreases for increasing P_s or decreasing Q. In the region near normal phonation, the approximation is still relatively good, with $R \sim 0.8$. As expected, the approximation is better for increasing x_0, since the effect of colliding folds is not included in the flapping model.

ii.   Isofrequency curves

One basic perceptual property of the voice is the pitch, identified with the fundamental frequency f_0 of the vocal folds oscillation. The production of different pitch contours is central to language, as they affect the semantic content of speech, carrying accent and intonation information. Although experimental data on pitch control is scarce, it was reported that it is actively controlled by the laryngeal muscles and the subglottal pressure. In particular, when the vocalis or interarytenoid muscle activity is inactive, a raise of the subglottal pressure produces an upraising of the pitch [16].

Compatible with these experimental results, we performed a theoretical analysis using P_s as a single control parameter for pitch. In the upper panels of Fig. 5, we show isofrequency curves in the range of normal speech for our model of Eqs. (1) to (10). Following the ideas developed in [22] for the avian case, we compare the behavior of the fundamental frequency with respect to pressure P_s in the two most usual cases presented in the literature: the cubic [1, 7] and the linear [10, 14] restitutions. In the lower panels of Fig. 5, we show the isofrequency curves that result from replacing the cubic restitution by a linear restitution $K_i(x_i) = k_i x_i + \Theta(\frac{x_i + x_0}{x_0})3k_i(x_i + x_0)$.

Although the curves $f_0(P_s)$ are not affected by the type of restitution at the very beginning of oscillations, the changes become evident for higher values of P_s, with positive slopes for the cubic case and negative for the linear case. This result suggests that a nonlinear cubic restitution force is a

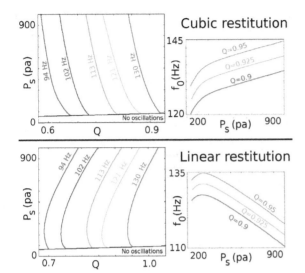

Figure 5: Relationship between pitch and restitution forces. Left panels: isofrequency curves in the plane (Q, P_s). Right panels: Curves $f_0(P_s)$ for $Q{=}0.9$, $Q{=}0.925$ and $Q{=}0.95$. In the upper panels, we used the model with the cubic nonlinear restitution of Eq. (2). In the lower panels, we show the curves obtained with a linear restitution, $K_i(x_i) = k_i x_i + \Theta(\frac{x_i + x_0}{x_0}) 3 k_i (x_i + x_0)$.

good model for the elastic properties of the oscillating tissue.

IV.   Conclusions

In this paper, we have analyzed a complete two-mass model of the vocal folds integrating collisions, nonlinear restitution and dissipative forces for the tissue and jets and viscous losses of the air-stream. In a framework of growing interest for detailed modeling of voice production, the aspects studied here contribute to understanding the role of the different physical terms in different dynamical behaviors.

We calculated the bifurcation diagram, focusing in two regimes: the oscillation onset and normal phonation. Near the parameters of normal phonation, a saddle repulsor bifurcation takes place that modifies the shape of the limit cycle, contributing to the spectral richness of the glottal flow, which is central to the production of voiced sounds. With respect to the oscillation onset, we showed how jets and viscous losses intervene in the hysteresis phenomenon.

Many different models for the restitution properties of the tissue have been used across the literature, including linear and cubic functional forms. Yet, its specific role was not reported. Here we showed that the experimental relationship between subglottal pressure and pitch is fulfilled by a cubic term.

Acknowledgements - This work was partially funded by UBA and CONICET.

[1] K Ishizaka, J L Flanagan, *Synthesis of voiced sounds from a two-mass model of the vocal cords*, Bell Syst. Tech. J. **51**, 1233 (1972).

[2] I R Titze, *The physics of smallamplitude oscillation of the vocal folds*, J. Acoust. Soc. Am. **83**, 1536 (1988).

[3] M A Trevisan, M C Eguia, G Mindlin, *Nonlinear aspects of analysis and synthesis of speech time series data*, Phys. Rev. E **63**, 026216 (2001).

[4] Y S Perl, E M Arneodo, A Amador, F Goller, G B Mindlin, *Reconstruction of physiological instructions from Zebra finch song*, Phys. Rev. E **84**, 051909 (2011).

[5] E M Arneodo, Y S Perl, F Goller, G B Mindlin, *Prosthetic avian vocal organ controlled by a freely behaving bird based on a low dimensional model of the biomechanical periphery*, PLoS Comput. Biol. **8**, e1002546 (2012).

[6] B H Story, I R Titze *Voice simulation with a bodycover model of the vocal folds*, J. Acoust. Soc. Am. **97**, 1249 (1995).

[7] J C Lucero, L Koening *Simulations of temporal patterns of oral airflow in men and women using a two-mass model of the vocal folds under dynamic control*, J. Acoust. Soc. Am. **117**, 1362 (2005).

[8] X Pelorson, X Vescovi, C Castelli, E Hirschberg, A Wijnands, A P J Bailliet, H M

A Hirschberg, *Description of the flow through in-vitro models of the glottis during phonation. Application to voiced sounds synthesis*, Acta Acust. **82**, 358 (1996).

[9] M E Smith, G S Berke, B R Gerratt, *Laryngeal paralyses: Theoretical considerations and effects on laryngeal vibration*, J. Speech Hear. Res. **35**, 545 (1992).

[10] I Steinecke, H Herzel *Bifurcations in an asymmetric vocalfold model*, J. Acoust. Soc. Am. **97**, 1874 (1995).

[11] N J C Lous, G C J Hofmans, R N J Veldhuis, A Hirschberg, *A symmetrical two-mass vocal-fold model coupled to vocal tract and trachea, with application to prosthesis design*, Acta Acust. United Ac. **84**, 1135 (1998).

[12] T Baer, *Vocal fold physiology*, University of Tokyo Press, Tokyo, (1981).

[13] T Ikeda, Y Matsuzak, T Aomatsu, *A numerical analysis of phonation using a two-dimensional flexible channel model of the vocal folds*, J. Biomech. Eng. **123**, 571 (2001).

[14] J C Lucero, *Dynamics of the two-mass model of the vocal folds: Equilibria, bifurcations, and oscillation region*, J. Acoust. Soc. Am. **94**, 3104 (1993).

[15] X Pelorson, A Hirschberg, R R van Hassel, A P J Wijnands, Y Auregan, *Theoretical and experimental study of quasisteadyflow separation within the glottis during phonation. Application to a modified twomass model*, J. Acoust. Soc. Am. **96**, 3416 (1994).

[16] T Baer, *Reflex activation of laryngeal muscles by sudden induced subglottal pressure changes*, J. Acoust. Soc. Am. **65**, 1271 (1979).

[17] J C Lucero, *A theoretical study of the hysteresis phenomenon at vocal fold oscillation onset-offset*, J. Acoust. Soc. Am. **105**, 423 (1999).

[18] I Titze, *Principles of voice production*, Prentice Hall, (1994).

[19] J Guckenheimer, P Holmes, *Nonlinear oscillations, dynamical systems and bifurcations of vector fields*, Springer, (1983).

[20] E Doedel, *AUTO: Software for continuation and bifurcation problems in ordinary differential equations*, AUTO User Manual, (1986).

[21] J Sitt, A Amador, F Goller, G B Mindin, *Dynamical origin of spectrally rich vocalizations in birdsong*, Phys. Rev. E **78**, 011905 (2008).

[22] A Amador, F Goller, G B Mindlin, *Frequency modulation during song in a suboscine does not require vocal muscles*, J. Neurophysiol. **99**, 2383 (2008).

3

KPZ: Recent developments via a variational formulation

Horacio S. Wio,[1][*] Roberto R. Deza,[2][†] Carlos Escudero,[3][‡] Jorge A. Revelli[4][§]

Recently, a variational approach has been introduced for the paradigmatic Kardar–Parisi–Zhang (KPZ) equation. Here we review that approach, together with the functional Taylor expansion that the KPZ nonequilibrium potential (NEP) admits. Such expansion becomes naturally truncated at third order, giving rise to a nonlinear stochastic partial differential equation to be regarded as a gradient-flow counterpart to the KPZ equation. A dynamic renormalization group analysis at one-loop order of this new mesoscopic model yields the KPZ scaling relation $\alpha + z = 2$, as a consequence of the exact cancelation of the different contributions to vertex renormalization. This result is quite remarkable, considering the lower degree of symmetry of this equation, which is in particular not Galilean invariant. In addition, this scheme is exploited to inquire about the dynamical behavior of the KPZ equation through a path-integral approach. Each of these aspects offers novel points of view and sheds light on particular aspects of the dynamics of the KPZ equation.

I. Introduction

Although readers whose careers span mostly on the 21th century might not care about this, back in the sixties (when transistors and lasers had already been invented) *equilibrium* critical phenomena were still a puzzle. In fact, although a sense of "universality" had been gained in 1950 through a field theory based on the innovative concept of *order parameter* [1,2], its predicted critical exponents were

---
[*]E-mail: wio@ifca.unican.es
[†]E-mail: deza@mdp.edu.ar
[‡]E-mail: cel@icmat.es
[§]E-mail: revelli@famaf.unc.edu.ar

[1] IFCA (UC-CSIC), Avda. de los Castros s/n, E-39005 Santander, Spain.

[2] IFIMAR (UNMdP-CONICET), Funes 3350, 7600 Mar del Plata, Argentina.

[3] Depto. Matemáticas & ICMAT (CSIC-UAM-UC3M-UCM), Cantoblanco, E-28049 Madrid, Spain.

[4] FaMAF-IFEG (CONICET-UNC), 5000 Córdoba, Argentina.

almost as a rule wrong. It was not until the seventies that a far more sophisticated field-theory approach [3] brought order home: equilibrium universality classes are determined solely by the dimensionalities of the order parameter and the ambient space. Since then, one of Statistical Physics' "holy grials" has been to conquer a similar achievement for *non-equilibrium* critical phenomena [4]. In such a (still unaccomplished) enterprise, a valuable field-theoretical tool has been in the last quarter of century the Kardar–Parisi–Zhang (KPZ) equation [5–7].

The KPZ equation [5–7] has become a paradigm for the description of a vast class of nonequilibrium phenomena by means of stochastic fields. The field $h(x,t)$, whose evolution is governed by this stochastic nonlinear partial differential equation, describes the height of a fluctuating interface in the context of surface-growth processes in which it was originally formulated. From a theoretical point of view, the KPZ equation has many interesting properties, for instance, its close relationship with the Burgers equation [8] or with a diffusion equation with multiplicative noise, whose field $\phi(x,t)$ can be interpreted as the restricted partition function of the directed polymer problem [9]. Many of the efforts put in investigating the behavior of its solutions were focused on obtaining the scaling laws and critical exponents in one or more spatial dimensions [11–17]. However, other questions of great interest are the development of suitable algorithms for its numerical integration [18,19], the construction of particular solutions [20–23], the crossover behavior between different regimes [10,24–26], as well as related ageing and pinning phenomena [27–29].

Among all the classical theoretical developments concerning this equation [6,7], two have recently drawn our attention. One was the scaling relation $\alpha + z = 2$, which is expected to be exact for the KPZ equation in any dimension. The exactness of this relation has been traditionally attributed to the Galilean invariance of the KPZ equation. Nevertheless, the assumed central role of this symmetry has been challenged in this as well as in other nonequilibrium models from both a theoretical [30–33] and a numerical [34–36] point of view. The second one is the generally accepted lack of existence of a suitable functional allowing to formulate the KPZ equation as a gradient flow. In fact, a variational approach to the closely related Sun-Guo-Grant [37] and Villain-Lai-Das Sarma [38,39] equations was developed in [40,41] by means of a geometric construction. In [46], a Lyapunov functional (with an explicit density) was found for the deterministic KPZ equation. Also, a *nonequilibrium potential* (NEP), a functional that allows the formal writing of the KPZ equation as a (stochastically forced) exact gradient flow, was introduced.

In this work we shortly review the consistency constraints imposed by the nonequilibrium-potential structure on *discrete* representations of the KPZ equation and show that they lead to explicit breakdown of Galilean invariance, despite the fact that the obtained numerical results are still those of the KPZ universality class. A Taylor expansion of the previously introduced NEP has (in terms of *fluctuations*) an explicit density and a thought-provoking structure [47], and leads to an equation of motion (for *fluctuations*, in the continuum) with exact gradient-flow structure, but different from the KPZ one. This equation has a lower degree of symmetry: it is neither Galilean invariant

nor even translation invariant. Its scaling properties are studied by means of a dynamic renormalization group (DRG) analysis, and its critical exponents fulfill at one-loop order the same scaling relation $\alpha + z = 2$ as those of the KPZ equation, despite the aforementioned lack of Galilean invariance. The concern with stability leads us to suggest the introduction of an equation related to the Kuramoto-Sivashinsky one, also with exact gradient-flow structure. We close this article exposing some novel developments based on a path-integral-like approach.

II. Brief review of the nonequilibrium potential scheme

Loosely speaking, the notion of NEP is an extension to nonequilibrium situations of that of equilibrium thermodynamic potential. In order to introduce it, we consider a general system of nonlinear stochastic equations (admitting the possibility of *multiplicative noises*)

$$\dot{q}^{\nu} = K^{\nu}(q) + g_i^{\nu}(q)\,\xi_i(t), \qquad \nu = 1,\ldots,n; \quad (1)$$

where repeated indices are summed over. Equation (1) is stated in the sense of Itô. The $\{\xi_i(t)\}$, $i = 1,\ldots,m \leq n$ are mutually independent sources of Gaussian white noise with typical strength γ.

The Fokker–Planck equation corresponding to Eq. (1) takes the form

$$\frac{\partial P}{\partial t} = -\frac{\partial}{\partial q^{\nu}}\,K^{\nu}(q)\,P + \frac{\gamma}{2}\frac{\partial^2}{\partial q^{\nu}\,\partial q^{\mu}}\,Q^{\nu\mu}(q)\,P \quad (2)$$

where $P(q,t;\gamma)$ is the probability density of observing $q = (q_1,\ldots,q_n)$ at time t for noise intensity γ, and $Q^{\nu\mu}(q) = g_i^{\nu}(q)\,g_i^{\mu}(q)$ is the matrix of transport coefficients of the system, which is symmetric and non-negative. In the long time limit ($t \to \infty$), the solution of Eq. (2) tends to the stationary distribution $P_{\mathrm{st}}(q)$. According to [42–44], the NEP $\Phi(q)$ associated to Eq. (2) is defined by

$$\Phi(q) = -\lim_{\gamma \to 0} \gamma \ln P_{\mathrm{st}}(q,\gamma). \quad (3)$$

In other words,

$$P_{\mathrm{st}}(q)\,d^n q = Z(q)\exp\left[-\frac{\Phi(q)}{\gamma} + \mathcal{O}(\gamma)\right] d\Omega_q,$$

where $\Phi(q)$ is the NEP of the system and the prefactor $Z(q)$ is defined as the limit

$$\ln Z(q) = \lim_{\gamma \to 0} \left[\ln P_{\text{st}}(q, \gamma) + \frac{1}{\gamma} \Phi(q) \right].$$

Here $d\Omega_q = d^n q / \sqrt{G(q)}$ is the invariant volume element in the q-space and $G(q)$ is the determinant of the contravariant metric tensor (for the Euclidean metric it is $G = 1$). It was shown [42] that $\Phi(q)$ is the solution of a Hamilton–Jacobi-like equation (HJE)

$$K^\nu(q) \frac{\partial \Phi}{\partial q^\nu} + \frac{1}{2} Q^{\nu\mu}(q) \frac{\partial \Phi}{\partial q^\nu} \frac{\partial \Phi}{\partial q^\mu} = 0,$$

and $Z(q)$ is the solution of a linear first-order partial differential equation depending on $\Phi(q)$ (not shown here).

Equation (3) and the normalization condition ensure that Φ is bounded from below. Furthermore, it follows that

$$\frac{d\Phi(q)}{dt} = K^\nu(q) \frac{\partial \Phi(q)}{\partial q^\nu} = -\frac{1}{2} Q^{\nu\mu}(q) \frac{\partial \Phi}{\partial q^\nu} \frac{\partial \Phi}{\partial q^\mu} \le 0,$$

i.e., Φ is a Lyapunov functional for the dynamics of the system when fluctuations are neglected. Under the deterministic dynamics, $\dot{q}^\nu = K^\nu(q)$, Φ decreases monotonically and takes a minimum value on attractors. In particular, Φ must be constant on all extended attractors (such as limit cycles or strange attractors) [42].

An alternative way to look into this problem is due to Ao [45]. The interesting feature of this approach is that it resorts neither to $P_{st}(q)$ nor to the small-noise limit, thus being applicable in principle to more general situations.

III. Variational approach for KPZ

The Kardar–Parisi–Zhang (KPZ) equation reads

$$\frac{\partial h(x,t)}{\partial t} = \nu \nabla^2 h(x,t) + \frac{\lambda}{2} \left[\nabla h(x,t) \right]^2 + \xi(x,t), \quad (4)$$

where $\xi(x,t)$ is a Gaussian white noise, of zero mean ($\langle \xi(x,t) \rangle = 0$) and correlation $\langle \xi(x,t)\xi(x',t') \rangle = 2\gamma \delta(x - x')\delta(t - t')$. As it is well known, this nonlinear differential equation describes the fluctuations of a growing interface with a surface tension given by ν; λ is proportional to the average growth velocity and arises because the surface slope is paralleled transported in such a growth process.

Lyapunov functional

The deterministic KPZ equation—obtained by setting $\gamma = 0$—is exactly solvable by means of the Hopf–Cole transformation ($\phi(x,t) = e^{\frac{\lambda}{2\nu} h(x,t)}$), which maps the nonlinear KPZ equation onto the (deterministic) linear diffusion equation [5]

$$\frac{\partial \phi(x,t)}{\partial t} = \nu \nabla^2 \phi(x,t). \quad (5)$$

Also, the multiplicative reaction-diffusion (RD) equation

$$\frac{\partial \phi(x,t)}{\partial t} = \nu \nabla^2 \phi(x,t) + \phi(x,t)\xi(x,t), \quad (6)$$

which is associated to the directed polymer problem [6,7,9] results, using the inverse transformation ($h(x,t) = \frac{2\nu}{\lambda} \ln \phi(x,t)$) to be mapped into the *complete* KPZ equation (4).

The deterministic part of Eq. (6) (i.e., Eq. (5)), can be written as

$$\frac{\partial \phi(x,t)}{\partial t} = -\frac{\delta \mathcal{F}[\phi(x,t)]}{\delta \phi(x,t)}, \quad (7)$$

where $\mathcal{F}[\phi(x,t)]$ is the Lyapunov functional of the deterministic RD problem given by

$$\mathcal{F}[\phi(x,t)] = \frac{\nu}{2} \int \left[\nabla \phi(x,t) \right]^2 \, dx.$$

Applying to this functional the above indicated inverse transformation we get [46]

$$\mathcal{F}[h] = \frac{\lambda^2}{8\nu} \int e^{\frac{\lambda}{\nu} h(x,t)} \left[\nabla h(x,t) \right]^2 \, dx, \quad (8)$$

that allows the KPZ equation to be written as

$$\frac{\partial}{\partial t} h(x,t) = -\Gamma[h] \frac{\delta \mathcal{F}[h]}{\delta h(x,t)} + \xi(x,t). \quad (9)$$

One can check the Lyapunov property $\dot{\mathcal{F}}[h] \le 0$, with the motility $\Gamma[h]$ given by

$$\Gamma[h] = \left(\frac{2\nu}{\lambda} \right)^2 e^{-\frac{\lambda}{\nu} h(x,t)},$$

and that its minimum is achieved by constant functions. Hence we have a Lyapunov functional for the deterministic KPZ equation that displays simple dynamics: the asymptotic stability of constant solutions indicates an approach to constant profiles at long times, for arbitrary initial conditions. Despite this simplicity in the deterministic case, the stochastic situation is far from trivial and gives rise to self-affine fractal profiles. In particular, the existence of this Lyapunov functional provides no a priori intuition on the stochastic dynamics.

The nonequilibrium potential

An alternative functional was also proposed in [46]. By starting from the functional Fokker-Planck equation, we look for the stationary solution (in fact steady-state solution), and after some integration by parts, it is possible to arrive to another form of Lyapunov functional

$$\Phi[h] = \frac{\nu}{2} \int \mathrm{d}x \, (\nabla h)^2 - \frac{\lambda}{2} \int \mathrm{d}x \int_{h_{\mathrm{ref}}}^{h(x,t)} \mathrm{d}\psi \, (\nabla\psi)^2 \,. \tag{10}$$

It is somehow inspired in the analytical form of "model A", according to the classification of critical phenomena in [48]. Here, the interpretation of the integral in the 2nd term on the rhs is $\int \mathrm{d}x \int_{h_{\mathrm{ref}}}^{h(x,t)} \mathrm{d}\psi = \sum_j \triangle x \int_{h_{\mathrm{ref},j}}^{h_j} d\psi_j$. According to this definition, the KPZ equation can be formally written as a stochastically forced gradient flow

$$\frac{\partial}{\partial t} h(x,t) = -\frac{\delta\Phi[h]}{\delta h(x,t)} + \xi(x,t). \tag{11}$$

The functional so defined fulfills the Lyapunov condition $\dot{\Phi}[h] = -\left(\frac{\delta\Phi[h]}{\delta h(x,t)}\right)^2 \leq 0$ as well, and could be identified as the *nonequilibrium potential* (NEP) for the KPZ case [53,54].

We will not pursue here the development of rigorous result concerning the functional (10). Our present interest falls in the calculation of quantities of physical interest rather than in building a completely rigorous mathematical theory. It is worth remarking that, as indicated in [46] and above, such a form has a discrete definition. It is also interesting to point out that analogous functionals involving functional integrals which are not carried out explicitly were obtained for the problem of interface fluctuations in random media [49–52].

NEP expansion

We now proceed to formally Taylor expand the NEP defined in Eq. (10) around a given reference

(or initial) state, denoted by h_0

$$\begin{aligned} \Phi[h] &= \frac{\nu}{2} \int \mathrm{d}x \, (\nabla h)^2 \\ &\quad - \frac{\lambda}{2} \int \mathrm{d}x \int_{h_0}^{h(x,t)} \mathrm{d}\psi \, (\nabla\psi)^2 \\ &\approx \Phi[h_0] + \delta\Phi[h_0] + \frac{1}{2}\delta^2\Phi[h_0] \\ &\quad + \frac{1}{6}\delta^3\Phi[h_0] + \cdots. \end{aligned} \tag{12}$$

The successive terms in the expansion of $\Phi[h]$ are

$$\begin{aligned} \delta\Phi[h_0] &= -\int \mathrm{d}x \left[\nu\nabla^2 h_0 + \frac{\lambda}{2}(\nabla h_0)^2 \right]\delta h, \\ \delta^2\Phi[h_0] &= -\int \mathrm{d}x \, \delta h \left(\nu\nabla^2 + \lambda\nabla h_0 \cdot \nabla\right)\delta h, \\ \delta^3\Phi[h_0] &= -\lambda\int \mathrm{d}x \, \delta h \, (\nabla\delta h)^2. \end{aligned} \tag{13}$$

Clearly, for higher order $(n \geq 4)$ terms we have

$$\delta^n\Phi[h_0] \equiv 0, \tag{14}$$

indicating that this formal expansion has a natural cut-off after the third order.

It is worth indicating that in this computation—as in all the other computations within this work—boundary terms vanish provided one of the following types of boundary conditions is assumed: homogeneous Dirichlet boundary conditions, homogeneous Neumann boundary conditions, periodic boundary conditions or an infinite space with the derivatives of δh vanishing as they approach an infinite distance from the origin.

The reference state h_0 is arbitrary (i.e., any initial condition), but it is particularly useful to take it as one that makes $\delta\Phi[h_0] = 0$, that is: a solution to the stationary counterpart of the deterministic KPZ equation. The complete set of solutions is $h_0 = c$, where c is an arbitrary constant (arbitrary up to the application of the boundary conditions, whenever this consideration applies), what physically corresponds to a flat interface. Hence we have $(\delta h = h - h_0)$

$$\Phi[h] = \Phi[h_0] + \frac{1}{2}\delta^2\Phi[h_0,\delta h] + \frac{1}{6}\delta^3\Phi[h_0,\delta h]. \tag{15}$$

The equation for fluctuations

From here we can define an effective NEP, which drives the dynamics of the fluctuations δh and has an explicit density. Clearly, it corresponds to the last two terms in Eq. (15). To simplify the notation we adopt $u(x,t) := \delta h(x,t)$, and so the NEP reads

$$
\begin{aligned}
\mathcal{I}[u] &= \int dx \left[\frac{\nu}{2} - \frac{\lambda}{6} u(x,t) \right] (\nabla u)^2 \qquad (16) \\
&= -\int dx\, u(x,t) \left[\nu\nabla^2 u + \frac{\lambda}{6}(\nabla u)^2 \right].
\end{aligned}
$$

The deterministic equation for u results

$$
\begin{aligned}
\frac{\partial u}{\partial t} &= -\frac{\delta \mathcal{I}[u]}{\delta u}, \\
\frac{\partial u}{\partial t} &= \left(\nu - \frac{\lambda u}{3} \right)\nabla^2 u - \frac{\lambda}{6}(\nabla u)^2. \qquad (17)
\end{aligned}
$$

Clearly, patterns like $u_0 = $ constant are stationary solutions of Eq. (17): for all of them $\mathcal{I}[u] = 0$, indicating that all such states have the same "energy". Finally, let us remark that although the formal Taylor expansion becomes naturally truncated at third order, the deterministic KPZ equation is not recovered. We call the stochastic version of this new equation "KPZW".

There is a remarkable difference between both equations (KPZ and Eq. (17)). It arises due to the fact that in the first case we have a **fixed** equation for h and for **any** initial condition, while in the second case we have a **fixed initial condition** ($u = 0$) with a **variable equation** whose coefficients depend on h_o! (it is an equation for the departure from the given initial condition). The question of the relevance of this aspect to ageing problems (as discussed for instance in [29]) arises naturally. This point, worth to be analyzed, will be the subject of further work.

Non-local kernel

In previous works [36, 46] it was indicated that the following functional, including a nonlocal contribution,

$$
\begin{aligned}
\mathcal{F}[h] &= \int_\Omega \left\{ \left(\frac{\lambda^2}{8\nu} \right)(\nabla h)^2 + e^{-\frac{\lambda}{2\nu}h(\mathbf{x},t)} \right. \\
&\quad \left. \times \int_\Omega d\mathbf{x}' G(\mathbf{x},\mathbf{x}')e^{\frac{\lambda}{2\nu}h(\mathbf{x}',t)} \right\} e^{\frac{\lambda}{\nu}h(\mathbf{x},t)} d\mathbf{x},
\end{aligned}
$$
(18)

leads, after functional derivation, to a generalized KPZ equation

$$
\begin{aligned}
\partial_t h(\mathbf{x},t) &= \nu\nabla^2 h(\mathbf{x},t) + \frac{\lambda}{2}[\nabla h(\mathbf{x},t)]^2 \\
&\quad - e^{-\frac{\lambda}{2\nu}h(\mathbf{x},t)}\int_\Omega d\mathbf{x}' G(\mathbf{x},\mathbf{x}')e^{\frac{\lambda}{2\nu}h(\mathbf{x}',t)} \\
&\quad + \xi(\mathbf{x},t). \qquad (19)
\end{aligned}
$$

It was also shown that if the nonlocal kernel has translational invariance ($G(\mathbf{x},\mathbf{x}') = G(\mathbf{x}-\mathbf{x}')$), and also, if it is of (very) "short" range, it can be expanded as

$$
G(\mathbf{x} - \mathbf{x}') = \sum_{n=0}^{\infty} A_{2n}\delta^{(2n)}(\mathbf{x} - \mathbf{x}'), \qquad (20)
$$

with $\delta^{(n)}(\mathbf{x}-\mathbf{x}') = \nabla_{\mathbf{x}'}^n\delta(\mathbf{x}-\mathbf{x}')$, and where symmetry properties were taken into account. Exploiting this form of the kernel, and considering different approximation orders, it is possible to recover contributions having the same form as the ones arising in several previous works, where scaling properties, symmetry arguments, etc., have been used to discuss the possible contributions to a general form of the kinetic equation [55–57]. Such different contributions are tightly related to several of other previously studied equations, like the Sun–Guo–Grant equation [37], as well as others [55,58].

We will not pursue this aspect here, but we will briefly refer again to it in a forthcoming section.

IV. Discretization issues, symmetry violation and all that

In this section we will review aspects related to two main symmetries associated with the 1D KPZ equation: Galilean invariance and the fluctuation–dissipation relation. On the one hand, Galilean invariance has been traditionally linked to the exactness of the relation $\alpha + z = 2$ among the critical exponents, in any spatial dimensionality (the roughness exponent α, characterizing the surface morphology in the stationary regime, and the dynamic exponent z, indicating the correlation length scaling as $\xi(t) \sim t^{1/z}$). However, this interpretation has been criticized in this and other nonequilibrium models [31,32,59]. On the other hand, the second symmetry essentially tells us that in 1D, the nonlinear (KPZ) term is not operative at long times.

Even when recognizing the interesting analytical properties of the KPZ equation, it is clear that investigating the behavior of its solutions requires the (stochastic) numerical integration of a discrete version. Such an approach has been used ,e.g., to obtain the critical exponents in one and more spatial dimensions [10–15,60]. Although a pseudo-spectral spatial discretization scheme has been recently introduced [18, 61], real-space discrete versions of Eq. (4) are still used for numerical simulations [62,63]. One reason is their relative ease of implementation and of interpretation in the case of non-homogeneous substrates, for example a quenched impurity distribution [64].

Consistency

Here, we use the standard, nearest-neighbor discretization prescription as a benchmark to elucidate the constraints to be obeyed by any spatial discretization scheme, arising from the mapping between the KPZ and the diffusion equation (with multiplicative noise) through the Hopf–Cole transformation.

The standard spatially discrete version of Eq. (6) is

$$\dot{\phi}_j = \frac{\nu}{a^2}\left(\phi_{j+1} - 2\phi_j + \phi_{j-1}\right) + \frac{\lambda\sqrt{\gamma}}{2\nu}\phi_j\xi_j, \quad (21)$$

with $1 \leq j \leq N \equiv 0$, because of the assumed periodic b.c. (the implicit sum convention is not meant in any of the discrete expressions). Here a is the lattice spacing. Then, using the discrete version of Hopf–Cole transformation $\phi_j(t) = \exp\left[\frac{\lambda}{2\nu}h_j(t)\right]$, we get

$$\dot{h}_j = \frac{2\nu^2}{\lambda a^2}\left(e^{\delta_j^+ a} + e^{\delta_j^- a} - 2\right) + \sqrt{\gamma}\,\xi_j, \quad (22)$$

with $\delta_j^\pm \equiv \frac{\lambda}{2\nu a}(h_{j\pm1} - h_j)$. By expanding the exponentials up to terms of order a^2, and collecting equal powers of a (observe that the zero-order contribution vanishes) we retrieve

$$\begin{aligned}\dot{h}_j &= \frac{\nu}{a^2}\left(h_{j+1} - 2h_j + h_{j-1}\right)\\&+ \frac{\lambda}{4\,a^2}\left[(h_{j+1} - h_j)^2 + (h_j - h_{j-1})^2\right]\\&+ \sqrt{\gamma}\,\xi_j.\end{aligned} \quad (23)$$

As we can see, the first and second terms on the r.h.s. of Eq. (23) are *strictly* related by virtue of the

Hopf-Cole transformation. In other words, the discrete form of the Laplacian in Eq. (21) constrains the discrete form of the nonlinear term in the transformed equation. Later we return, in another way, to the tight relation between the discretization of both terms. Known proposals [60] fail to comply with this natural requirement.

An important feature of the Hopf–Cole transformation is that it is *local*, i.e., it involves neither spatial nor temporal transformations. An effect of this feature is that the discrete form of the Laplacian is the same, regardless of whether it is applied to ϕ or h.

The aforementioned criterion dictates the following discrete form for $\mathcal{F}[\phi]$ (the one just before Eq. (8)), thus a Lyapunov function for any finite N

$$\begin{aligned}\mathcal{F}[\phi] &= \frac{\nu}{2}\sum_{j=1}^N a\left((\partial_x\phi)^2\right)_j\\&= \frac{\nu}{4a}\sum_{j=1}^N\left[(\phi_{j+1} - \phi_j)^2 + (\phi_j - \phi_{j-1})^2\right].\end{aligned} \quad (24)$$

It is a trivial task to verify that the Laplacian is $(\partial_x^2\phi)_j = -a^{-1}\partial_{\phi_j}\mathcal{F}[\phi]$. Now, the obvious fact that this functional can also be written as $\mathcal{F}[\phi] = \frac{\nu}{2a}\sum_{j=1}^N(\phi_{j+1} - \phi_j)^2$ illustrates a fact that for a more elaborate discretization requires explicit calculations: the Laplacian does not *uniquely* determine the Lyapunov function [34–36].

Equation (22) has also been written in [14], although with different goals than ours. Their interest was to analyze the strong coupling limit via mapping to the directed polymer problem.

An accurate consistent discretization

Since the proposals of [60] already involve next-to-nearest neighbors, one may seek for a prescription that minimizes the numerical error. An interesting choice for the Laplacian is [65]

$$\frac{1}{12\,a^2}\left[16(\phi_{j+1} + \phi_{j-1}) - (\phi_{j+2} + \phi_{j-2}) - 30\,\phi_j\right], \quad (25)$$

which has the associated discrete form for the KPZ term

$$(\partial_x\phi)^2 = \frac{1}{24\,a^2}\Big\{16\left[(\phi_{j+1}-\phi_j)^2\right.$$
$$\left.+(\phi_j-\phi_{j-1})^2\right]$$
$$-\left[(\phi_{j+2}-\phi_j)^2+(\phi_j-\phi_{j-2})^2\right]\Big\}$$
$$+\mathcal{O}(a^4). \qquad (26)$$

Replacing this into the first line of Eq. (24), we obtain Eq. (25). Since this discretization scheme fulfills the consistency conditions, it is accurate up to $\mathcal{O}(a^4)$ corrections, and its prescription is not more complex than other known proposals, we expect that it will be the convenient one to use when high accuracy is required in numerical schemes [34–36].

Relation with the Lyapunov functional

In Sect. III we have indicated the form of the NEP for KPZ, and the way in which the functionals $\mathcal{F}[\phi]$ and $\mathcal{F}[h]$ are related [46]. According to the previous results, we can write the discrete version of Eq. (8) as

$$\mathcal{F}[h] = \frac{\lambda^2}{8\nu}\frac{1}{2\,a}\sum_j e^{\frac{\lambda}{\nu}h_j}\left[(h_{j+1}-h_j)^2\right.$$
$$\left.+(h_j-h_{j-1})^2\right].$$

Introducing this expression into $\partial_t h_j = \Gamma_j\frac{\delta\mathcal{F}[h]}{\delta h_j}$, and through a simple algebra, we obtain Eq. (23). This reinforces our previous result, and moreover indicates that the discrete variational formulation naturally leads to a consistent discretization of the KPZ equation.

The fluctuation–dissipation relation

This relation is, together with Galilean invariance, a fundamental symmetry of the one-dimensional KPZ equation. It is clear that both symmetries are recovered when the continuum limit is taken in any reasonable discretization scheme. Thus, an accurate enough partition must yield suitable results.

The stationary probability distribution for the KPZ problem in 1D is known to be [6,7]

$$\mathcal{P}_{\text{stat}}[h] \sim \exp\left\{-\frac{\nu}{2\gamma}\int dx\,(\partial_x h)^2\right\}.$$

For the discretization scheme in Eq. (23), this is

$$\sim \exp\left\{\frac{\nu}{2\varepsilon}\frac{1}{2a}\sum_j\left[(h_{j+1}-h_j)^2+(h_j-h_{j-1})^2\right]\right\}. \qquad (27)$$

Inserting this expression into the stationary Fokker–Planck equation, the only surviving term has the form

$$\frac{1}{2a^3}\sum_j\quad\left[(h_{j+1}-h_j)^2+(h_j-h_{j-1})^2\right]$$
$$\times[h_{j+1}-2h_j+h_{j-1}]. \qquad (28)$$

The continuum limit of this term is $\int dx\,(\partial_x h)^2\,\partial_x^2 h$, that is identically zero [6,7]. A numerical analysis of Eq. (28) indicates that it is several orders of magnitude smaller than the value of the exponents' pdf [in Eq. (27)], and typically behaves as $\mathcal{O}(1/N)$, where N is the number of spatial points used in the discretization. Moreover, it shows an even faster approach to zero if expressions with higher accuracy [like Eqs. (25) and (26)] are used for the differential operators. In addition, when the discrete form of $(\partial_x h)^2$ from [60] is used together with its consistent form for the Laplacian, the fluctuation–dissipation relation **is not** exactly fulfilled. This indicates that the problem with the fluctuation–dissipation theorem in $1+1$, discussed in [18,60] can be just circumvented by using more accurate expressions.

Galilean invariance

This invariance means that the transformation

$$x\to x-\lambda vt,\quad h\to h+vx,\quad F\to F-\frac{\lambda}{2}v^2, \quad (29)$$

where v is an arbitrary constant vector field, leaves the KPZ equation invariant. The equation obtained using the classical discretization

$$\partial_x h\to\frac{1}{2\,a}(h_{j+1}-h_{j-1}), \qquad (30)$$

is invariant under the discrete Galilean transformation

$$ja\to ja-\lambda vt,\quad h_j\to h_j+vja,\quad F\to F-\frac{\lambda}{2}v^2. \qquad (31)$$

However, the associated equation is known to be numerically unstable [14], at least when a is not small enough. Besides, Eq. (23) is not invariant under the discrete Galilean transformation. In fact, the transformation $h \to h + vja$ yields an excess term which is compatible with the gradient discretization in Eq. (30); however, this discretization does not allow to recover the quadratic term in Eq. (23), indicating that this finite-difference scheme is not Galilean-invariant.

Since Eq. (21) is invariant under the transformation indicated in Eq. (31), it is the nonlinear Hopf–Cole transformation (within the present discrete context) which is responsible for the loss of Galilean invariance. Note that these results are independent of whether we consider this discretization scheme or a more accurate one.

Galilean invariance has always been associated with the exactness of the one-dimensional KPZ exponents, and with a relation that connects the critical exponents in higher dimensions [68]. If the numerical solution obtained from a finite-difference scheme as Eq. (23), which is not Galilean invariant, *yields the well known critical exponents*, this will be an indicative that Galilean invariance is not strictly necessary to get the KPZ universality class. The numerical results presented in [34–36] clearly show that this is the case.

We will not discuss here the simulation procedure but only indicate that to make the simulations we introduced a discrete representation of $h(x,t)$ along the substrate direction x with lattice spacing $a = 1$, and that a standard second-order Runge–Kutta algorithm (with periodic boundary conditions) was employed (see [66]). In [34–36] it was shown that all the cases (consistent or not) exhibit the same critical exponents. Moreover, we want to note that the discretization used in Refs. [60], which also violates Galilean invariance, yields the same critical exponents too. Additionally, stochastic differential equations which are not explicitly Galilean invariant have been shown to obey the relation $\alpha + z = 2$ ([33], see also next section). Hence, our numerical analysis indicates that there are discrete schemes of the KPZ equation which, even not obeying Galilean invariance, show KPZ scaling.

The moral from the present analysis is clear: due to the locality of the Hopf–Cole transformation, the discrete forms of the Laplacian and the nonlinear (KPZ) term cannot be chosen independently; more-

over, the prescriptions should be the same, regardless of the fields they are applied to. For further details we refer to [34–36].

V. Renormalization-group analysis for Fluctuations

In section III we have built a gradient flow counterpart of the deterministic KPZ equation. In this section we consider the corresponding stochastically forced gradient flow

$$\partial_t u = -\frac{\delta \mathcal{I}}{\delta u} + \xi(x,t), \qquad (32)$$

with the density indicated in Eq. (16). We obtain the KPZW equation, which is the following SPDE

$$\partial_t u = \nu \nabla^2 u - \frac{\lambda}{6}(\nabla u)^2 - \frac{\lambda}{3} u \nabla^2 u + \xi(x,t). \quad (33)$$

Our present goal will be to analyze the scaling behavior of the fluctuations of the solution to this equation.

Since Eq. (33) is nonlinear, we focus on a perturbative technique. We choose the dynamic renormalization group as employed in [67, 68]. Employing this method, we find at one-loop order the following flow equations [47]

$$\frac{d\lambda}{d\ell} = \lambda(\alpha + z - 2), \qquad (34)$$

$$\frac{d\nu}{d\ell} = \nu\left(z - 2 - \frac{1}{36}\frac{\lambda^2 D}{\nu^3}K_d\frac{1-d}{d}\right), \quad (35)$$

$$\frac{d\gamma}{d\ell} = \gamma\left(z - d - 2\alpha + \frac{K_d}{72}\frac{\lambda^2\gamma}{\nu^3}\right), \qquad (36)$$

where $K_d = S_d/(2\pi)^d$, $S_d = 2\pi^{d/2}/\Gamma(n/2)$ is the surface area of the $d-$dimensional unit sphere, and Γ is the gamma function. We find that the coupling constant $\bar{g} := K_d\lambda^2\gamma/\nu^3$ obeys the one-loop differential equation

$$\frac{d\bar{g}}{d\ell} = (2-d)\bar{g} + \frac{6-5d}{72d}\bar{g}^2, \qquad (37)$$

revealing that the critical dimension of this model is $d_c = 2$ as could be anticipated by means of power counting. For $d > 2$ the coupling constant approaches zero exponentially fast in the scale ℓ; for $d = 2$, this approach is algebraic. So for these dimensions one expects the large-scale space-time properties of Eq. (33) to be dominated by its linear counterpart (up to marginal corrections in $d = 2$). In $d = 1$, the coupling constant runs to infinity for finite ℓ, suggesting the presence of a non-perturbative fixed point (as the one in the KPZ equation for $d = 2$).

The values of the critical exponents which yield scale invariance can be formally calculated by identifying with zero the right hand sides of Eqs. (34)–(36). We get

$$\alpha = \frac{2(2-d)(1-d)}{6-5d}, \tag{38}$$

$$z = \frac{12 - 10d - 2(2-d)(1-d)}{6-5d}, \tag{39}$$

which in particular obey the relation $\alpha + z = 2$ in any dimensionality, despite the fact that Eq. (33) does not obey any sort of Galilean invariance. We note that in both $d = 1$ and $d = 2$ it is $\alpha = 0$ and $z = 2$, whereas α becomes negative in higher dimensions. Hence, in all dimensions, the exponent α indicates that the interface is either flat or at most marginally rough. The values for $d = 1$ make both diffusion and nonlinearity in Eq. (33) invariant under the scale transformation $\{x, t, u\} \to \{bx, b^z t, b^\alpha u\}$, as far as $b > 1$. In this case, the noise grows with the scale (a fact that might explain the growth of the coupling constant in the renormalization group flow). In $d = 2$, the exponents are those of the linear equation. An interesting result is that for $d = 0$, the exponents become those of the KPZ equation: $\alpha = 2/3$ and $z = 4/3$, although this limit is highly singular for Eq. (37). Of course, these results have been obtained by means of a perturbative dynamic renormalization group and could be modified by non-perturbative contributions. One possible path to study such a possibility could be to adapt some *non perturbative renormalization group* techniques used for KPZ [69] to the present KPZW case.

Among all the results in this section, we would like to highlight the one given by Eq. (34). We recall that the RG analysis of the KPZ equation yields non-renormalization of the vertex and renormalization of propagator and noise. Our variational equation yields exactly the same result. Vertex non-renormalization at one-loop order is expressed by Eq. (34). The origin of this result is analogous to that of its equivalent in the KPZ equation: three non-vanishing Feynman diagrams contribute to vertex renormalization, but they cancel out each other [5] (a fact that has been traditionally attributed to the Galilean invariance of the KPZ equation). Here we have shown that the same result appears in a SPDE that is not even invariant under the translation $u \to u$+constant.

VI. Stability

We have carried out the NEP expansion about a constant solution of the KPZ equation and found that constants are still solutions to KPZW (Eq. (17)). In this section we will study the linear stability of such solutions. We start considering the solution

$$u(x,t) = c + \epsilon v(x,t), \tag{40}$$

where c is an arbitrary constant and ϵ is the small parameter. Substituting in Eq. (17), we find

$$\partial_t v = \frac{3\nu - \lambda c}{3} \nabla^2 v, \tag{41}$$

at first order in ϵ. So v obeys a diffusion equation whose diffusion constant depends on c. For $c < 3\nu/2\lambda$, the diffusion constant is positive and correspondingly the constant solution is linearly stable. For $c > 3\nu/2\lambda$, the diffusion constant is negative and consequently the constant solution is unstable. Furthermore, in this case the problem becomes linearly ill posed.

Since for large values of c, the problem becomes linearly ill posed, numerical solutions are not available. In order to solve this disadvantage, we could include a higher order term in our problem. We concentrate on the gradient flow

$$\partial_t u = -\frac{\delta \mathcal{J}}{\delta u} + \xi(x,t), \tag{42}$$

with density

$$\mathcal{J}[u] = \frac{\nu}{2} \int \mathrm{d}x \, (\nabla u)^2 - \frac{\lambda}{6} \int \mathrm{d}x \, u(\nabla u)^2 + \frac{\mu}{2} \int \mathrm{d}x \, (\nabla^2 u)^2, \tag{43}$$

leading to the following equation

$$\partial_t u = \nu \nabla^2 u - \frac{\lambda}{6} (\nabla u)^2 - \frac{\lambda}{3} u \nabla^2 u - \mu \nabla^4 u + \xi(x,t). \tag{44}$$

Note that the deterministic counterpart of this fourth-order equation can be considered as a variational version of the Kuramoto-Sivashinsky equation. It is worth remarking that this ad hoc construction resembles the one that, as indicated in [46], could more formally be obtained by considering the expansion of a nonlocal, short range interaction.

The regime of linear stability/instability of this equation is identical to that of Eq. (33) but in this case, the problem is always linearly well posed. Furthermore, the term proportional to μ is presumably irrelevant in the large spatiotemporal scale (as simple power counting of the linear terms reveals) so the results of the previous RG analysis could possibly hold for this case too. Anyway, due to the presence of the deterministic instability, further analysis are needed in order to assure this (note that both linear terms in the equation are stabilizing and that this instability has its origin in the vertex structure).

VII. Crossover: a path integral point of view

Another recently discussed related aspect [70] is based in a path-integral Monte Carlo-like method for the numerical evaluation of the mean rugosity and other typical averages whose approach, which radically differs from one introduced before [71], exploits some of our previous results [34–36]. Here we limit ourselves to quote the temporally (μ) and spatially (j) discrete form of the "stochastic action"

$$
\begin{aligned}
\mathbf{S}[h] &= \frac{1}{2\tau} \sum_{j,\mu} \{h_{j,\mu+1} - h_{j,\mu} \\
&\quad -\tau[\alpha\mathbf{L}_{j,\mu+1} + (1-\alpha)\mathbf{L}_{j,\mu}]\}^2 \\
&\quad -2\nu\alpha Nt \\
&\quad -\tau\alpha\frac{\lambda}{2}\sum_{j,\mu}[h_{j+1,\mu} - 2h_{j,\mu} + h_{j-1,\mu}],
\end{aligned}
$$

(45)

and briefly discuss the obtained numerical results. τ is the time step, $0 < \alpha < 1$ a time-discretization parameter meant to be fixed for explicit calculation

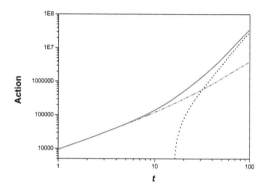

Figure 1: Crossover-like behavior from EW to KPZ regime for $\lambda = 1$, on a lattice of 1028 sites ($\nu = D = 1$). Red solid line: KPZ action; blue dash-dotted line: EW action; black-dotted line: difference.

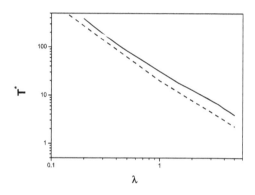

Figure 2: Same data as the previous figure. Solid line: time T^* vs. λ; dashed line: trend for $T^* \sim \lambda^{-1.35}$, included for comparison.

[72,73], and $\mathbf{L}_{j,\mu}$ the "stochastic Lagrangian"

$$
\begin{aligned}
\mathbf{L}_{j,\mu} &= \nu (h_{j+1,\mu} - 2h_{j,\mu} + h_{j-1,\mu}) \\
&\quad +\frac{\lambda}{4}\left[(h_{j+1,\mu} - h_{j,\mu})^2 \right. \\
&\quad \left. +(h_{j,\mu} - h_{j-1,\mu})^2\right].
\end{aligned}
$$

(46)

Figure 1 shows the crossover-like behavior from the Edwards–Wilkinson (EW) regime to the KPZ one. We take as estimator of such a transition the time at which the difference (dotted black line) between KPZ (red solid curve) and EW actions (blue dash-dotted line) crosses the EW one (it grossly coincides with the time at which the asymptotes cross). This estimator numerically agrees neither with the results in [10] (where a value of $\phi \sim 4$ was found) nor with the one in [24] (with $\phi \sim 3$, but corresponding to a 2D case). In Fig. 2 we have plotted the dependence of this estimator on λ. For comparison, we have also included the trend for $\lambda^{-\phi}$ with $\phi = 1.35$ (dotted line). Preliminary results for $\lambda > 7$ seem to indicate a marked change in the value of ϕ, maybe a hint that the system is entering a *strong coupling* region [71].

Short-time propagator

Our aim here is to work out a variant of the method introduced in [70] by exploiting the first form of Lyapunov functional found in [46], namely Eq. (10), that leads us to Eq. (11), the full KPZ equation.

Whereas Eq. (45) is valid whatever the value of τ, we now seek for a simpler expression valid for $\tau \ll 1$. This idea parallels in some sense other studies in the literature [71], but here we exploit the functional $\mathcal{F}[h]$ of Eq. (8). We denote as $\{h\} = (h_{1,\mu}, h_{2,\mu}, \ldots, h_{j,\mu}, \ldots, h_{N,\mu})$ the interface configuration at time μ. The transition pdf between patterns h_0 at t_0 and h_f at t_f can be written as

$$
P(\{h_f\}, t_f | \{h_0\}, t_0) = \\
\int \mathcal{D}[h] \exp\left(-\frac{1}{\gamma} \int_{t_0}^{t_f} \mathcal{L}[h, \dot{h}]\right), \quad (47)
$$

with

$$
\mathcal{L}[h, \dot{h}] = \frac{1}{2} \int_0^L dx \left[\left(\partial_t h + \Gamma[h] \frac{\delta}{\delta h} \mathcal{F}[h]\right)^2 \right.\\
\left. + \alpha \frac{\delta}{\delta h}\left(\Gamma[h] \frac{\delta}{\delta h} \mathcal{F}[h]\right)\right], \quad (48)
$$

whose discrete form is given by Eq. (46). A key observation is the (temporally and spatially) "diagonal" character of Eq. (9), highlighted in its discrete version

$$
\dot{h}_j(t) = -\Gamma_j \frac{\delta \mathcal{F}}{\delta h_j} + \sqrt{\gamma} \, \xi_j(t). \quad (49)
$$

Guided by Eq. (46), we propose the following form of $P(\{h_f\}, t_f | \{h_0\}, t_0)$ for $\tau \ll 1$, or short-time

propagator (STP)

$$
P(h_f, \tau | h_0, 0) = \\
\int_{h_0}^{h_f} \mathcal{D}[h] e^{\left[-\frac{1}{2\gamma} \int_0^\tau ds \int_0^L dx \left(\partial_t h + \Gamma \frac{\delta \mathcal{F}}{\delta h}\right)^2\right]} \\
\approx \exp\left\{-\frac{\tau}{2\gamma} \int_0^L dx \right. \\
\left. \left[\left(\frac{h_f - h_0}{\tau} + \frac{1}{2}\left[\Gamma_f \frac{\delta \mathcal{F}}{\delta h_f} + \Gamma_0 \frac{\delta \mathcal{F}}{\delta h_0}\right]\right)^2\right]\right\}.
$$
(50)

Here, for simplicity, we have chosen a discretization with $\alpha = 0$. As it is well known [72, 73], the Jacobian of the transformation from the noise variable to the height variable depends on α. With this choice, the Jacobian results equal to 1.

Incidentally, the form in Eq. (50) coincides with the discretization used in [71] for determining the least-action trajectory. The "quasi-Gaussian" character of this STP is better evidenced in the following approximate form

$$
P(h_f, t_f = \tau | h_0, t_0 = 0) \sim e^{\left[-\frac{1}{2\gamma\tau} \int_0^L dx (h_f - h_0)^2\right]} \\
\times \left\{1 - \frac{1}{2\gamma} \int_0^L dx \left[(h_f - h_0)\frac{1}{2}\left(\Gamma_f \frac{\delta \mathcal{F}}{\delta h_f}\right.\right.\right. \\
\left.\left.\left. + \Gamma_0 \frac{\delta \mathcal{F}}{\delta h_0}\right) + \mathcal{O}(\tau)\right]\right\}
$$
(51)

where the exponential term has been separated out since it is of order τ^{-1}, whereas the following two are of order τ^0 and τ^1, respectively (of lesser weight and negligible respectively, in the limit $\tau \to 0$). It is worth remarking that the term that could come from the Jacobian is also of order τ^1.

It is easy to check that we can recover the known FPE from the proposed form of STP (adopting $\alpha = 0$ for simplicity). We will not reiterate this calculation here. An immediate result of this form is that at very short times, behavior of the Edwards–Wilkinson type is obtained

$$
\sqrt{\langle h^2 \rangle} \approx \tau^{\frac{1}{2}}.
$$

VIII. Conclusions

Herein, in addition to reviewing some recent results [34–36, 47, 70], we have furthered the study in [46], where it was shown that the deterministic KPZ equation admits a Lyapunov functional, and a (formal) definition of a *nonequilibrium potential* was introduced. We have carried out a Taylor expansion of such a nonequilibrium potential, what led us to a different equation of motion than the KPZ one, the KPZW which is an exact gradient flow and has an explicit density. In particular, it has a lower degree of symmetry: it is neither Galilean invariant, nor even translational invariant. The critical exponents determining its scaling properties were obtained through a one-loop dynamic renormalization group analysis. These exponents fulfill the same scaling relation as the KPZ equation, $\alpha + z = 2$, traditionally attributed to the Galilean invariance of the latter. The fact that the same scaling relation arises in a SPDE (i.e., the KPZW) that is not only non-Galilean invariant but even non-invariant under the translation $u \rightarrow u+$constant supports recent theoretical and numerical results indicating that Galilean invariance does not necessarily play the relevant role previously assumed in defining the universality class of the KPZ equation and different nonequilibrium models [30–36].

We have, moreover, analyzed the stability properties of the solutions to the present equation, finding the threshold condition for the appearance of diffusive instabilities, which indicates that in this case the problem becomes linearly ill posed. After considering the simplest way to correct such an ill-posed problem, we have met a kind of Kuramoto–Sivashinsky equation, resembling the one that, as indicated in [46], could be obtained by considering a nonlocal, short range interaction. This equation has an exact gradient flow structure with an explicit density. Furthermore, when subject to stochastic forcing, its scaling properties could be formally described by the same critical exponents because the stabilizing term is irrelevant in the large scale from a dimensional analysis viewpoint.

Exploiting some elements of a path integral description of the problem, we have also shown what seems to be a simple form of viewing and studying the crossover from the EW to the KPZ regimes.

The present review-like study aims to open new points of view on, as well as alternative routes to study, the KPZ problem. Among the many aspects to be further studied, an interesting one is to test the (kind) of stability of the recently found exact solutions [20–23] by exploiting the indicated form of the NEP.

Acknowledgements - Financial support from MINECO (Spain) is especially acknowledged, through Projects PRI-AIBAR-2011-1323 (which enabled international cooperation), FIS2010-18023 (HSW and CE) and RYC-2011-09025 (CE). Also acknowledged is the support from CONICET, UNC (JAR) and UNMdP (RRD) of Argentina. Collaboration with M.S. de la Lama and E. Korutcheva during different stages of this research is highly appreciated.

---

[1] L D Landau, E M Lifshitz, *Statistical physics*, Butterworth Heinemann, Oxford, 1980.

[2] L P Kadanoff, W Götze, D Hamblen, R Hecht, E A S Leis, V V Palciauskas, M Rayl, J Swift, D Aspnes, J Kane, *Static phenomena near critical points: Theory and experiment*, Rev. Mod. Phys. **39**, 395 (1967).

[3] K G Wilson, J Kogut, *The renormalization group and the ϵ expansion*, Phys. Rep. **12**, 75 (1974); K G Wilson, *The renormalization group: Critical phenomena and the Kondo problem*, Rev. Mod. Phys. **47**, 773 (1975).

[4] J Marro, R Dickman, *Nonequilibrium phase transitions in lattice models*, Cambridge U. Press, Cambridge, UK (1999); M Henkel, H Hinrichsen, S Lübeck, *Nonequilibrium phase transitions - I*, Springer, Berlin (2008); M Henkel, M Pleimling, *Nonequilibrium phase transitions - II*, Springer, Berlin (2010); G Ódor, *Universality in nonequilibrium lattice systems*, World Scientific, Singapore (2008).

[5] M Kardar, G Parisi, Y-C Zhang, *Dynamic scaling of growing interfaces*, Phys. Rev. Lett. **56**, 889 (1986).

[6] T Halpin-Healy, Y-C Zhang, *Kinetic roughening phenomena, stochastic growth, directed polymers and all that. Aspects of multidisciplinary statistical mechanics*, Phys. Rep. **254**, 215 (1995).

[7] A-L Barabási, H E Stanley, *Fractal concepts in surface growth*, Cambridge U. Press, Cambridge, UK (1995).

[8] V Gurarie, A Migdal, *Instantons in the Burgers equation,* Phys. Rev. E **54**, 4908 (1996).

[9] M Kardar, *Replica Bethe ansatz studies of two-dimensional interfaces with quenched random impurities,* Nucl. Phys. B **290**, 582 (1987).

[10] B M Forrest, R Toral, *Crossover and finite-size effects in the (1+1)-dimensional Kardar–Parisi–Zhang equation,* J. Stat. Phys. **70**, 703 (1993).

[11] M Beccaria, G Curci, *Numerical simulation of the Kardar–Parisi–Zhang equation,* Phys. Rev. E **50**, 4560 (1994).

[12] K Moser, D E Wolf, *Vectorized and parallel simulations of the Kardar–Parisi–Zhang equation in 3+ 1 dimensions,* J. Phys. A **27**, 4049 (1994).

[13] M Scalerandi, P P Delsanto, S Biancotto, *Time evolution of growth phenomena in the KPZ model,* Comput. Phys. Commun. **97**, 195 (1996).

[14] T J Newman, A J Bray, *Strong-coupling behaviour in discrete Kardar–Parisi–Zhang equations,* J. Phys. A **29**, 7917 (1996).

[15] C Appert, *Universality of the growth velocity distribution in 1+ 1 dimensional growth models,* Comput. Phys. Commun. **121-122**, 363 (1999).

[16] E Marinari, A Pagnani, G Parisi, *Critical exponents of the KPZ equation via multi-surface coding numerical simulations,* J. Phys. A **33**, 8181 (2000).

[17] T J Oliveira, S G Alves, S C Ferreira, *Kardar–Parisi–Zhang universality class in (2+1) dimensions: Universal geometry-dependent distributions and finite-time corrections,* Phys. Rev. E **87**, 040102 (2013).

[18] L Giada, A Giacometti, M Rossi, *Pseudospectral method for the Kardar–Parisi–Zhang equation,* Phys. Rev. E **65**, 036134 (2002).

[19] V G Miranda, F D A Aarão Reis, *Numerical study of the Kardar–Parisi–Zhang equation,* Phys. Rev. E **77**, 031134 (2008).

[20] T Sasamoto, H Spohn, *One-dimensional Kardar–Parisi–Zhang equation: an exact solution and its universality,* Phys. Rev. Lett. **104**, 230602 (2010).

[21] T Sasamoto, H Spohn, *The 1+1-dimensional Kardar–Parisi–Zhang equation and its universality class,* J. Stat. Mech. P11013 (2010).

[22] G Amir, I Corwin, J Quastel, *Probability distribution of the free energy of the continuum directed random polymer in 1+ 1 dimensions,* Commun. Pure Appl. Math. **64**, 466 (2011).

[23] P Calabrese, P Le Doussal, *Exact solution for the Kardar–Parisi–Zhang equation with flat initial conditions,* Phys. Rev. Lett. **106**, 250603 (2011).

[24] H Guo, B Grossmann, M Grant, *Crossover scaling in the dynamics of driven systems,* Phys. Rev. A **41**, 7082 (1990).

[25] C M Horowitz, E V Albano, *Relationships between a microscopic parameter and the stochastic equations for interface's evolution of two growth models,* Eur. Phys. J. B **31**, 563 (2003).

[26] F D A Aarão Reis, *Scaling in the crossover from random to correlated growth,* Phys. Rev. E **73**, 021605 (2006).

[27] S Bustingorry, L Cugliandolo, J L Iguain, *Out-of-equilibrium relaxation of the Edwards-Wilkinson elastic line,* J. Stat. Mech. P09008 (2007); S Bustingorry, *Aging dynamics of nonlinear elastic interfaces: The Kardar–Parisi–Zhang equation,* J. Stat. Mech. 10002 (2007).

[28] S Bustingorry, P LeDoussal, A Rosso, *Universal high-temperature regime of pinned elastic objects,* Phys. Rev. B **82**, 140201 (2010); S Bustingorry, A B Kolton, T Giamarchi, *Random-manifold to random-periodic depinning of an elastic interface,* Phys. Rev. B **82**, 094202 (2010).

[29] M Henkel, J D Noh, N Pleimling, *Phenomenology of aging in the Kardar–Parisi–Zhang equation,* Phys. Rev. E **85**, 030102 (2012).

[30] W D McComb, *Galilean invariance and vertex renormalization in turbulence theory,* Phys. Rev. E **71**, 037301 (2005).

[31] A Berera, D Hochberg, *Gauge symmetry and Slavnov-Taylor identities for randomly stirred fluids,* Phys. Rev. Lett. **99**, 254501 (2007).

[32] A Berera, D Hochberg, *Gauge fixing, BRS invariance and Ward identities for randomly stirred flows,* Nucl. Phys. B **814**, 522 (2009).

[33] M Nicoli, R Cuerno, M Castro, *Unstable nonlocal interface dynamics,* Phys. Rev. Lett. **102**, 256102 (2009).

[34] H S Wio, J A Revelli, R R Deza, C Escudero, M S de La Lama, *KPZ equation: Galilean-invariance violation, consistency, and fluctuation-dissipation issues in real-space discretization,* Europhys. Lett. **89**, 40008 (2010).

[35] H S Wio, J A Revelli, R R Deza, C Escudero, M S de La Lama, *Discretization-related issues in the Kardar–Parisi–Zhang equation: Consistency, Galilean-invariance violation, and fluctuation-dissipation relation,* Phys. Rev. E **81**, 066706 (2010).

[36] H S Wio, C Escudero, J A Revelli, R R Deza, M S de La Lama, *Recent developments on the Kardar–Parisi–Zhang surface-growth equation,* Phil. Trans. R. Soc. A **369**, 396 (2011).

[37] T Sun, H Guo, M Grant, *Dynamics of driven interfaces with a conservation law,* Phys. Rev. A **40**, 6763 (1989).

[38] J Villain, *Continuum models of crystal growth from atomic beams with and without desorption,* J. Phys. I (France) **1**, 19 (1991).

[39] Z-W Lai, S Das Sarma, *Kinetic growth with surface relaxation: Continuum versus atomistic models,* Phys. Rev. Lett. **66**, 2348 (1991).

[40] C Escudero, *Geometric principles of surface growth,* Phys. Rev. Lett. **101**, 196102 (2008).

[41] C Escudero, E Korutcheva, *Origins of scaling relations in nonequilibrium growth,* J. Phys. A: Math. Theor. **45**, 125005 (2012).

[42] R Graham, *Weak noise limit and nonequilibrium potentials of dissipative dynamical systems,* In: Instabilities and nonequilibrium structures, Eds. E Tirapegui, D Villaroel, D, Reidel Pub. Co., Dordrecht (1987).

[43] H S Wio, *Nonequilibrium potential in reaction-diffusion systems,* In: 4th Granada seminar in computational physics, Eds. P Garrido, J Marro, Pag. 135, Springer-Verlag, Berlin (1997).

[44] H S Wio, R R Deza, J M López, *Introduction to stochastic processes and nonequilibrium statistical physics,* Revised Edition, World Scientific, Singapore (2013).

[45] P Ao, *Potential in stochastic differential equations: novel construction,* J. Phys. A **37**, L25 (2004).

[46] H S Wio, *Variational formulation for the KPZ and related kinetic equations,* Int. J. Bif. Chaos **19**, 2813 (2009).

[47] C Escudero, E Korutcheva, H S Wio, R R Deza, J A Revelli, *KPZ equation as a gradient flow: Nonequilibrium-potential expansion and renormalization-group treatment of fluctuations,* unpublished.

[48] P Hohenberg, B Halperin, *Theory of dynamic critical phenomena,* Rev. Mod. Phys. **49**, 435 (1977).

[49] G Grinstein, S K Ma, *Surface tension, roughening, and lower critical dimension in the random-field Ising model,* Phys. Rev. B **28**, 2588 (1983).

[50] J Koplik, H Levine, *Interface moving through a random background,* Phys. Rev. B **32**, 280 (1985).

[51] R Bruinsma, G Aeppli, *Interface motion and nonequilibrium properties of the random-field Ising model,* Phys. Rev. Lett. **52**, 1547 (1984).

[52] D Kessler, H Levine, Y Tu, *Interface fluctuations in random media*, Phys. Rev. A **43**, 4551 (1991).

[53] H S Wio, S Bouzat, B Von Haeften, *Stochastic resonance in spatially extended systems: the role of far from equilibrium potentials*, Physica A **306**, 140 (2002).

[54] H S Wio, R R Deza, *Aspects of stochastic resonance in reaction-diffusion systems: The nonequilibrium-potential approach*, Eur. Phys. J.-Spec. Top. **146**, 111 (2007).

[55] H G E Hentschel, *Shift invariance and surface growth*, J. Phys. A: Math. Gen. **27**, 2269 (1994).

[56] S J Linz, M Raible, P Hänggi, *Stochastic field equation for amorphous surface growth*, In: Stochastic processes in physics, chemistry, and biology, Eds. J A Freund, T Pöschel, **557**, Pag. 473, Springer, Berlin (2000).

[57] J M López, M Castro, R Gallego, *Scaling of local slopes, conservation laws, and anomalous roughening in surface growth*, Phys. Rev. Lett. **94**, 166103 (2005).

[58] M Castro, J Muñoz-García, R Cuerno, M M García-Hernández, L Vázquez, *Generic equations for pattern formation in evolving interfaces*, New J. Phys. **9**, 102 (2007).

[59] E Hernández-García, T Ala-Nissila, M Grant, *Interface roughening with a time-varying external driving force*, Europhys. Lett. **21**, 401 (1993).

[60] C-H Lam, F G Shin, *Improved discretization of the Kardar–Parisi–Zhang equation*, Phys. Rev. E **58**, 5592 (1998); C-H Lam, F G Shin, *Formation and dynamics of modules in a dual-tasking multilayer feed-forward neural network*, Phys. Rev. E, **57**, 6506 (1998).

[61] R Gallego, M Castro, J M López, *Pseudospectral versus finite-difference schemes in the numerical integration of stochastic models of surface growth*, Phys. Rev. E **76**, 051121 (2007).

[62] S M A Tabei, A Bahraminasab, A A Masoudi, S S Mousavi, M R R Tabar, *Intermittency of height fluctuations in stationary state of the Kardar–Parisi–Zhang equation with infinitesimal surface tension in 1+1 dimensions*, Phys. Rev. E **70**, 031101 (2004).

[63] K Ma, J Jiang, C B Yang, *Scaling behavior of roughness in the two-dimensional Kardar–Parisi–Zhang growth*, Physica A **378**, 194 (2007).

[64] M S de la Lama, J M López, J J Ramasco, M A Rodríguez, *Activity statistics of a forced elastic string in a disordered medium*, J. Stat. Mech., P07009 (2009).

[65] M Abramowitz, I A Stegun, *Handbook of mathematical functions: With formulas, graphs, and mathematical tables*, Pag. 884, Dover, New Tork (1965).

[66] M San Miguel, R Toral, *Stochastic effects in physical systems*, In: Instabilities and nonequilibrium structures VI, Eds. E Tirapegui, J Martínez-Mardones, R Tiemann, Pag. 35, Kluwer Academic Publishers (2000).

[67] D Forster, D R Nelson, M J Stephen, *Large-distance and long-time properties of a randomly stirred fluid*, Phys. Rev. A **16**, 732 (1977).

[68] E Medina, T Hwa, M Kardar, Y-C Zhang, *Burgers equation with correlated noise: Renormalization-group analysis and applications to directed polymers and interface growth*, Phys. Rev. A **39**, 3053 (1989).

[69] L Canet, H Chate, B Delamotte, *General framework of the non-perturbative renormalization group for non-equilibrium steady states*, J. Phys. A **44**, 495001 (2011); L Canet, H Chate, B Delamotte, N Wschebor, *Nonperturbative renormalization group for the Kardar–Parisi–Zhang equation: General framework and first applications*, Phys. Rev. E **84**, 061128 (2011); Th Kloss, L Canet, N Wschebor, *Nonperturbative renormalization group for the stationary Kardar–Parisi–Zhang equation: Scaling functions and amplitude ratios in 1+1, 2+1, and 3+1 dimensions*, Phys. Rev. E **86**, 051124 (2012).

[70] H S Wio, R R Deza, J A Revelli, C Escudero, *A novel approach to the KPZ dynamics*, Acta Phys. Pol. B **44** 889 (2013).

[71] H C Fogedby, W Ren, *Minimum action method for the Kardar–Parisi–Zhang equation*, Phys. Rev. E **80**, 041116 (2009).

[72] F Langouche, D Roekaerts, E Tirapegui, *Functional integration and semiclassical expansions*, D. Reidel Pub. Co., Dordrecht (1982).

[73] H S Wio, *Path integrals for stochastic processes: An introduction*, World Scientific, Singapore (2013).

4

Study of the characteristic parameters of the normal voices of Argentinian speakers

E. V. Bonzi,[1,2*] G. B. Grad,[1†] A. M. Maggi,[3‡] M. R. Muñóz[3§]

The voice laboratory permits to study the human voices using a method that is objective and noninvasive. In this work, we have studied the parameters of the human voice such as pitch, formant, jitter, shimmer and harmonic-noise ratio of a group of young people. This statistical information of parameters is obtained from Argentinian speakers.

I. Introduction

The voice is a multidimensional phenomenon that must be evaluated using special tools for determining acoustic parameters. These parameters are: the pitch or voice tone, the timbre, considered as the personality of the voice that is particular of each person (determined by fundamental frequency, its harmonics and formants) and the degree of hoarseness.

During sustained vibration, the vocal fold will exhibit variations of fundamental frequency and amplitude; these phenomena are called "frequency perturbation" (jitter) and "amplitude perturbation" (shimmer). They reflect fluctuations in tension and biochemical characteristics of the vocal folds, as well as variation in their neural control and the physiological properties of the individuals voices.

The acoustic analysis is one of the major advances in the study of voice, increasing the accuracy of diagnosis in this area. Normal values as standards are important and necessary to guide voice professionals.

There are not many studies performed for the Latin languages [1–3]. However, there are several of them for the English language, such as those in Refs. [4–8].

In the same way, the software used for voice therapy is in general designed for other languages than Spanish. A comparison has been made, though, between the two vowel systems of English and Spanish (the variation spoken in Madrid, Spain), which triggered relatively large versus small vowel inventories [9]. That is the reason why we consider it is very important and necessary to produce more results for the Spanish speaking population.

We analyzed 72 audio files of female and male voices from an Argentinian Spanish speaking population to obtain the acoustical parameters using the Praat program [10]. Our data were compared to Bradlow [9], Hualde [11] and Casado Morente *et al.* [12]. The pitches measured were lower than expected and the First formant of the /a/ and /u/

*E-mail: bonzie@famaf.unc.edu.ar
†E-mail: grad@famaf.unc.edu.ar
‡E-mail: alicia.maggi@hotmail.com
§E-mail: eudaimonia13@hotmail.com

[1] Facultad de Matemática, Astronomía y Física, Universidad Nacional de Córdoba, Ciudad Universitaria, 5000 Córdoba, Argentina.

[2] Instituto de Física Enrique Gaviola (CONICET), Ciudad Universitaria, 5000 Córdoba, Argentina.

[3] Escuela de Fonoaudiología, Facultad de Ciencias Médicas, Universidad Nacional de Córdoba, Ciudad Universitaria, 5000 Córdoba, Argentina.

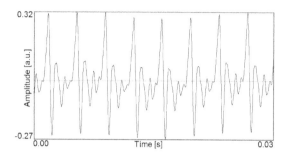

Figure 1: Wave shape of the /a/ sound.

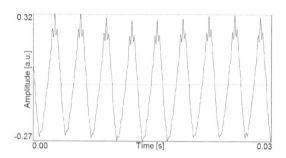

Figure 2: Wave shape of the /i/ sound.

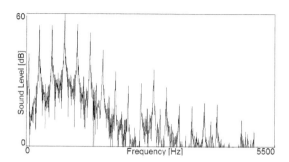

Figure 3: Harmonics of the /a/ vowel.

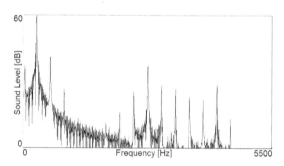

Figure 4: Harmonics of the /i/ vowel.

vowels is higher than the published data. Additionally, the Harmonic to Noise Ratio (HNR) values discriminated per vowel are presented.

II. Measurement methodology

Pitch, First and Second formants, Jitter, Shimmer and Harmonic to Noise Ratio (HNR) are the cornerstones of acoustic measurement of voice signals, and are often regarded as indices of the perceived quality of both normal and pathological voices [13].

In this work, we analyzed the audio files from the five Spanish vowels produced by 72 female and male individuals, in order to study the parameters previously mentioned. The individuals are Argentinian university students whose ages range between 20 and 30, coming from different regions without any special geographical distribution.

The voices were recorded using a Behringer C-1U (USB) cardioid microphone and a notebook.

The microphone was placed at a distance of 10 cm respect to the mouth of the subjects while they were pronouncing the vowels with an intensity and tone that was comfortable in an acoustically treated room. Each sound was sustained for,

at least, five seconds.

The Praat program, commonly used in linguistics for the scientific analysis of the human voice [10], was used to record, analyze the wav files and obtain all the parameters presented in this work. A sample rate of 44100 Hz was used to record the sound file.

The wave shapes of the sounds corresponding to /a/ and /i/ vowels are shown in Figs. 1 and 2. In Figs. 3 and 4, the harmonic components obtained by applying Fourier Transform to the respective vowel signal are shown.

Pitch

The pitch is a perceptual attribute of sound closely related to frequency, being this perception a subjective notion.

In psychoacoustics, the pitch is related to the fundamental frequency of vibration of the vocal cords, allowing the perception of the tone frequency.

Nevertheless, for Praat program [10], the pitch is coincident with the fundamental harmonic of the wave and we used this definition in this work.

This parameter depends on gender, being higher for women and lower for men.

Formants

The voice is created in the vocal cord, shaped as complex sound with harmonics and modified in the vocal tract by the resonating frequencies. Then, the amplitude of harmonics frequencies are enveloped forming a spectrum of energy, the peaks or maximum observed in these spectra are named "formants." Consequently, a formant is a concentration of acoustic energy around a particular frequency in the speech wave. There are several formants, each one at a different frequency corresponding to a resonance in the vocal tract, and especially the first two are related to the movement of the tongue. The high-low magnitude of the First one (F1) is inversely related to the up-down tongue position and the Second formant (F2) is related to the front tongue position.

Jitter and Shimmer

The naturalness factor of sustained vowels is attributed to a fundamental frequency and the signal amplitude. Still there are unwanted variations in time of the sound signal properties in the voice production.

While jitter indicates the variability or perturbation of fundamental frequency, shimmer refers to the same perturbation but, in this case, related to amplitude of sound wave, or intensity of vocal emission. Jitter is affected mainly by lack of control of vocal fold vibration and shimmer by reduction of glottic resistance and mass lesions in the vocal folds, which are related to the presence of noise at emission and breathiness [10, 14].

Harmonic to Noise Ratio - HNR

The amount of energy conveyed in the fundamental frequency (f_0) and its harmonics, divided by the energy in noise frequencies, is defined as the harmonic-to-noise ratio. Frequencies that are not integer multiples of f_0 are regarded as noise. This parameter is related to the perception of vocal roughness and hoarseness [10].

Normal voices have a low level of noise and high HNR. On the contrary, the degree of hoarseness increases the noise component and decreases HNR.

III. Results and Discussion

The measured data were processed statistically and the results are shown in the Tables 1, 4, 5, 6 and Figs. 5 and 6.

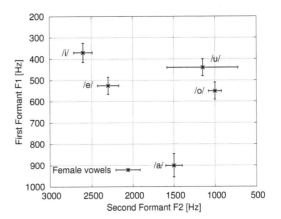

Figure 5: Female formant chart.

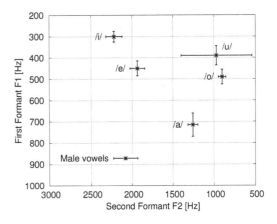

Figure 6: Male formant chart.

The pitches for female and male individuals are shown in Table 1. We used the minimum and maximum values to address the dispersion instead of the standard deviation because the data distribution was not normal. Our values are in general lower for both genders compared to the published data [9, 11, 12].

Tables 2 and 3 show the First and Second formants values and Figs. 5 and 6 show the chart of formants corresponding to female and male populations obtained in this work.

We have compared our male results with formant data of male Spanish speakers published by Bradlow [9].

In general, the First (F1) and Second (F2) formants values are comparable to the published ones.

In particular, the F1 formants for the /a/ and

	Female	Male
Maximum	314	196
Medium	225	128
Minimum	155	85

Table 1: Pitch values of female and male subjects in Hz.

/u/ vowels are higher than the reported ones, 12 and 21 %, respectively.

The Second formant, F2, for the /o/ vowel is lower than Bradlow by 12 %.

On the other hand, we cannot compare our female formant values with published results because we could not find results for female individuals in the literature. Comparing female versus male F1 formants, we observed that most of them are higher by 20 % but in the case of the /o/ vowel the difference is 11 %.

Comparing F2 formants, the female values are higher than the male ones, reaching almost the 25 % for /a/ and /i/ vowels.

Furthermore, the F2 of the /u/ vowel in our samples show an important scatter for both genders, female and male.

In the Tables 4 and 5, the obtained Jitter and Shimmer values for each vowel are shown. They are comparable to the Jitter and Shimmer averages obtained by Casado Morente et al. [12] in a study that involves a group of normal people. In our work, we have observed that the Jitter and the Shimmer values of the /a/ vowel are bigger than the corresponding ones of the other vowels.

Finally, the HNR results, see Table 6, are according to the average value presented by Casado Morente et al. [12]. However, we could not find in the bibliography the HNR values for each of the five Spanish vowels, so we had to make the comparison with the average of them. In the present work, we have found that the vowels show an increasing HNR value from /a/ to /u/, meaning that /u/ has better signal to noise ratio than the other vowels.

IV. Concluding remarks

The objective of this research was to measure acoustical properties of the Spanish voices of Argentinian speakers.

Vowels	F1 [Hz]	F2 [Hz]
/i/	370 ± 45	2600 ± 110
/e/	525 ± 40	2300 ± 130
/a/	900 ± 55	1500 ± 100
/o/	550 ± 40	1000 ± 80
/u/	440 ± 40	1150 ± 430

Table 2: First and Second formant of female.

Vowels	F1 [Hz]	F2 [Hz]
/i/	300 ± 25	2220 ± 100
/e/	450 ± 35	1935 ± 90
/a/	715 ± 55	1260 ± 60
/o/	490 ± 35	900 ± 45
/u/	390 ± 45	970 ± 430

Table 3: First and Second formant of male.

Vowels	Shimmer Local [%]	Jitter Local [%]
/a/	2.7 ± 1.1	0.31 ± 0.10
/e/	2.1 ± 0.7	0.28 ± 0.08
/i/	2.2 ± 0.6	0.29 ± 0.07
/o/	2.0 ± 0.7	0.26 ± 0.11
/u/	2.1 ± 0.7	0.27 ± 0.09

Table 4: Shimmer and Jitter of female subjects.

Vowels	Shimmer Local [%]	Jitter Local [%]
/a/	3.0 ± 0.9	0.36 ± 0.10
/e/	2.3 ± 0.8	0.33 ± 0.09
/i/	2.3 ± 0.7	0.28 ± 0.08
/o/	2.2 ± 0.8	0.29 ± 0.10
/u/	2.3 ± 0.9	0.25 ± 0.07

Table 5: Shimmer and Jitter of male subjects.

Vowels	Female	Male
/a/	21 ± 3	20 ± 2
/e/	20 ± 2	21 ± 2
/i/	22 ± 3	22 ± 2
/o/	25 ± 3	24 ± 3
/u/	25 ± 4	25 ± 3

Table 6: Harmonic to Noise Ratio of female and male subjects in dB.

These voice parameters are generally assessed subjectively by several authors. This form of perceptual analysis of voice has significant limitations and the subtle interpretative judgments of verbal classifications may not be accurate.

The differences we found in the parameters of the vowels measured in a group of people from Argentina compared to the parameters obtained from Spanish speaking people living in Spain suggests the region of study has an important influence in the results, as expected.

This kind of studies are very useful to compare the properties of normal and pathological voices of people from different regions.

It is necessary to test the same parameters in female Spanish speakers as well.

Such work should be performed in larger quantities and should be extended to other countries or regions of Latin America, especially where different ethnic groups can be found.

---

[1] W R Rodríguez, O Saz, E Lleida, *Análisis robusto de la voz infantil con aplicación en terapia de voz*, Areté **10**, 70 (2010).

[2] T Cervera, J L Miralles, J González-Álvarez, *Acoustical analysis of Spanish vowels produced by laryngectomized subjects*, J. Speech Lang. Hear Res. **44**, 988 (2001).

[3] J Muñoz, E Mendoza, M D Fresneda, G Carballo, P Lopez, *Acoustic and perceptual indicators of normal and pathological voice*, Folia Phoniatr. Logop. **55**, 102 (2003).

[4] H K Vorperian, R D Kent *Vowel acoustic space development in children: A synthesis of acoustic and anatomic data*, J. Speech Lang. Hear Res. **50**, 1510 (2007).

[5] S P Whiteside, *Sex-specific fundamental and formant frequency patterns in a cross-sectional study*, J. Acoust. Soc. Am. **110**, 464 (2001).

[6] P White, *Formant frequency analysis of childrens spoken and sung vowels using sweeping fundamental frequency production*, J. Voice, **13**, 570 (1999).

[7] R O Coleman, *Male and female voice quality and its relationship to vowel formant frequencies*, J. Speech Lang. Hear Res. **14**, 565 (1971).

[8] S Bennett, *Vowel formant frequency characteristics of preadolescent males and females*, J. Acoust. Soc. Am. **69** 231 (1981).

[9] A R Bradlow *A comparative acoustic study of English and Spanish vowels*, J. Acoust. Soc. Am. **97**, 1916 (1995).

[10] P Boersma, D Weenink, *Praat: doing phonetics by computer [Computer program]*, Version 5.3.51, retrieved 2 June 2013 from http://www.praat.org/.

[11] J I Hualde, *The sounds of Spanish*, Cambridge University Press, Cambridge (2005).

[12] J C Casado Morente, J A Adrián Torres, M Conde Jiménez, D Piédrola Maroto, V Povedano Rodríguez, E Muñoz Gomariz, E Cantillo Baños, A Jurado Ramos, *Estudio objetivo de la voz en población normal y en la disfonía por nódulos y pólipos vocales*, Acta Otorrinolaringol. Esp. **52**, 476 (2001).

[13] J Kreimana, B R Gerrattb, *Perception of aperiodicity in pathological voice*, J. Acoust. Soc. Am. **117**, 2201 (2005).

[14] H F Wertzner, S Schreiber, L Amaro, *Analysis of fundamental frequency, jitter, shimmer and vocal intensity in children with phonological disorders*, Rev. Bras. Otorrinolaringol. **71**, 582 (2005).

5

Critical phenomena in the spreading of opinion consensus and disagreement

A. Chacoma,[1] D. H. Zanette[1,2]*

We consider a class of models of opinion formation where the dissemination of individual opinions occurs through the spreading of local consensus and disagreement. We study the emergence of full collective consensus or maximal disagreement in one- and two-dimensional arrays. In both cases, the probability of reaching full consensus exhibits well-defined scaling properties as a function of the system size. Two-dimensional systems, in particular, possess nontrivial exponents and critical points. The dynamical rules of our models, which emphasize the interaction between small groups of agents, should be considered as complementary to the imitation mechanisms of traditional opinion dynamics.

I. Introduction

The remarkable regularities observed in many human social phenomena —which, in spite of the disparate behavior of individual human beings, emerge as a consequence of their interactions— have since long attracted the attention of physicists and applied mathematicians. Collective manifestations of human behavior have been mathematically modeled in a variety of socioeconomic processes, such as opinion formation, decision making, resource allocation, cultural and linguistic evolution, among many others, often using the tools provided by statistical physics [1]. The stylized nature of these models emphasizes the identification of the generic mechanisms at work in human interactions, as well as the detection of broadly significant fea-

*E-mail: zanette@cab.cnea.gov.ar

[1] Instituto Balseiro and Centro Atómico Bariloche, 8400 San Carlos de Bariloche, Río Negro, Argentina.

[2] Consejo Nacional de Investigaciones Científicas y Técnicas, Argentina.

tures in their macroscopic outcomes. They provide the key to a deep insight into the common elements that underlie those processes.

Models of opinion formation constitute a central paradigm in the mathematical description of social processes from the viewpoint of statistical physics. Starting in the seventies and eighties [2–5], much work —which we cannot aim at inventorying here, but which has been comprehensibly reviewed in recent literature [1]— has exploited the formal resemblance between opinion spreading and spin dynamics in order to apply well-developed statistical techniques to the analysis of such models.

The key mechanism driving most agent-based models of opinion formation is imitation. For instance, in the voter model —to which we refer several times in the present paper— the basic interaction event consists in an agent copying the opinion of another agent chosen at random from a specified neighborhood. At any given time, the opinion of each agent adopts one of two values, typically denoted as ± 1. The voter model can be exactly solved for populations of agents distributed over regular (hyper)cubic arrays in any dimension

[6]. For infinitely large populations, it is character-ized by the conservation of the average opinion. In one dimension, a finite population always reaches an absorbing state of full collective consensus, all agents sharing the same opinion. The probability of final consensus on either opinion coincides with the initial fraction of agents with that opinion, and the time needed to reach the absorbing state is of the order of the population size squared [1].

In this paper, we present an introductory anal-ysis of a class of models where opinion dynamics is driven by the spreading of consensus and dis-agreement, rather than by the dissemination of in-dividual opinions. The basic concept behind these models is that agreement of individual opinions in a localized portion of the population may promote the emergence of consensus in the neighborhood while, in contrast, local disagreement may inhibit the growth of, or even decrease, the degree of con-sensus in the surrounding region. In real social systems, the mechanism of consensus and disagree-ment spreading should be complementary to the direct transmission of opinions between individual agents. In our models, however, we disregard the latter to focus on the dynamical effects of the for-mer.

Since the degree of consensus can only be defined for two or more agents, the spreading of consensus and disagreement engages groups of agents rather than individuals. Such groups are, thus, the ele-mentary entities involved in the social interactions [7–11]. We stress that several other social phenom-ena — related, notably, to decision making [10] and resource allocation [12]— are also based on group interactions that cannot be reduced to two-agent events. In the class of models analyzed here, each interaction event is conceived to occur between two groups: an *active* group G and a *reference* group G'. As a result of the interaction, the agents in G change their individual opinions in such a way that the level of consensus in G approaches that of G'. This generic mechanism extends dynamical rules where the opinion of each single agent changes in response to the collective state of a reference group [1, 8, 13, 14]. The size and internal structure of the interacting groups, as well as the precise way in which opinions are modified in the active group with respect to the reference group, defines each model in this class. For the sake of concrete-ness, we limit the analysis to systems where, as in

the voter model, individual opinions can adopt two values (± 1). In the next section, we analyze the case where both the active group and the reference group are formed by two agents, and the popula-tion is structured as a one-dimensional array. In this case, the system admits stationary absorbing states of full consensus and maximal disagreement, with simple scaling laws with the population size. In Section III., we study a two-dimensional version of the same kind of model with larger groups, where nontrivial critical phenomena —not present in the one-dimensional case— emerge. Results and per-spectives are summarized in the final section.

II.   Two-agent   groups   on   one-dimensional arrays

We begin by considering the simple situation where each of the two groups involved in each interaction event is formed by just two agents. The situation within each group, thus, is one of either full con-sensus (when the two agents bear the same opinion, either $+1$ or -1) or full disagreement (when their opinions are different). We take a population where agents are distributed on a one-dimensional array, and consecutively labeled from 1 to N. Periodic boundary conditions are applied at the ends. At each time step, we choose four contiguous agents, say, $i-1$ to $i+2$. The central pair i, $i+1$ acts as the reference group G'. If they are in disagreement, the agents $i-1$ and $i+2$ respectively adopt the opinions opposite to those of i and $i+1$ with probability p_D, while with the complementary probability $1 - p_D$ nothing happens. If, on the other hand, i and $i+1$ agree with each other, $i-1$ and $i+2$ copy the common opinion in G' with probability p_C, while with probability $1 - p_C$ nothing happens. In this way, both consensus and disagreement spread from G' outwards, to the left and right. The probabili-ties p_C and p_D control the relative frequency with which consensus and disagreement are effectively transmitted. The left panel of Fig. 1 illustrates the states of the four consecutive agents in the two possible outcomes of the interaction (up to opinion inversions).

It is not difficult to realize that, for $p_D = p_C = 1$, our one-dimensional array is equivalent to two in-tercalated subpopulations —respectively occupying even and odd sites— each of them evolving accord-

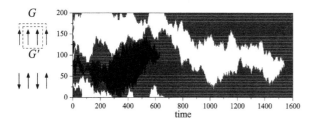

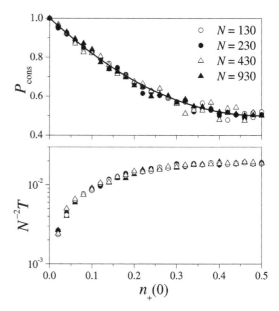

Figure 1: Left: The two possible outcomes of the interaction, up to opinion inversions, for four consecutive agents along the one-dimensional array. The active and the reference groups, G and G', are respectively formed by the outermost and innermost agents. Right: Time evolution of a 200-agent array with $n_+(0) = 0.5$ and $p_D = p_C = 1$. Black and white dots correspond, respectively, to opinions $+1$ and -1. At time $t = 1534$, an absorbing state of maximal disagreement is reached.

Figure 2: Numerical results for consensus and disagreement spreading on a one-dimensional array with $p_D = p_C = 1$, obtained from 10^3 realizations for each parameter set (see text for details). Upper panel: Probability of reaching full consensus, P_{cons}, as a function of the initial fraction of agents with opinion $+1$, $n_+(0)$, for four values of the population size N. Lower panel: Total time T needed to reach the final absorbing state, normalized by the squared population size N^2. Since both P_{cons} and $N^{-2}T$ are symmetric with respect to $n_+(0) = 1/2$, only the lower half of the horizontal axis is shown.

ing to the voter model. The dynamical rules are reduced in this case to binary interactions between agents. In fact, whatever the opinions in group G' at each interaction event, agent $i-1$ and $i+2$ respectively copy the opinions of $i+1$ and i. Now, since the voter model always leads a finite population to an absorbing state of full consensus, the final state of our system can be one of full consensus on either opinion, or a state of maximal disagreement where opposite opinions alternate over the sites of the one-dimensional array. In the latter, the two neighbors of each agent with opinion $+1$ have opinion -1 and vice versa. The right panel of Fig. 1 shows the evolution of a 200-agent array for $n_+(0) = 0.5$ and $p_D = p_C = 1$, black and white dots respectively corresponding to opinions $+1$ and -1. At any given time, the population is divided into well-defined domains either of consensus in one of the opinions or disagreement. Note that the domain boundaries show the typical diffusive motion found in stochastic coarsening processes [1, 15].

Taking into account that, in the voter model, the probability of ending with full consensus on opinion $+1$ is given by the initial fraction of agents with that opinion, $n_+(0)$, and assuming that the initial distribution of opinions is homogeneous over the array, the probability that our system ends in a state of full consensus on either opinion is $P_{\text{cons}} = n_+^2(0) + n_-^2(0) = 1 - 2n_+(0) + 2n_+^2(0)$. Note

that this coincides with the probability that, in the initial state, any two contiguous agents are in consensus. Moreover, we know that the time needed to reach an absorbing state in the one-dimensional voter model is proportional to N^2, a result that should also hold in our case.

The upper panel of Fig. 2 shows numerical results for the probability of final full consensus P_{cons}, determined as the fraction of realizations that ended in full consensus out of 10^3 runs, as a function of $n_+(0)$ and for several population sizes N. The curve is the analytic prediction given above. The result is analogous to the probability of final consensus found in Sznajd-type models [13]. The lower panel shows the total time T needed to reach the final absorbing state (of either consensus or disagree-

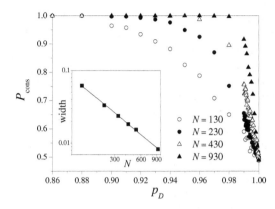

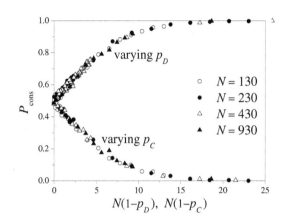

Figure 3: Probability of reaching full consensus, P_{cons}, as a function of the probability p_D, with $p_C = 1$ and for four values of the system size N. Results were obtained averaging over 10^3 realizations for each parameter set. Insert: Width of the variation range of P_{cons} as a function of N. The straight line has slope -1.

Figure 4: Probability of reaching full consensus, P_{cons}, as a function of $N(1-p_D)$ when varying p_D with $p_C = 1$, and as a function of $N(1-p_C)$ when varying p_C with $p_D = 1$.

ment), averaged over 10^3 realizations and normalized by N^2. As expected, both P_{cons} and $N^{-2}T$ are independent of the population size.

When $p_D \neq p_C$, the two intercalated subpopulations cannot be considered independent of each other any more. If $p_D < p_C$, for instance, an opinion prevailing in one of the subpopulations will invade the other subpopulation faster than the opposite opinion, thus favoring the establishment of collective consensus. To analyze this asymmetric situation, we first fix $p_C = 1$ and let p_D vary in $(0, 1)$, so that the spreading of consensus is more probable than that of disagreement. The main plot in Fig. 3 shows numerical results for P_{cons}, measured as explained above, as a function of p_D and for four values of N. In all the realizations, $n_+(0) = 0.5$, and the two opinions are homogeneously distributed over the population. As p_D decreases below 1, the probability of reaching full consensus grows rapidly, approaching $P_{\text{cons}} = 1$. As N grows, moreover, the change in P_{cons} is more abrupt. Fitting of a sigmoidal function to the data of P_{cons} vs. p_D near $p_D = 1$ makes it possible to assign a width to the range where P_{cons} changes between 1 and 0.5. The insert of Fig. 3 shows this width as a function of the system size N in a log-log plot. The slope of the linear fitting is -1.00 ± 0.02.

Therefore, the width is inversely proportional to N.

The facts that $P_{\text{cons}} = 0.5$ for $p_D = 1$ and for all N, and that the width of the range where P_{cons} changes decreases as N^{-1}, make it possible to conjecture the existence of a function $\Phi(u)$, with $\Phi(0) = 0.5$ and $\Phi(u) \to 1$ for large u, such that $P_{\text{cons}} = \Phi[N(1 - p_D)]$. To test this hypothesis, we have plotted our numerical data for P_{cons} against $N(1-p_D)$ in Fig. 4. The results are those in the upper half of the plot ("varying p_D"). The collapse of the data for different N on the same curve confirms the conjecture.

Analogous results were obtained when fixing $p_D = 1$ and p_C was varied. Now, P_{cons} drops to 0 in a narrow interval for p_C just below 1, indicating the prevalence of disagreement. Again, the width of the interval is proportional to N^{-1}. The results in the lower half of Fig. 4 ("varying p_C") illustrate the collapse of the corresponding values of P_{cons} when plotted against $N(1 - p_C)$.

In our numerical realizations with $p_D \neq p_C$, we have also recorded the average time T needed to reach the final absorbing state. Figure 5 shows results for $N^{-2}T$ in the case where $p_C = 1$ and p_D changes (cf. lower panel of Fig. 2). In contrast with the case with $p_D = p_C = 1$, rescaling of the time T with N^2 leaves a remnant discrepancy between results for different population sizes N. Specifically, for $p_D < 1$, T grows faster than N^2. Moreover, T

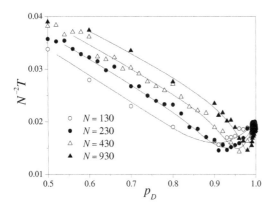

Figure 5: Total time T needed to reach the final absorbing state, normalized by the squared population size N^2, as a function of the probability p_D ($p_C = 1$). Bézier curves have been plotted as a guide to the eye.

is nonmonotonic as a function of p_D, exhibiting a minimum which shifts towards $p_D = 1$ as N grows. The same dependence with N and p_C is observed when we fix $p_D = 1$ and let p_C vary.

Summarizing our results for a one-dimensional population with two-agent groups, we can say that the possibility that both consensus and disagreement spread over the system makes it possible to find absorbing collective states of either full consensus, with all the agents having the same opinion, or maximal disagreement, where opposite opinions alternate between consecutive neighbor agents. For large populations, the relative prevalence of collective consensus and disagreement is controlled by how the probabilities p_D and p_C compare with each other. Our results suggest that, in the limit $N \to \infty$, the condition $p_C > p_D$ univocally leads to full consensus and vice versa. For smaller sizes, however, the system can approach full consensus even when $p_D > p_C$, and vice versa —presumably due to finite-size fluctuations.

III. Larger groups on two-dimensional arrays

A two-dimensional version of the above model, where agents occupy the $N = L \times L$ sites of a regular square lattice with periodic boundary con-

ditions, can be defined as follows. The reference group G' at each interaction event is a randomly chosen 2×2-agent block. The corresponding active group G is formed by the eight nearest neighbors to the agents in G' which are not in turn members of the reference group. The active group, thus, surrounds G'. Of the sixteen possible opinion configurations of the reference group, two correspond to full consensus —with the four agents sharing the same opinion— and six correspond to maximal disagreement —with two agents in each opinion. The remaining eight configurations correspond to partial consensus, with only one agent disagreeing with the other three. The dynamical rules are the following: (1) if G' is in full consensus, all the agents in G copy the common opinion in G'; (2) if G' is in maximal disagreement, each agent in G adopts the opinion opposite to that of the nearest neighbor in G'; (3) otherwise, nothing happens. Hence, both consensus and disagreement spread outwards from the reference group. Probabilities p_D and p_C for the spreading of disagreement and consensus are introduced exactly as above. The left part of Fig. 6 shows, up to rotations and opinion inversions, the three possible outcomes of a single interaction event.

The states of full collective consensus —with all the agents in the population having the same opinion— and of maximal collective disagreement —with the two opinions alternating site by site along each direction over the lattice— are absorbing states, in correspondence with the one-dimensional case. However, for $p_D = p_C = 1$, the system cannot be reduced anymore to a collection of sublattices governed by the voter model. The definition of G and G' establish now correlations between the opinion changes in the active group at each interaction event. Moreover, some opinion configurations in the reference group induce evolution in the active group, while others do not. Figure 6 shows, to its right, four snapshots of a 120×120-agent population, along a realization starting with $n_+(0) = 0.35$ and $p_D = p_C = 1$. Note the formation of consensus clusters at rather early stages, and the final prevalence of disagreement. The line boundaries between disagreement regions are also worth noticing.

Following the same lines as for the one-dimensional array, we study first the probability P_{cons} of reaching full collective consensus as a func-

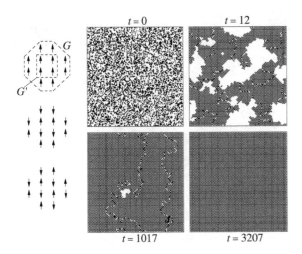

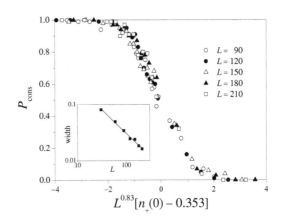

Figure 7: Numerical results for the probability of reaching full consensus, P_{cons}, on a two-dimensional lattice with $p_D = p_C = 1$, obtained from 100 realizations for each parameter set. Collapse for several system sizes L is obtained plotting P_{cons} against $L^{0.83}[n_+(0) - 0.353]$. Insert: Scaling of the width of the transition zone of P_{cons}, determined from fitting a sigmoidal function, as a function of the size L. The straight line has slope -0.83.

Figure 6: Left: The three possible outcomes of the interaction, up to $\pm 90°$ rotations and opinion inversions, on the two-dimensional lattice. The active and the reference groups, G and G', are respectively formed by the outermost and innermost agents. Right: Four snapshots of a population with $L = 120$, for $n_+(0) = 0.35$ and $p_D = p_C = 1$, including the initial condition and two intermediate states. At time $t = 3207$, an absorbing state of maximal disagreement has been reached. Black and white dots correspond, respectively, to opinions $+1$ and -1.

tion of the initial fraction of agents with opinion $+1$, $n_+(0)$, in the case $p_D = p_C = 1$. Opinions are homogeneously distributed all over the population. For very small $n_+(0)$, as expected, we find $P_{\text{cons}} \approx 1$. However, in sharp contrast with the one-dimensional case (see Fig. 3), P_{cons} remains close to its maximal value until $n_+(0) \approx 0.35$, where it drops abruptly to $P_{\text{cons}} \approx 0$. The width of the transition zone decreases as a nontrivial power of the system size, $\sim L^{0.83 \pm 0.04}$, as illustrated in the insert of Fig. 7. Our best estimate for the critical value of $n_+(0)$ at which P_{cons} drops is $n_+^{\text{crit}} = 0.353 \pm 0.001$. The main plot in the figure shows the collapse of numerical measurements of P_{cons} as a function of $n_+(0)$ for different sizes L, averaged over 100 realizations, when plotted against the rescaled shifted variable $L^{0.83}[n_+(0) - 0.353]$.

These results suggest that, for very large populations, the probability of reaching full consensus jumps discontinuously from $P_{\text{cons}} = 1$ to 0 at

$n_+(0) = n_+^{\text{crit}}$. Compare this with the smooth, size-independent behavior of the one-dimensional case. Note also that n_+^{crit} is close to, but does not coincide with, $n_+(0) = 1/3$. At this latter value, in the initial condition with homogeneously distributed opinions, the probability of finding a 2×2-agent block in full consensus becomes lower than that of maximal disagreement as $n_+(0)$ grows.

In the above simulations, we have also measured the average total time T needed to reach the final absorbing state. Results are shown in Fig. 8. Again in contrast with the one-dimensional case, T exhibits a remarkable change in its scaling with the system size as $n_+(0)$ overcomes the critical value n_+^{crit}.

Going now to the dependence of P_{cons} on the probability of disagreement spreading p_D —with $p_C = 1$ and $n_+(0) = 0.5$— it qualitatively mirrors that of the one-dimensional case, shown in Fig. 3. Namely, as p_D decreases from 1, P_{cons} grows from 0 to 1 in an interval whose width decreases with the population size. In the two-dimensional system, however, the transition takes place at a critical probability p_D^{crit} that can be

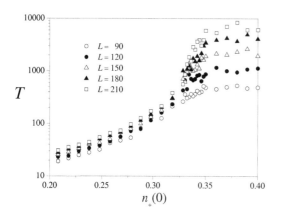

Figure 8: Total time T needed to reach the final absorbing state in a two-dimensional lattice, as a function of $n_+(0)$, for different sizes L.

clearly discerned from $p_D = 1$. Our estimate is $p_D^{\mathrm{crit}} = 0.984 \pm 0.002$. Moreover, the scaling of the transition width with the population size exhibits a nontrivial exponent, decreasing as $L^{-0.93\pm0.05}$. Collapse of the rescaled numerical results for various sizes, obtained from averages of 100 realizations, are shown in Fig. 9, where we plot P_{cons} as a function of $L^{0.93}(0.984-p_D)$ (cf. Fig. 4). The insert displays the power-law dependence of the width on the size L. Analogous results are obtained if the probability of consensus spreading p_C is varied, with $p_D = 1$.

Finally, we have found that the transition in P_{cons} as a function of the disagreement probability p_D shows a dependence on the initial fraction of agents with opinion $+1$. To characterize this effect in a way that highlights the relative prevalence of disagreement and consensus, we have measured the value of p_D at which the probability of getting full collective consensus reaches $P_{\mathrm{cons}} = 0.5$, as a function of $n_+(0)$. The parameter plane $(n_+(0), p_D)$, thus, becomes divided into regions where a final state of full consensus is more probable than that of maximal disagreement, and vice versa. Results for a 120×120-agent population are presented in Fig. 10.

In summary, while spreading of consensus and disagreement on a two-dimensional lattice bears superficial qualitative similarity with the one-

dimensional case, the probability that the population reaches full collective consensus in two dimen-

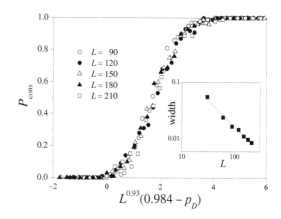

Figure 9: Collapse of numerical results for the probability of reaching full consensus, P_{cons}, on a two-dimensional lattice with $p_C = 1$ and $n_+(0) = 0.5$, for several system sizes L when plotted against $L^{0.93}(0.984-p_D)$. Insert: Scaling of the width of the transition zone of P_{cons} as a function of the size L. The straight line has slope -0.93.

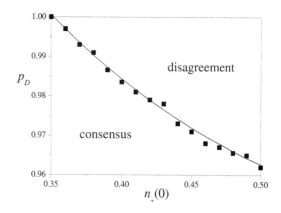

Figure 10: Zones of relative prevalence of full consensus and maximal disagreement in a two-dimensional lattice with $L = 120$, plotted on the parameter plane $(n_+(0), p_D)$. Symbols stand for numerical results, and the curve serves as a guide to the eye.

sions exhibits a quite different dependence on the system size, on the initial conditions, and on the spreading probabilities. In particular, our results reveal the existence of critical phenomena involving scaling laws with nontrivial exponents.

IV. Conclusion

In this paper, we have considered the emergence of collective opinion in a population of interacting agents where, instead of imitation between individual agents, opinions are transmitted through the spreading of local consensus and disagreement toward their neighborhoods. The basic interacting units in this mechanism are not individual agents but rather small groups of agents which mutually compare their internal degrees of consensus and modify their opinions accordingly. In this sense, it extends the basic mechanism underlying such models as the majority-rule and Sznajd-like dynamics [1, 8, 13], where the opinion of each individual agent changes in response to the collective state of a reference group. It is expected that in real social systems the dissemination of individual opinions through agent-to-agent imitation on one side, and the spreading of consensus and disagreement by group interaction on the other, are complementary mechanisms simultaneously shaping the overall opinion distribution. Here, in order to gain insight on the specific effects of the second class, we have focused on models solely driven by the spreading of consensus and disagreement. The combined effects of the two mechanisms is a problem open to future work.

Our numerical simulations concentrated on two-opinion models evolving on one- and two-dimensional arrays [14]. In both cases, absorbing states with all the population bearing the same opinion (full consensus) and with half of the population in each opinion (maximal disagreement) are possible final states for the system. Maximal disagreement states are characterized by alternating opinions between neighbor sites along the arrays.

A relevant quantity to characterize the behavior is the probability of reaching full consensus, as a function of the initial condition —i.e., the initial fraction of the population with each opinion— and of the relative probabilities of consensus and disagreement spreading. The total time needed to reach the final absorbing state, averaged over realizations, has also been measured as a characterization of the dynamics. We have found that, in several cases, these quantities display critical phenomena when the control parameters are changed, with power-law scaling laws as functions of the system size, pointing to the presence of discontinuities in the limit of infinitely large populations. It is interesting to remark that the scaling laws are rather simple for one-dimensional arrays, but involve nontrivial exponents and critical points in the case of two-dimensional systems.

Within the same one- and two-dimensional models analyzed here, an aspect that deserves further exploration is the dynamics and mutual interaction of the opinion domains that develop since the first stages of evolution (Figs. 1 and 6). However, the most interesting extension of the present analysis should progress along the direction of considering more complex social structures. The interplay between the dynamical rules of consensus and disagreement spreading and the topology of the interaction pattern underlying the population might bring about the emergence of new kinds of collective self-organization phenomena.

Acknowledgements - We acknowledge enlightening discussions with Eduardo Jagla. Financial support from ANPCyT (PICT2011-545) and SECTyP UNCuyo (Project 06/C403), Argentina, is gratefully acknowledged.

---

[1] C Castellano, S Fortunato, V Loreto, *Statistical physics of social dynamics*, Rev. Mod. Phys. **81**, 591 (2009).

[2] W Weidlich, *The statistical description of polarization phenomena in society*, Br. J. Math. Stat. Psychol. **24**, 251 (1971).

[3] R Holley, T Liggett, *Ergodic theorems for weakly interacting infinite systems and the voter model*, Ann. Probab. **3**, 643 (1975).

[4] S Galam, Y Gefen, Y Shapir, *A mean behavior model for the process of strike*, J. Math. Sociol. **9**, 1 (1982).

[5] S Galam, *Majority rule, hierarchical structures and democratic totalitarism: A statistical approach*, J. Math. Psychol. **30**, 426 (1986).

[6] S Redner, *A guide to first-passage processes*, Cambridge University Press, Cambridge (2001).

[7] K Starkey, Ch Barnatt, S Tempest, *Beyond networks and hierarchies: Latent organization in the UK television industry*, Org. Sci. **11**, 299 (2000).

[8] P Krapivsky, S Redner, *Dynamics of majority rule in two-state interacting spin systems*, Phys. Rev. Lett. **90**, 238701 (2003).

[9] J Johnson, *Multidimensional events in multilevel systems*, In: The Dynamics of Complex Urban Systems, Eds. S Albeverio et al., Pag. 311, Physica-Verlag, Heidelberg (2008).

[10] D H Zanette, *Beyond networks: Opinion formation in triplet-based populations*, Phil. Trans. R. Soc. A **367**, 3311 (2009).

[11] D H Zanette, *A note on the consensus time of mean-field majority-rule dynamics*, Pap. Phys. **1**, 010002 (2009).

[12] D G Hernández, D H Zanette, *Evolutionary dynamics of resource allocation in the Colonel Blotto game*, J. Stat. Phys. **151**, 623 (2013).

[13] K Sznajd-Weron, J Sznajd, *Opinion evolution in closed community*, Int. J. Mod. Phys. C **11**, 1157 (2000).

[14] D Stauffer, A O Sousa, S M de Oliveira, *Generalization to square lattice of Sznajd sociophysics model*, Int. J. Mod. Phys. C **11**, 1239 (2000).

[15] A Bray, *Theory of phase-ordering kinetics*, Adv. Phys. **43**, 357 (1994).

6

Jamming transition in a two-dimensional open granular pile with rolling resistance

C. F. M. Magalhães,[1*]   A. P. F. Atman,[2,3†]   G. Combe,[5]   J. G. Moreira[4‡]

We present a molecular dynamics study of the jamming/unjamming transition in two-dimensional granular piles with open boundaries. The grains are modeled by viscoelastic forces, Coulomb friction and resistance to rolling. Two models for the rolling resistance interaction were assessed: one considers a constant rolling friction coefficient, and the other one a strain dependent coefficient. The piles are grown on a finite size substrate and subsequently discharged through an orifice opened at the center of the substrate. Varying the orifice width and taking the final height of the pile after the discharge as the order parameter, one can devise a transition from a jammed regime (when the grain flux is always clogged by an arch) to a catastrophic regime, in which the pile is completely destroyed by an avalanche as large as the system size. A finite size analysis shows that there is a finite orifice width associated with the threshold for the unjamming transition, no matter the model used for the microscopic interactions. As expected, the value of this threshold width increases when rolling resistance is considered, and it depends on the model used for the rolling friction.

I. Introduction

Granular materials are ubiquitous either in nature —desert dunes, beach sand, soil, etc.— or in indus-

*E-mail: cfmm@unifei.edu.br
†E-mail: atman@dppg.cefetmg.br
‡E-mail: jmoreira@fisica.ufmg.br

[1] Universidade Federal de Itajubá - Campus de Itabira, Rua Irmã Ivone Drumond, 200, 35900-000 Itabira, Brazil.

[2] Departamento de Física e Matemática, Centro Federal de Educação Tecnológica de Minas Gerais, Av. Amazonas, 7675, 30510-000 Belo Horizonte, Brazil.

[3] Instituto Nacional de Ciência e Tecnologia Sistemas Complexos, 30510-000 Belo Horizonte, Brasil.

[4] Universidade Federal de Minas Gerais, Caixa Postal 702, 30161-970 Belo Horizonte, Brasil.

[5] UJF-Grenoble 1, Grenoble-INP, CNRS UMR 5521, 3SR Lab. Grenoble F-38041, France.

trial processes as mineral extraction and processing, or food, construction and pharmaceutical industries [1–3]. In fact, any particulate matter made of macroscopic solid elements can be classified as granular material. The vast phenomenology exhibited by these systems combined with an incomplete understanding about the microscopic physical mechanisms responsible for the macroscopic behavior of these materials have motivated the increasing interest of the physics community in the past years [4,5].

Although materials of this class are not sensitive to thermal fluctuations, they can be found at gas, liquid or solid phases [6]. The transition between solid and liquid phases in granular matter, which is commonly referred to as jamming/unjamming transition, has been extensively studied from both theoretical and experimental perspectives [4,6–11]. Currently, a great effort is being made to understand the nature of this transition, which is still a subject of debate [12]. The jamming/unjamming transition is not a specific property of granular matter, being observed in many kinds of materials, such as foams [13], emulsions [14], colloids [15], gels [16], and also in usual molecular liquids [17] —glass transition. Liu and Nagel [9] proposed a general phase diagram as an attempt to unify the several approaches to study jamming/unjamming in disordered materials. This work has motivated several theoretical, experimental and numerical investigations, but a comprehensive understanding of this transition is still lacking.

O'Hern et al. [18, 19] have performed numerical simulations of granular materials approaching to jamming in two and three dimensions. They have explored the packing fraction axis of the general phase diagram proposed by Liu and Nagel and have demonstrated, by means of finite-size analysis, the existence of a unique critical point in which the system jams in the thermodynamic limit. The authors have also shown some evidence that this point is an ordinary critical point, indicating that the jamming transition would be a second order phase transition. These results were corroborated later by experiments (c.f. Majmudar et al. [10]) and by simulations (c.f. Manna and Khakhar [20]). In Ref. [20], the authors have revealed evidence of self-organized criticality (SOC) by measuring the internal avalanches resulting from the opening of an orifice at the bottom of granular piles.

Experimental investigation of the jamming transition in granular materials under gravitational field has been conducted in a variety of ways, addressing the role of many parameters, like the grain shape, the friction coefficient, and the system geometry. However, a common feature of all these approaches is the analysis of the granular flow through bottlenecks.

The jamming of three-dimensional piles seems to be settled after the work of Zuriguel et al. [21]. They have demonstrated experimentally, for piles composed of different kinds of grains, the divergence on the mean internal avalanche size. It means that, as the outlet size approaches a critical value, the internal avalanche increases without limit and a permanent flow is established. This critical outlet is insensitive to the density, stiffness and roughness of the grains, but shows a significant dependence on the grain shape. For spherical grains, a critical outlet width $w_c \sim 4.94d$ was obtained, for cylindrical grains $w_c \sim 5.03d$ and for rice grains $w_c \sim 6.15d$.

Nevertheless, the jamming transition in two-dimensional piles is still a question under debate. In order to address it, To et al. [22] have carried out experiments using two-dimensional hoppers in order to find a critical outlet size for jamming events. The jamming probability J has presented a rapid decay from $J = 1$ to $J = 0$ close to the aperture width $w \sim 3.8d$, signaling a possible phase transition. The authors discussed, based on a restricted random walk model, the connection between jamming and the arch formation mechanism. Nevertheless, the point needs further investigation, especially at the limit of high hopper angle. There exist several works [23–25] focused on the mechanisms of arch formation, but none had explored its relation to jamming probability.

Janda et al. [26] have made some progress by simulating discharges of two-dimensional silos. They have improved the definition of jamming so that the internal avalanche size is considered. This modification addresses the extremely long relaxation times associated with jamming. Within the framework of a probabilistic theory concerning the arch formation, the authors have tested two hypothesis for the internal avalanche behavior: a functional form that predicts a divergence in the mean size, and a functional form where the mean size exists for all values of the orifice width w. Since the latter one is more compatible with the arch formation model, the authors claimed that "no critical opening size exists beyond which there is not jamming" [26].

The results mentioned so far are related to fully confined systems. Recently, a simulation study on discharges of granular piles with open boundaries [27] provided new insights on the problem. The piles are composed by homogeneous disks interacting via elastic and frictional forces. Using finite size analysis, the authors have shown that a catastrophic regime, in which the pile is completely destroyed by the opening of the orifice, is well defined. At the limit of infinitely large systems, the catastrophic regime coincides with the unjammed phase, since it implies a divergent internal avalanche. Hence, the results indicate the existence of the jamming transition. It is important to note, however, that the pile geometry could probably play a role in the causes for this distinct behavior, due to the absence of the Janssen effect, but further investigations are necessary to confirm this point.

In the present work, the investigation of jamming in 2D open systems is extended in order to consider a rolling resistance term in the grain interaction model, following the prescription adopted by Chevalier *et al.* [28]. The main objective of this study is the verification of rolling resistance influence on jamming. Many factors contribute to the appearance of rolling friction, including microsliding, plastic deformation, surface adhesion, grain shape, etc., but mainly, it is due to the contact deformation [29]. Here, it will be taken into account only the effect due to the contact deformation by implementing the micromechanical model proposed by Jiang *et al.* [30]. The rolling friction produces a resistance to roll which, among other effects, is responsible for granting more stability to granular piles [31], and for the occurrence of different types of failure modes in granular matter [32,33]. Rolling friction was also used to model a system of polygonal grains by making a correspondence between the rolling stiffness and the number of sides of the polygon [34]. It was demonstrated that it is an essential ingredient to reproduce experimental compression tests in mixtures of two-dimensional circular and rectangular grains [35]. These facts suggest that jamming could be affected by rolling friction. Nevertheless, most studies on granular materials based on computer simulations do not deal with it.

The paper is organized as follows: after a review in the Introduction, the next section is concerned with the methodology. Then, we present the results and a brief discussion. Finally, the last section gathers the conclusions and some perspectives.

II. Methods

The jamming transition is assessed by means of discharges of granular piles, simulated using the molecular dynamics method [36] with the Velocity-Verlet algorithm [37]. In a few words, the molecular dynamics consists in integrating numerically the equations of motion that governs the system dynamics. The system is constituted by N free grains governed by Newton's second law and by a finite and horizontally aligned substrate made of fixed grains. The free grains, which will form the pile, are homogeneous bi-dimensional disks that are free to translate and rotate around their center, and whose radii are uniformly distributed around an average value d, a small polydispersity of 5% was imposed in order to avoid crystallization effects. All spatial quantities will be expressed in terms of d. Since the grains have all the same mass density, their masses are proportional to their respective areas. Normalized by the mass of the heaviest grain, the masses are given by $m_i = d_i^2/d_{max}^2$, where d_{max} stands for the diameter of the largest grain. The finite substrate of length L is composed by fixed grains of the same kind but with a smaller and fixed diameter $d_s = 0.1d$. These grains are aligned horizontally and are equally spaced in order to form a grid without gaps.

The grains are subject to a uniform gravitational field orthogonal to the substrate line, and to short range binary interactions. Besides the viscoelastic and coulomb friction interactions used in past works [25,27,38–42], two grains in contact are also subject to a rolling resistance moment due to the finite contact length l_c. The rolling resistance is introduced through a micromechanical model of the contact line between two grains [30]. This model treats the contact as an object formed by a set of springs and dashpots connecting the borders of the two grains. As one grain rolls over the other, the springs in one side of the contact line contract while the springs in the opposite side stretch. This configuration generates an unbalanced force distribution and a consequent moment with respect to the grain center (see Fig. 1). This moment grows linearly as the grain rolls, until the rolling displacement δ_r reaches some threshold value at which the springs located near one end of the contact line break up and new ones emerge at the other end. At that time, the moment saturates at some value that depends on the properties of the grain. The rolling displacement is defined by $\delta_r = \sum(\omega_i - \omega_j)\Delta t$, where ω_i refers to the angular velocity of grain i, Δt is molecular dynamics time step, and the sum runs over time during the whole existence of contact. Based on these assumptions, the authors have derived an analytic expression for the rolling resistance moment as a function of rolling displacement. They have also proposed a simplified version - the one used in this study - in order to improve numerical computations:

$$\tau_{rr} = \begin{cases} -k_r\delta_r, & k_r|\delta_r| \leq \mu_r f_{el} \\ -\dfrac{\delta_r}{|\delta_r|}\mu_r f_{el}, & k_r|\delta_r| > \mu_r f_{el}, \end{cases} \quad (1)$$

where k_r is the rolling stiffness, μ_r is the coefficient of rolling resistance, and f_{el} is the compressive elastic force, normal to contact line. As can be noted, the expression for the rolling resistance momentum possesses a striking resemblance with the Coulomb static friction force, and as a matter of fact, the rolling resistance interaction is implemented in the molecular dynamics algorithm in the same way as the static friction. The model predicts that the rolling stiffness and the coefficient of rolling resistance are related to the contact length l_c by the equations $k_r = k_n l_c^2$ and $\mu_r = \mu l_c$, where k_n and μ are respectively the normal elastic constant and friction coefficient. The values used for these parameters were the same as in Ref. [27], $\mu = 0.5$ and $k_n = 1000$ in normalized unities (see [40] for further details). Two cases were investigated: systems in which the rolling resistance parameters k_r and μ_r vary according to the above-mentioned equations, and systems with fixed rolling resistance parameters, assuming that all contacts have the same deformation value.

The simulation procedure consists of two steps: (1) the formation of the granular pile with open boundaries, by deposing grains from rest, under gravity, over the substrate until an stationary state is reached; (2) the discharge itself, which consists in opening an orifice of a given width at the center of the substrate. In the first step, the initial positions of the free grains are randomly sorted along a horizontal line located at height L from the substrate - the releasing height is equal to the substrate length. To avoid initial overlapping of grains, a 50% filling ratio was imposed to each line of grains released, and the time interval between successive rows is the inverse of the frequency f. Each row was released after the predecessor had fallen a distance equivalent to the maximum grain diameter. This deposition protocol mimics a dense rain of grains. During deposition, the grains may leave the system through the lateral boundaries, so that the total number of grains in the pile fluctuates as the process evolves. The release of grains ceases when the number of grains in the pile reaches a stationary value. The deposition phase ends only when a

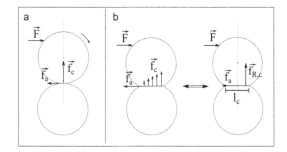

Figure 1: In panel (a), a perfectly rigid grain rolling over another one is exhibited. The contact force is concentrated on one point and is aligned through the grain center, thus, not generating momentum with respect to the center. Panel (b) shows the same situation but with deformable grains. Now, the contact force is non-uniformly distributed over a segment of length l_c (contact length). The resultant force $\vec{F}_{R,c}$ is dislocated from the center line and an opposing moment with respect to the center appears.

mechanical equilibrium state is attained, and the configuration is recorded for later analysis. The equilibrium state must satisfy the following criteria: mechanical stability, absence of slipping contacts, vertical and horizontal force balance, and vanishingly small kinetic energy [43]. In the second step, the configuration recorded is loaded and an outlet of width w is opened at the center of the substrate, allowing the grains to flow through it. As the grains pass through the outlet, they are removed from the system, in order to improve computational efficiency. The simulation runs until a new equilibrium state is reached, which happens either due to the formation of an arch above the outlet or after the pile has been completely discharged. The remaining pile configuration is then recorded.

The average height h of the resulting pile after discharge, measured from the center of the substrate, was taken as an order parameter to distinguish between the two regimes. If h does not change significantly with respect to the original height, it means that an arch does readily clog the flux after the orifice opening, but if $h = 0$, it means that the pile has collapsed, and a jammed state was not attained for the corresponding orifice diameter and substrate length. As the orifice width w approaches

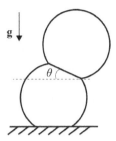

Figure 2: Diagram of a possible equilibrium configuration if rolling resistance is considered. The grain on top has only one contact and, even so, it can sustain a stable position. The maximum value of the angle θ depends on the friction coefficient and on the rolling resistance parameters.

a certain threshold value w_t, the system suffers a transition from jammed to unjammed state. This threshold is defined as the orifice width for which the height fluctuations is maximum. The connection to the jamming transition occurs when the substrate length diverges since the collapse of an infinite substrate pile implies a continuous flowing state. Then, the jamming transition can be characterized by a critical aperture width, as defined by the following expression

$$w_c = \lim_{L \to \infty} w_t(L) \ . \tag{2}$$

III. Results and Discussion

Two different models of rolling resistance interaction were tested in the simulations: the original model proposed by Jiang *et al.* [30] mentioned earlier, with a rolling constant and a coefficient of rolling resistance that depends on the contact length ($k_r = k_n l_c^2$ and $\mu_r = \mu l_c$) and a simpler derived version, in which these parameters are constant over time assuming that $l_c = 0.05\ d$ for all contacts. This fixed l_c model assumes a mean contact length equivalent to 5% of the average grain diameter, a scenario which could be associated to a system composed by polygonal grains, for example, a extreme rolling resistance regime.

For either models, the numerical simulations described in the last section were carried out for

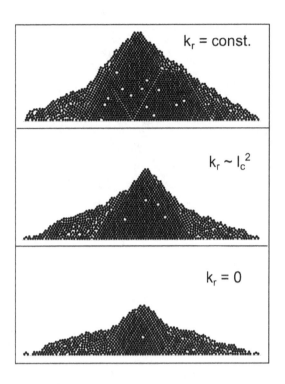

Figure 3: Images of the pile equilibrium states for the three grain interaction models before the discharge step. The piles were grown over a $L = 100d$ substrate and are representative samples from each type of pile. As shown in the figure, the top image represents a pile composed by grains with a fixed k_r rolling resistance interaction, the image in the middle is a pile of grains with a $k_r \sim l_c^2$ rolling resistance interaction, and the bottom image is a pile of grains without rolling resistance.

various system sizes and orifice widths. Figure 3 shows equilibrium configurations of typical piles with $L = 100d$ for the two tested rolling resistance models and for the absence of rolling resistance case. It can be seen that the inclusion of the rolling resistance term modifies significantly the macroscopic features of the pile. It provides more stability to the structure, which is reflected by a steeper free surface, a fact also observed elsewhere [31]. Indeed, it should be noted that the rolling resistance makes possible some otherwise very unstable two-grain configuration, as exemplified in Fig. 2.

The behavior of the order parameter h as function of w is presented in Fig. 4 for the three cases.

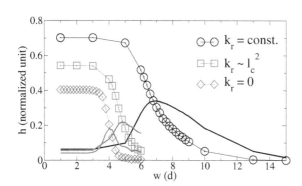

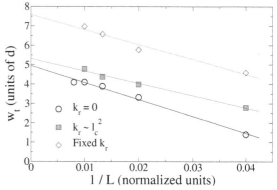

Figure 4: Order parameter h as a function of the orifice width w for all types of pile. The graphs were generated from the simulation data of $L = 100d$ piles. While the symbols represent the parameter h itself, the lines indicate the corresponding fluctuations. The thick line is related to the fixed k_r curve, the medium thickness line to the $k_r \sim l_c2$ curve, and the dashed line to $k_r = 0$ curve.

Note that the transition region translates to the right as the rolling resistance becomes more important, which is an expected result since the increasing of stability allows the formation of larger arches. Figure 5 exhibits the dependence of w_t (fluctuation maximum) on the system size for the three models considered and, again, the curves are dislocated along the w_t axis. Nevertheless, they all share the same functional aspect, which is an evidence that the rolling resistance does not change the nature of the jamming transition, only the critical value of the threshold width. Despite the tendency to a heavy tail distribution observed for large L and w, in all scenarios, the data suggest the existence of the jamming transition. It means that the features described in Ref. [27] to characterize the transition were also observed in both scenarios tested for the rolling resistance. The fitting values were $w_c = (5.3 \pm 0.1)d$ for contact length dependent k_r model and $w_c = (7.6 \pm 0.2)d$ for the fixed k_r one, while for the model without rolling resistance, $w_c = (5.0 \pm 0.1)d$ [27]. These results indicate that the rolling resistance only slightly changes the critical width when included in a system of disks, expressed by the contact length model. But, in other

Figure 5: Graphs of the threshold orifice width w_t as a function of the reciprocal system size $1/L$ for models with and without rolling resistance. As the caption indicates, the symbols represent the data obtained from numerical simulations and the lines are fitting curves, which provide the respective values of w_c.

approaches, as in the fixed length model, it can alter significantly the critical aperture width.

The probability density functions (PDF) of h for the absent rolling friction and for the fixed rolling friction models are exhibited in Fig. 6. For each model, the PDFs are obtained for three characteristic values of w: $w = 1.0d$, $w \sim w_t$, and $w > w_t$. It can be noted that the PDFs have an approximately Gaussian peak around the initial height for all values of w, meaning that there is always a certain amount of samples that are simply not disturbed by the outlet. Apart from these samples, the great majority is completely discharged for $w > w_t$. This result is in consonance with that obtained earlier in a different context [25]. In that study, it was shown that for large w, all blocked events occurred after only few grains had passed through the outlet. These facts indicate that the initial conditions may play an important role in jamming experiments. However, this issue needs further investigation.

IV.   Conclusions

Evidence of the jamming transition was observed in molecular dynamics simulations of open granular piles with rolling resistance, in consonance with

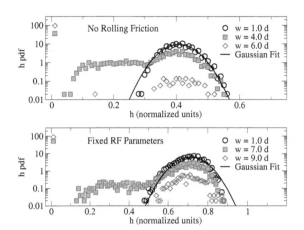

Figure 6: Probability density functions of the order parameter h for different values of w in the case of absent rolling friction grains (top) and fixed rolling friction grains (bottom).

the results found in granular piles without this interaction. This result strengthens the expectation that the transition exists in real granular piles and is probably affected by the system geometry. We observe that when there was rolling resistance, the piles built were more stable, denoted for the large mean height of the samples, and also more robust against perturbations, since the critical aperture width increased. In future works, we plan to present a detailed study of the arching statistics for the two approaches considered here.

Acknowledgements - We are grateful for CNPq, FAPEMIG Brazilian funding agencies. APFA and GC thanks to CEFET-MG by the international interchange which made this interaction possible.

[1] S J Antony, W Hoyle, Y Ding, *Granular materials: Fundamentals and applications* The Royal Society of Chemistry, Cambridge (2004).

[2] J Duran, *Sands, powders and grains*, Springer, Berlin (1997).

[3] T Halsey, A Mehta, *Challenges in granular physics*, World Scientific Publishing, New Jersey (2002).

[4] A Yu , K Dong , R Yang, S Luding, *Powders and grains 2013: Proceedings of the 7th international conference on micromechanics of granular media*, AIP Series. Vol. 1542, Sydney, Australia (2013).

[5] GdR MiDi, Eur. Phys. J. E **14**, 341 (2004).

[6] H M Jaeger, S R Nagel, *Granular solids, liquids, and gases*, Rev. Mod. Phys. **68**, 1259 (1996).

[7] P Cixous, E Kolb, N Gaudouen, J-C Charme, *Jamming and unjamming by penetration of a cylindrical intruder inside a 2 dimensional dense and disordered granular medium*, In: Powders and grains 2009, Proceedings of the 6th international conference on micromechanics of granular media **1145**, 539 (2009).

[8] A P F Atman, P Claudin, G Combe, R Mari, *Mechanical response of an inclined frictional granular layer approaching unjamming*, Europhys. Lett. **101**, 44006 (2013).

[9] A J Liu, S R Nagel, *Jamming is not just cool any more*, Nature **396**, 21 (1998).

[10] T S Majmudar, M Sperl, S Luding, R P Behringer, *Jamming transition in granular systems*, Phys. Rev. Lett. **98**, 058001 (2007).

[11] C Mankoc, A Janda, R Arévalo, J M Pastor, I Zuriguel, A Garcimartín and D Maza. *The flow rate of granular materials through an orifice*, Granul. Matter **9**, 407 (2007).

[12] A P F Atman, P Claudin, G Combe, G H B Martins, *Mechanical properties of inclined frictional granular layers*, Granul. Matter **16**, 1 (2014).

[13] G Katgert, M van Hecke, *Jamming and geometry of two-dimensional foams*, Europhys. Lett. **92**, 34002 (2010).

[14] N D Denkov, S Tcholakova, K Golemanov, A Lips, *Jamming in sheared foams and emulsions, explained by critical instability of the films between neighboring bubbles and drops*, Phys. Rev. Lett. **103**, 118302 (2009).

[15] A Kumar, J Wu, *Structural and dynamic properties of colloids near jamming transition*, Colloid. Surf. A **247**, 145151 (2004).

[16] A Fluerasu, A Moussaid, A Madsen, A Schofield, *Slow dynamics and aging in colloidal gels studied by x-ray photon correlation spectroscopy*, Phys. Rev. E **76**, 010401 (2007).

[17] B Duplantier, T C Halsey, V Rivasseau, *Glasses and grains: Poincaré Seminar 2009*, Springer, Basel (2011).

[18] C S O'Hern, S A Langer, A J Liu, S R Nagel, *Random packings of frictionless particles*, Phys. Rev. Lett. **88**, 075507 (2002).

[19] C O'Hern, L E Silbert, A J Liu, S R Nagel, *Jamming at zero temperature and zero applied stress: The epitome of disorder*, Phys. Rev. E **68**, 011306 (2003).

[20] S S Manna, D V Khakhar, *Internal avalanches in a granular medium*, Phys. Rev. E **58**, R6935 (1998).

[21] I Zuriguel, A Garcimartín, D Maza, L A Pugnaloni, J M Pastor, *Jamming during the discharge of granular matter from a silo*, Phys. Rev. E **71**, 051303 (2005).

[22] K To, P-Y Lai, H K Pak, *Jamming of granular flow in a two-dimensional hopper*, Phys. Rev. Lett. **86**, 71 (2001).

[23] A Garcimartín, I Zuriguel, L A Pugnaloni, A Janda, *Shape of jamming arches in two-dimensional deposits of granular materials*, Phys. Rev. E **82**, 031306 (2010).

[24] A Drescher, A J Waters, C A Rhoades, *Arching in hoppers .2. Arching theories and critical outlet size*, Powder Technol. **84**, 177 (1995).

[25] C F M Magalhães, A P F Atman, J G Moreira, *Segregation in arch formation*, Eur. J. Phys. E **35**, 38 (2012).

[26] A Janda, I Zuriguel, A Garcimartín, L A Pugnaloni, D Maza, *Jamming and critical outlet size in the discharge of a two-dimensional silo*, Europhys. Lett. **84**, 44002 (2008).

[27] C F M Magalhães, J G Moreira, A P F Atman, *Catastrophic regime in the discharge of a granular pile*, Phys. Rev. E **82**, 051303 (2010).

[28] B Chevalier, G Combe, P Villard, *Experimental and discrete element modeling studies of the trapdoor problem: Influence of the macro-mechanical frictional parameters*, Acta Geotech. **7**, 15 (2012).

[29] J Ai, J-F Chen, J M Rotter, J Y Ooi, *Assessment of rolling resistance models in discrete element simulations*, Powder Technol. **206**, 269 (2011).

[30] M J Jiang, H-S Yu, D Harris, *A novel discrete model for granular material incorporating rolling resistance*, Comput. Geotech. **32**, 340357 (2005).

[31] Y C Zhou, B D Wright, R Y Yang, B H Xu, A B Yu, *Rolling friction in the dynamic simulation of sandpile formation*, Physica A **269**, 536 (1999).

[32] K Iwashita, M Oda, *Rolling resistance at contacts in simulation of shear band development by DEM*, J. Eng. Mech. **124**, 285292 (1998).

[33] X Li, X Chu, Y T Feng, *A discrete particle model and numerical modeling of the failure modes of granular materials*, Eng. Computation. **22**, 894 (2005).

[34] N Estrada, E Azéma, F Radjai, A Taboada, *Identification of rolling resistance as a shape parameter in sheared granular media*, Phys. Rev. E **84**, 011306 (2011).

[35] E-M Charalampidou, G Combe, G Viggiani, J Lanier, *Mechanical behavior of mixtures of circular and rectangular 2D particles*, In: Powders and grains 2009: Proceedings of the 6th international conference on micromechanics of granular media, AIP Conf. Proc., Vol. 1145, Pag. 821, (2009).

[36] D C Rapaport, *The art of molecular dynamics simulation*, Cambridge University Press, Cambridge (2004).

[37] W C Swope, H C Andersen, P H Berens, K R Wilson, *Computer simulation method for the*

calculation of equilibrium constants for the formation of physical clusters of molecules: Application to small water clusters, J. Chem. Phys. **76**, 637 (1982).

[38] C Goldenberg, A P F Atman, P Claudin, G Combe, I Goldhirsch, *Scale separation in granular packings: Stress plateaus and fluctuations*, Phys. Rev. Lett. **96**, 168001 (2006).

[39] S F Pinto, A P F Atman, M S Couto, S G Alves, A T Bernardes, H F V Resende, E C Souza, *Granular fingers on jammed systems: New fluidlike patterns arising in grain-grain invasion experiments*, Phys. Rev. Lett. **99**, 068001 (2007).

[40] A P F Atman, P Claudin, G Combe, *Departure from elasticity in granular layers: Investigation of a crossover overload force*, Comput. Phys. Commun. **180**, 612 (2009).

[41] P A Cundall, O D L Strack, *A discrete numerical model for granular assemblies*, Geotechnique **29**, 47 (1979).

[42] R D Mindlin, *Compliance of elastic bodies in contact*, J. Appl. Mech. **71**, 259 (1949).

[43] A P F Atman, P Brunet, J Geng, G Reydellet, G Combe, P Claudin, R P Behringer, E Clement, *Sensitivity of the stress response function to packing preparation*, J. Phys.: Cond. Matter **17**, S2391 (2005).

[44] H Hertz, *On the contact of elastic solids*, J. Reine Angew. Math. **92**, 156 (1881).

[45] M P Allen, D J Tildesley, *Computer simulation of liquids*, Clarendon Press, Oxford (1987).

[46] O O'Sullivan, *Computing quaternions*, In: The art of numerical manipulation, Eds. A Q Rista, M Nadola, Pag. 132, North Holland, Amsterdam (2003).

7

Sequential evacuation strategy for multiple rooms toward the same means of egress

D. R. Parisi,[1,2*] P. A. Negri[2,3†]

This paper examines different evacuation strategies for systems where several rooms evacuate through the same means of egress, using microscopic pedestrian simulation. As a case study, a medium-rise office building is considered. It was found that the standard strategy, whereby the simultaneous evacuation of all levels is performed, can be improved by a sequential evacuation, beginning with the lowest floor and continuing successively with each one of the upper floors after a certain delay. The importance of the present research is that it provides the basis for the design and implementation of new evacuation strategies and alarm systems that could significantly improve the evacuation of multiple rooms through a common means of escape.

I. Introduction

A quick and safe evacuation of a building when threats or hazards are present, whether natural or man-made, is of enormous interest in the field of safety design. Any improvement in this sense would increase evacuation safety, and a greater number of lives could be better protected when fast and efficient total egress is required.

Evacuation from real pedestrian facilities can have different degrees of complexity due to the particular layout, functionality, means of escape, occupation and evacuation plans. During the last two decades, modeling and simulation of pedestrian

*E-mail: dparisi@itba.edu.ar
†E-mail: pnegri@uade.edu.ar

[1] Instituto Tecnológico de Buenos Aires, 25 de Mayo 444, 1002 Ciudad Autónoma de Buenos Aires, Argentina.

[2] Consejo Nacional de Investigaciones Científicas y Técnicas, Av. Rivadavia 1917, 1033 Ciudad Autónoma de Buenos Aires, Argentina.

[3] Universidad Argentina de la Empresa, Lima 754, 1073 Ciudad Autónoma de Buenos Aires, Argentina.

movements have developed into a new approach to the study of this kind of system. Basic research on evacuation dynamics has started with the simplest problem of evacuation from a room through a single door. This "building block" problem of pedestrian evacuation has extensively been studied in the bibliography, for example, experimetally [1, 2], or by using the social force model [3–5], and cellular automata models [6–8], among many others.

As a next step, we propose investigating the egress from multiple rooms toward a single means of egress, such as a hallway or corridor. Examples of this configuration are schools and universities where several classrooms open into a single hallway, cinema complexes, museums, office buildings, and the evacuation of different building floors via the same staircase. The key variable in this kind of system is the timing (simultaneity) at which the different occupants of individual rooms go toward the common means of egress. Clearly, this means of egress has a certain capacity that can be rapidly exceeded if all rooms are evacuated simultaneously and thus, the total evacuation time can be suboptimal. So, it is valid to ask in what order the different rooms should be evacuated.

The answer to this question is not obvious. Depending on the synchronization and order in which the individual rooms are evacuated, the hallway can be saturated in different sectors, which could hinder the exit from some rooms and thus, the corresponding flow rate of people will be limited by the degree of saturation of the hallway. This is because density is a limitation for speed. The relationship between density and velocity in a crowd is called "fundamental diagram of pedestrian traffic" [9–14]. Therefore, the performance of the egress from each room will depend on the density of people in the hallway, which is difficult to predict from analytical methods. This type of analysis is limited to simple cases such as simultaneous evacuation of all rooms, assuming a maximum degree of saturation on the stairs. An example of an analytical resolution for this simple case can be seen in Ref. [12], on chapter 3-14, where the egress from a multistory building is studied.

From now on we will analyze a 2D version of this particular case: an office building with 7 floors being evacuated through the same staircase, which is just an example of the general problem of several rooms evacuating through a common means of egress.

i. Description of the evacuation process

The evacuation process comprises two periods:

- E_1, reaction time indicating the time period between the onset of a threat or incident and the instant when the occupants of the building begin to evacuate.

- E_2, the evacuation time itself is measured from the beginning of the egress, when the first person starts to exit, until the last person is able leave the building.

E_1 can be subdivided into: time to detect danger, report to building manager, decision-making of the person responsible for starting the evacuation, and the time it takes to activate the alarm. These times are of variable duration depending on the usage given to the building, the day and time of the event, the occupants training, the proper functioning of the alarm system, etc. Because period E_1 takes place before the alarm system is triggered, it

must be separated from period E_2. The duration of E_1 is the same for the whole building. In consequence, for the present study only the evacuation process itself described as period E_2 is considered. The total time of a real complete evacuation will be necessarily longer depending on the duration of E_1.

ii. Hypothesis

This subsection defines the scope and conditions that are assumed for the system.

1. The study only considers period E_2 (the evacuation process itself) described in subsection I. i. above.

2. All floors have the same priority for evacuation. The case in which there is a fire at some intermediate floor is not considered.

3. The main aspect to be analyzed is the movement of people who follow the evacuation plan. Other aspects of safety such as types of doors, materials, electrical installation, ventilation system, storage of toxic products, etc., are not included in the present analysis.

4. After the alarm is triggered on each floor, the egress begins under conditions similar to those of a fire drill, namely:

   - People walk under normal conditions, without running.

   - If high densities are produced, people wait without pushing.

   - Exits are free and the doors are wide open.

   - The evacuation plan is properly signaled.

   - People start to evacuate when the alarm is activated on their own floor, following the evacuation signals.

   - There is good visibility.

II. Simulations

i. The model

The physical model implemented is the one described in [15], which is a modification of the social force model (SFM) [3]. This modification allows a better approximation to the fundamental diagram of Ref. [12], commonly used in the design of pedestrian facilities.

The SFM is a continuous-space and force-based model that describes the dynamics considering the forces exerted over each particle (p_i). Its Newton equation reads

$$m_i \mathbf{a}_i = \mathbf{F}_{Di} + \mathbf{F}_{Si} + \mathbf{F}_{Ci}, \qquad (1)$$

where $\mathbf{a}_i$ is the acceleration of particle p_i. The equations are solved using standard molecular dynamics techniques. The three forces are: "Driving Force" ($\mathbf{F}_{Di}$), "Social Force" ($\mathbf{F}_{Si}$) and "Contact Force" ($\mathbf{F}_{Ci}$). The corresponding expressions are as follows

$$\mathbf{F}_{Di} = m_i \frac{(v_{di}\, \mathbf{e}_i - \mathbf{v}_i)}{\tau}, \qquad (2)$$

where m_i is the particle mass, $\mathbf{v}_i$ and v_{di} are the actual velocity and the desired velocity magnitude, respectively. $\mathbf{e}_i$ is the unit vector pointing to the desired target (particles inside the corridors or rooms have their targets located at the closest position over the line of the exit door), τ is a constant related to the time needed for the particle to achieve v_d.

$$\mathbf{F}_{Si} = \sum_{j=1,j \neq i}^{N_{\mathrm{p}}} A \, \exp\left(\frac{-\epsilon_{ij}}{B}\right)\, \mathbf{e}_{ij}^n, \qquad (3)$$

with N_{p} being the total number of pedestrians in the system, A and B are constants that determine the strength and range of the social interaction, $\mathbf{e}_{ij}^n$ is the unit vector pointing from particle p_j to p_i; this direction is the "normal" direction between two particles, and ϵ_{ij} is defined as

$$\epsilon_{ij} = r_{ij} - (R_i + R_j), \qquad (4)$$

where r_{ij} is the distance between the centers of p_i and p_j and R is their corresponding particle radius.

$$\mathbf{F}_{Ci} = \qquad (5)$$

$$\sum_{j=1,j \neq i}^{N_{\mathrm{p}}} \left[(-\epsilon_{ij}\ k_n)\, \mathbf{e}_{ij}^n + (v_{ij}^t\ \epsilon_{ij}\ k_t)\, \mathbf{e}_{ij}^t\right]\, g(\epsilon_{ij}),$$

where the tangential unit vector ($\mathbf{e}_{ij}^t$) indicates the corresponding perpendicular direction, k_n and k_t

are the normal and tangential elastic restorative constants, v_{ij}^t is the tangential projection of the relative velocity seen from $p_j (\mathbf{v}_{ij} = \mathbf{v}_i - \mathbf{v}_j)$, and the function $g(\epsilon_{ij})$ is: $g = 1$ if $\epsilon_{ij} < 0$ or $g = 0$ otherwise.

Because this version of the SFM does not provide any self-stopping mechanism for the particles, it cannot reproduce the fundamental diagram of pedestrian traffic as shown in Ref. [15]. In consequence, the modification consists in providing virtual pedestrians with a way to stop pushing other pedestrians. This is achieved by incorporating a semicircular respect area close to and ahead of the particle (p_i). While any other pedestrian is inside this area, the desired velocity of pedestrians (p_i) is set equal to zero ($v_{di} = 0$). For further details and benefits of this modification to the SFM, we refer the reader to Ref. [15].

The kind of model used allows one to define the pedestrian characteristics individually. Following standard pedestrian dynamics bibliography (see, for example, [3–5, 15]), we considered independent and uniform distributed values between the ranges: pedestrian mass $m \,\epsilon$ [70 kg, 90 kg]; shoulder width $d \,\epsilon$ [48 cm, 56 cm]; desired velocity $v_d \,\epsilon$ [1.1 m/s, 1.5 m/s]; and the constant values are: $\tau = 0.5$ s, $A = 2000$ N, $B = 0.08$ m, $k_n = 1.2\ 10^5$ N/m, $k_t = 2.4\ 10^5$ kg/m/s.

Beyond the microscopic model, pedestrian behavior simply consists in moving toward the exit of the room and then toward the exit of the hallway, following the evacuation plan.

From the simulations, all the positions and velocities of the virtual pedestrians were recorded every 0.1 second. From these data, it is possible to calculate several outputs; in the present work we focused on evacuation times.

ii. Definition of the system under study

As a case study, we have chosen that of a medium-rise office building with $N = 7$, N being the number of floors. This system was studied analytically in Chapter 3-14 in Ref. [12], only for the case of simultaneous evacuation of all floors.

The building has two fire escapes in a symmetric architecture. At each level, there are 300 occupants. Exploiting the symmetric configuration, we will only consider the egress of 150 persons toward one of the stairs. Thus, on each floor, 150 people

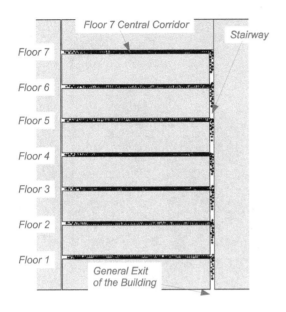

Figure 1: Schematic of the two-dimensional system to be simulated. Each black dot indicates one person.

are initially placed along the central corridor that is 1.2 m wide and 45 m long. In total, 1050 pedestrians are considered for simulating the system.

For the sake of simplicity, we define a two-dimensional version of a building where the central corridors of all the floors and the staircase are considered to be on the same plane as shown in Fig. 1.

The central corridors can be identified with the "rooms" of the general problem described in section I. and the staircase is the common means of egress. The effective width of the stairway is 1.4 m. The central corridors of each floor are separated by 10.66 m. This separation arises from adding the horizontal distance of the steps and the landings between floors in the 3D system [12]. So the distance between two floors in the 2D version of the problem is of the same length as the horizontal distance that a person should walk, also between two floors, along the stairway in the 3D building.

iii.   Evacuation strategies

The objective of proposing a strategy in which different floors start their evacuation at different times

is to investigate whether this method allows an improvement over the standard procedure, which is the simultaneous evacuation of all floors.

The parameters to be varied in the study are the following:

a The order in which the different levels are evacuated. In this sense, we study two procedures: a.1) "Bottom-Up": indicates that the evacuation begins on the lowest (1st) floor and then follows in order to the immediately superior floors. a.2) "Top-Down" indicates that the evacuation begins on the top floor (7th, in this case), and continues to the next lower floor, until the 1st floor is finally evacuated.

b The time delay dt between the start of the evacuation of two consecutive floors. This could be implemented in a real system through a segmented alarm system for each floor, which triggers the start of the evacuation in an independent way for the corresponding floor.

The initial time, when the first fire alarm is triggered in the building, is defined as T_0.

The instant $t_0^f {}_{\{BU,TD,SE\}}$ indicates the time when the alarm is activated on floor f. Subindices $\{BU, TD, SE\}$ are set if the time t belongs to the Bottom-Up, Top-Down, or Simultaneous Evacuation strategies, respectively.

The Bottom-Up strategy establishes that the 1st floor is evacuated first: $t_0^1 {}_{BU} = T_0$. Then the alarm on the 2nd floor is triggered after dt seconds, $t_0^2 {}_{BU} = t_0^1 {}_{BU} + dt$, and so on in ascending order up to the 7th floor . In general, the time when the alarm is triggered on floor f can be calculated as:

$$t_0^f {}_{BU} = T_0 + dt \times (f - 1). \tag{6}$$

The Top-Down strategy begins the building evacuation on the top floor (7th, in this case): $t_0^7 {}_{TD} = T_0$. After a time dt, the evacuation of the floor immediately below starts, and so on until the evacuation of the 1st floor:

$$t_0^f {}_{TD} = T_0 + dt \times (N - f). \tag{7}$$

Simultaneous Evacuation is the special case in which $dt = 0$ and thus, it considers the alarms on all the floors to be triggered at the same time:

$$t_0^f {}_{SE} = T_0|_{f=1,2,...,7}. \tag{8}$$

III. Results

This section presents the results of simulations made by varying the strategy and the time delay between the beginning of the evacuation of the different levels.

Each configuration was simulated five times, and thus, the mean values and standard deviations are reported. This is consistent with reality, because if a drill is repeated in the same building, total evacuation times will not be exactly the same.

i. Metrics definition

Here we define the metrics that will be used to quantify the efficiency of the evacuation process of the system under study.

It is called *Total Evacuation Time (TET)*, starting at T_0, when everyone in the building ($150 \times 7 = 1050$ persons) has reached the exit located on the ground floor (see Fig. 1), which means that the building is completely evacuated.

The fth *Floor Evacuation Time* (FET_f) refers to the time elapsed since initiating the evacuation of floor f until its 150 occupants reach the staircase. It must be noted that this evacuation time does not consider the time elapsed between the access to the staircase and the general exit from the building, nor does it consider as starting time the time at which the evacuation of some other level or of the building in general begins. It only considers the beginning of the evacuation of the current floor. Average Floor Evacuation Time (FET) is the average of the seven FET_f.

From these definitions, it follows that $TET > FET_f$ for any floor (even the lowest one).

ii. Simultaneous evacuation strategy

In general, the standard methodology consists in evacuating all the floors having the same priority at the same time.

Under these conditions, the capacity of the stairs saturates quickly, and so all floors have a slow evacuation. Figure 2 shows a snapshot from one simulation of this strategy. Here, the profile of the queues at each level can be observed. The differences in the length of queues are due to differences in the temporal evolution of density in front of each door.

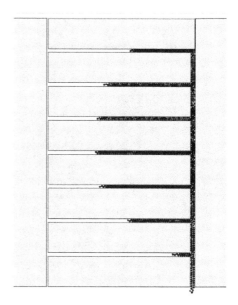

Figure 2: Snapshot taken at 73 seconds since the start of the simultaneous evacuation, where the queues of different lengths can be observed on each floor.

In this evacuation scheme, the first level that can be emptied is the 1st floor (105 ± 6 s) and the last one is the 6th floor (259 ± 3 s).

The Total Evacuation Time (TET) of the building for this configuration is 316 ± 8 s, and the mean Floor Evacuation Time (FET) is 195 ± 55 s.

For reference, the independent evacuation of a single floor toward the stairs was also simulated. It was found that the evacuation time of only one level toward the empty stair is 65 ± 4 s.

iii. Bottom-Up strategy

Figure 3(a) shows the evacuation times for different time delays dt following the Bottom-Up strategy.

It can be seen that the Total Evacuation Time (TET) remains constant for time delays (dt) up to 30 seconds. Therefore, TET is the same as the simultaneous evacuation strategy ($dt = 0$ s) in this range. It is worth noting that 30 seconds is approximately one half of the time needed to evacuate a floor if the staircase were empty.

Furthermore, the mean Floor Evacuation Time (FET) declines as dt increases, reaching the

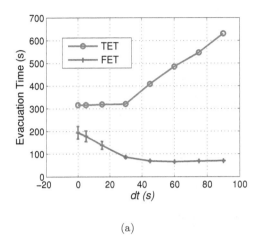 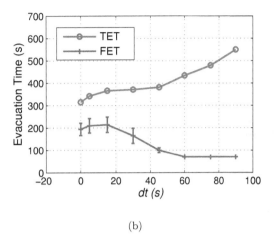

(a)   (b)

Figure 3: *TET* and *FET*, obtained from simulations for different phase shifts (*dt*) following sequential evacuation: (a) Bottom-Up strategy, (b) Top-Down strategy. The symbols and error bars indicate one standard deviation.

asymptotic value for 65 seconds, which is the evacuation time of a single floor considering the empty stairway. As expected, if the levels are evacuated one at a time, with a time delay greater than the duration of the evacuation time of one floor, the system is at the limit of decoupled or independent levels. In these cases, *TET* increases linearly with *dt*.

Since *TET* is the same for $dt < 30$ s and *FET* is significantly improved (it is reduced by half) for $dt = 30$ s, this phase shift can be taken as the best value, for this strategy, to evacuate this particular building.

This result is surprising because the *TET* of the building is not affected by systematic delays (*dt*) at the start of the evacuation of each floor if $dt \leq 30$ s, which reaches up to 180 seconds for the floor that further delays the start of the evacuation.

More details can be obtained by looking at the discharge curves corresponding to one realization of the building egress simulation. The evacuation of the first 140 pedestrians (93%) of each floor is analyzed by plotting the occupation as a function of time in Fig. 4 for three time delays between the relevant range $dt \, \epsilon[0, 30]$. For $dt = 0$ [Fig. 4(a)] there is an initial transient of about 10 seconds in which every floor can be evacuated toward a free part of the staircase before reaching the congestion

due to the evacuation of lower levels. After that, it can be seen that the egress time of different floors has important variations, the lower floors (1st and 2nd) being the ones that evacuate quicker and intermediate floors such as 5th and 6th the ones that take longer to evacuate. After an intermediate situation for $dt = 15$ s [Fig. 4(b)] we can observe the population profiles for the optimum phase shift of $dt = 30$ in Fig. 4(c). There, it can be seen that the first 140 occupants of different floors evacuate uniformly and very little perturbation from one to another is observed.

In the curves shown in Fig. 4, the derivative of the population curve is the flow rate, meaning that low slopes (almost horizontal parts of the curve such as the one observed in Fig. 4(a) for the 5th floor between 40 and 100 s) can be identified with lower velocities and higher waiting time for the evacuating people. Because of the fundamental diagram, we know that lower velocities indicate higher densities. In consequence, we can say that the greater the slope of the population curves, the greater the comfort of the evacuation (more velocity, less waiting time, less density). Therefore, it is clear that the situation displayed in Fig. 4(c) is much more comfortable than the one in Fig. 4(a).

In short, for the Bottom-Up strategy, the time delay $dt = 30$ s minimizes the perturbation among

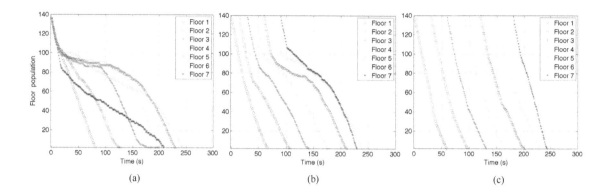

Figure 4: Time evolution of the number of pedestrians in each floor up to 3 m before the exit to the staircase. (a) for the simultaneous evacuation ($dt = 0$); (b) for delay of $dt = 15$ s and (c) for $dt = 30$ s.

evacuating pedestrians from successive levels; it reduces FET to one half of the simultaneous strategy ($dt = 0$ s); it maintains the total evacuation time (TET) at the minimum and, overall, it exploits the maximum capacity of the staircase maintaining each pedestrian's evacuation time at a minimum. This result is highly beneficial for the general system and for each floor, because it can avoid situations generating impatience due to waiting for gaining access to the staircase.

iv.   Top-Down strategy

Figure 3(b) shows the variation of TET and FET, as a function of the time delay dt, for the Top-Down strategy.

It must be noted that TET increases monotonously for all dt, which is sufficient to rule out this evacuation scheme.

In addition, for $dt < 15$ s, FET also increased, peaking at $dt = 15$ s. It can be said that for the system studied, the Top-Down strategy with a time delay of $dt = 15$ s leads to the worst case scenario.

For 15 s $< dt < 45$ s, there is a change of regime in which FET decreases and TET stabilizes.

For values of $dt > 45$ s, FET reaches the limit of independent evacuation of a single floor (see section III.ii.). And the TET of the building increases linearly due to the increasing delays between the start of the evacuation of the different floors.

In summary, the Top-Down Strategy does not present any improvement with respect to the standard strategy of simultaneous evacuation of all floors ($dt = 0$).

IV.   Conclusions

In this paper, we studied the evacuation of several pedestrian reservoirs ("rooms") toward the same means of egress ("hallway"). In particular, we focused on an example, namely, a multistory building in which different floors are evacuated toward the staircase. We studied various strategies using computer simulations of people's movement.

A new methodology, consisting in the sequential evacuation of the different floors (after a time delay dt) is proposed and compared to the commonly used strategy in which all the floors begin to evacuate simultaneously.

For the system under consideration, the present study shows that if a strategy of sequential evacuation of levels begins with the evacuation of the 1st floor and, after a delay of 30 seconds (in this particular case, 30 s is approximately one half of the time needed to evacuate only one floor if the staircase were empty), it follows with the evacuation of the 2nd floor and so on (Bottom-Up strategy), the quality of the overall evacuation process improves. From the standpoint of the evacuation of the building, TET is the same as that for the reference state. However, if FET is considered, there is a significant improvement since it falls to about half. This will make each person more comfortable during an evacuation, reducing the waiting time and thus, the probability of causing anxiety that may bring undesirable consequences.

So, one important general conclusion is that a sequential Bottom-Up strategy with a certain phase shift can improve the quality of the evacuation of a building of medium height.

On the other hand, the simulations show that the sequential Top-Down strategy is unwise for any time delay (dt). In particular, for the system studied, the value $dt = 15$ s leads to a very poor evacuation since the TET is greater than that of the reference, and it maximizes FET (which is also higher than the reference value at $dt = 0$). In consequence, the present study reveals that this would be a bad strategy that should be avoided.

The perspectives for future work are to generalize this study to buildings with an arbitrary number of floors (tall buildings), seeking new strategies. We also intend to analyze strategies where some intermediate floor must be evacuated first (e.g., in case of a fire) and then the rest of the floors.

The results of the present research could form the basis for developing new and innovative alarm systems and evacuation strategies aimed at enhancing the comfort and security conditions for people who must evacuate from pedestrian facilities, such us multistory buildings, schools, universities, and other systems in which several "rooms" share a common means of escape.

Acknowledgements - This work was financially supported by Grant PICT2011 - 1238 (ANPCyT, Argentina).

---

[1] T Kretz, A Grnebohm, M Schreckenberg, *Experimental study of pedestrian flow through a bottleneck*, J. Stat. Mech. P10014 (2006).

[2] A Seyfried, O Passon, B Steffen, M Boltes, T Rupprecht, W Klingsch, *New insights into pedestrian flow through bottlenecks*, Transport. Sci. **43**, 395 (2009).

[3] D Helbing, I Farkas, T Vicsek, *Simulating dynamical features of escape panic*, Nature **407**, 487 (2000).

[4] D R Parisi, C Dorso, *Microscopic dynamics of pedestrian evacuation*, Physica A **354**, 608 (2005).

[5] D R Parisi, C Dorso, *Morphological and dynamical aspects of the room evacuation process*, Physica A **385**, 343 (2007).

[6] A Kirchner, A Schadschneider, *Simulation of evacuation processes using a bionics-inspired cellular automaton model for pedestrian dynamics*, Physica A **312**, 260 (2002).

[7] C Burstedde, K Klauck, A Schadschneider, J Zittartz, *Simulation of pedestrian dynamics using a two-dimensional cellular automaton*, Physica A **295**, 507 (2001).

[8] W Song, X Xu, B H Wang, S Ni, *Simulation of evacuation processes using a multi-grid model for pedestrian dynamics*, Physica A **363**, 492 (2006).

[9] U Weidmann, *Transporttechnik der eussgänger, transporttechnische eigenschaften des fussgängerverkehrs*, Zweite, Ergänzte Auflage, Zürich, 90 (1993).

[10] J Fruin, *Pedestrian planning and design*, The Metropolitan Association of Urban Designers and Environmental Planners, New York (1971).

[11] A Seyfried, B Steffen, W Klingsch, M Boltes, *The fundamental diagram of pedestrian movement revisited*, J. Stat. Mech. P10002 (2005).

[12] P J Di Nenno (Ed.), *SFPE Handbook of fire protection engineering*, Society of Fire Protection Engineers and National Fire Protection Association (2002).

[13] D Helbing, A Johansson, H Al-Abideen, *Dynamics of crowd disasters: An empirical study*, Phys. Rev. E **75**, 046109 (2007).

[14] http://www.asim.uni-wuppertal.de/database-new/data-from-literature/fundamental-diagrams.html, accessed November 27, 2014.

[15] D R Parisi, B M Gilman, H Moldovan, *A modification of the social force model can reproduce experimental data of pedestrian flows in normal conditions*, Physica A **388**, 3600 (2009).

8

Adapting a Fourier pseudospectral method to Dirichlet boundary conditions for Rayleigh–Bénard convection

I. C. Ramos,[1]   C. B. Briozzo[1*]

We present the adaptation to non–free boundary conditions of a pseudospectral method based on the (complex) Fourier transform. The method is applied to the numerical integration of the Oberbeck–Boussinesq equations in a Rayleigh–Bénard cell with no-slip boundary conditions for velocity and Dirichlet boundary conditions for temperature. We show the first results of a 2D numerical simulation of dry air convection at high Rayleigh number ($R \sim 10^9$). These results are the basis for the later study, by the same method, of wet convection in a solar still.

I.   Introduction

Experimental observations [1] show that the onset of a turbulent convective flux can significantly enhance the efficiency of a basin-type solar still, but until now a theoretical explanation is lacking. Any adequate hydrodynamical simulation must incorporate the effects of moisture and condensation. Recent works [2, 3] show that this can be achieved through a Boussinesq-like approximation, which simplifies considerably the problem. However, realistic simulations are still demanding, given the need to resolve fine flux details and cope with Rayleigh numbers up to $\sim 10^9$ [4].

Spectral methods [5] are well suited for this kind of tasks, and have many attractive features: they are simple to implement, show much better resolution and accuracy properties than finite difference or finite volume methods, and are highly efficient in large-scale simulations [6]. Fourier-based pseu-

dospectral methods are the simplest and fastest, since the discretized spatial differential operators are local, nonlinear terms can be computed through Fast Fourier Transform (FFT) convolutions, and solving the Poisson equation originating from the incompressibility (divergence-free) condition is almost trivial. Nevertheless, they usually work only for free (in fact, periodic) boundary conditions (BCs).

The presence of non-free (*e.g.*, Dirichlet or Neumann) BCs introduces additional complications. For example, two-dimensional Rayleigh-Bénard convection with laterally periodic BCs can be treated by using a spectral Galerkin–Fourier technique in the horizontal coordinate and a collocation-Chebyshev method in the vertical one [7], but vertical derivatives must then be computed by matrix multiplication. On a grid with N horizontal and M vertical points, this needs the solution of a linear system of dimension M for each of the N horizontal Fourier modes, at each time-integration step.

Another complication with non-free BCs arises from the need to fulfill numerically the divergence-free condition, leading mainly to two different groups of methods (see Ref. [6] and references

*E-mail: briozzo@famaf.unc.edu.ar

[1] Facultad de Matemática, Astronomía y Física, Universidad Nacional de Córdoba, X5000HUA Córdoba, Argentina.

therein). In a first group the velocity field is written in terms of scalar potentials such that the divergence-free condition is satisfied by construction, *e.g.*, in the 2D streamfunction-vorticity formulation or the 3D decomposition into toroidal and poloidal velocity potentials. In these methods, pressure is not present in the equations, but they lead to systems of higher-order partial differential equations with coupled BCs. In a second group, a primitive variable formulation of the equations is adopted and projection methods [8] are used to decouple velocity and pressure. These methods use a specific splitting of the equation system based on the chosen time-integration scheme, and determine pressure by projecting an appropriate velocity field onto a divergence-free space, leading to predictor-corrector algorithms. Besides the problem of correctly specifying the pressure BCs [9, 10], these methods require solving a Poisson equation for the pressure at each time-integration step. On a $\mathcal{N} = N \times M$ grid, the best Fourier-based Fast Poisson Solvers (FPS) have operation counts $\mathcal{O}(\mathcal{N}\log_2 \mathcal{N})$ for the lowest (second) order discretization (and significantly worse for higher orders) [11,12], and those using GMRES are $K\mathcal{O}(\mathcal{N})$ with $K \gtrsim 100$ [13–15].

In this work, we will show how a Fourier-based pseudospectral method can be adapted to simple non-free (but periodic) BCs without losing its more appealing features. This is a first step towards building a pseudospectral simulation of wet air convection inside a basin-type solar still, and must be considered just as a proof of concept.

II. System

We consider 2D dry air convection in a Rayleigh–Bénard cell of width $L = 1\,\text{m}$ and height $H = 0.5\,\text{m}$, close to room temperature and with temperature differences $\Delta T = T_h - T_c$ up to $\sim 65\text{K}$ between the hot lower ($T = T_h$) and cold upper ($T = T_c$) plates (roughly the parameters of a real still [1]). Discarding thermal fluctuations and the heat generated by viscous dissipation, and assuming an incompressible fluid across which all thermodynamical parameters change little, the dynamics is given by [16]

$$\rho(\partial_t + \mathbf{u} \cdot \nabla)\mathbf{u} = -\nabla P + \eta \nabla^2 \mathbf{u} - \rho g \hat{\mathbf{z}}, \quad (1)$$

$$\rho c_p(\partial_t + \mathbf{u} \cdot \nabla)T = K\nabla^2 T, \quad (2)$$

$$\nabla \cdot \mathbf{u} = 0. \quad (3)$$

In these equations, the dynamical variables are the density ρ, the velocity $\mathbf{u}$, the temperature T, and the pressure P. The parameters are the shear (or dynamic) viscosity η, the constant pressure specific heat capacity c_p, the thermal conductivity K, and the gravitatioinal acceleration g.

The Boussinesq approximation [17] consists in discarding the dependence of η, c_p, and K on temperature and density, keeping

$$\rho = \bar{\rho}[1 - \alpha(T - \bar{T})] \quad (4)$$

in the buoyancy term ($-\rho g \hat{\mathbf{z}}$ in Eq. 1) but otherwise setting $\rho = \bar{\rho}$ elsewhere. Here, $\bar{T} = \frac{1}{2}(T_h + T_c)$ is a reference temperature, $\bar{\rho}$ is a reference density (that of air at normal temperature and pressure), and α is the thermal expansion coefficient. Dropping the bars signifying reference quantities, absorbing some constants into the pressure gradient term, and defining the viscous diffusivity (or kinematic viscosity) $\nu = \eta/\rho$ and the thermal diffusivity $\kappa = K/(\rho c_p)$, we obtain the (dimensional) Oberbeck–Boussinesq equations [17]

$$(\partial_t + \mathbf{u} \cdot \nabla)\mathbf{u} = -\nabla(P/\rho) + \nu \nabla^2 \mathbf{u} - \alpha g T \hat{\mathbf{z}}, \quad (5)$$

$$(\partial_t + \mathbf{u} \cdot \nabla)T = \kappa \nabla^2 T, \quad (6)$$

$$\nabla \cdot \mathbf{u} = 0. \quad (7)$$

Assuming perfect thermal contact with the lower ($z = 0$) and upper ($z = H$) plates, these equations admit the stationary conductive solution

$$\mathbf{u} = 0,$$

$$T_0(z) = \bar{T} - \frac{\Delta T}{H} z',$$

$$P_0(z) = \bar{P} + \rho g \alpha \left(-\bar{T} z' + \frac{\Delta T}{2H} z'^2\right), \quad (8)$$

where $\bar{P}$ is a reference (*e.g.* normal atmospheric) pressure, and $z' = z - H/2$.

We now scale lengths with the cell height H, times with the characteristic vertical thermal diffusion time $t_c = H^2/\kappa$, and temperatures with the temperature difference ΔT. The nondimensional lengths, times, and velocities are then

$$\mathbf{r}' = \frac{1}{H}\mathbf{r}, \qquad t' = \frac{\kappa}{H^2}t, \qquad \mathbf{u} = \frac{\kappa}{H}\mathbf{u}'. \qquad (9)$$

We also define the nondimensional temperature θ and pressure P' by

$$\theta = \frac{1}{\Delta T}\left(T - T_0(z)\right),$$
$$P' = \frac{H^2}{\kappa\nu}\left(P - P_0(z)\right). \qquad (10)$$

Substituting into Eqs. (5)–(7), discarding the primes for simplicity, and absorbing all pure gradient terms into the pressure, we get the dimensionless Oberbeck–Boussinesq equations [17]

$$\sigma^{-1}(\partial_t + \mathbf{u}\cdot\nabla)\mathbf{u} = -\nabla P + \theta\hat{\mathbf{z}} + \nabla^2\mathbf{u}, \qquad (11)$$
$$(\partial_t + \mathbf{u}\cdot\nabla)\theta = Ru_z + \nabla^2\theta, \qquad (12)$$
$$\nabla\cdot\mathbf{u} = 0, \qquad (13)$$

where

$$R = \frac{g\alpha\Delta T H^3}{\kappa\nu} \qquad (14)$$

is the Rayleigh number and $\sigma = \nu/\kappa$ is Prandtl's number ($\simeq 0.7$ for dry air). The BCs we adopt are periodic in the horizontal direction, and homogeneous Dirichlet for both velocity $\mathbf{u}$ (no-slip BCs) and temperature θ (perfect thermal contact) on the lower and upper plates.

Note that Eq. (13) is not a differential equation but a constitutive relationship, expressing the incompressibility of the flux. In fact, the pressure term in Eq. (11) is computed by enforcing Eq.(13), which gives

$$\nabla^2 P = -\sigma^{-1}\Sigma_{i,j}\partial_i\partial_j(u_iu_j) + \partial_z\theta, \qquad (15)$$

where $i,j = x,z$. In primitive variable integration schemes, this Poisson equation must be solved with adequate BCs at each time-step [6,9], to insure $\nabla\cdot(\partial_t\mathbf{u}) = 0$.

III.   Helmholtz decomposition

Given a vector field $\mathbf{f}$, twice continuously differentiable, Helmholtz's Theorem [18] states that it can be decomposed as

$$\mathbf{f} = \mathbf{f}_\parallel + \mathbf{f}_\perp. \qquad (16)$$

Here, $\mathbf{f}_\parallel$ and $\mathbf{f}_\perp$ are the longitudinal (or irrotational) and the transverse (or solenoidal) components of the field [8], respectively, with

$$\nabla\times\mathbf{f}_\parallel = 0, \qquad\qquad \nabla\cdot\mathbf{f}_\perp = 0. \qquad (17)$$

We are now going to rewrite Eq. (11) in the form

$$\partial_t\mathbf{u} = \sigma\left(\mathbf{f} - \nabla P\right) \qquad (18)$$

with

$$\mathbf{f} = -\sigma^{-1}(\mathbf{u}\cdot\nabla)\mathbf{u} + \theta\hat{\mathbf{z}} + \nabla^2\mathbf{u}. \qquad (19)$$

Then, if $\nabla\cdot\mathbf{u} = 0$ initially, the incompressibility condition Eq. (13) requires

$$\nabla\cdot(\mathbf{f} - \nabla P) = 0. \qquad (20)$$

This amounts to requiring the field $\mathbf{f} - \nabla P$ to be purely transverse, that is

$$(\mathbf{f} - \nabla P)_\parallel = \mathbf{f}_\parallel - \nabla P = 0, \qquad (21)$$

where we used $(\nabla P)_\parallel \equiv \nabla P$, since the pressure gradient is purely longitudinal. This shows that the *only* effect of the term ∇P in Eq. (18) is to cancel the longitudinal component of $\mathbf{f}$. Equation (18) can then be set as

$$\partial_t\mathbf{u} = \sigma\mathbf{f}_\perp, \qquad (22)$$

with no explicit reference to a pressure field.

In 2D and in free space (that is, disregarding surface terms), the longitudinal and transverse components of $\mathbf{f}$ can be computed as the projections

$$\tilde{\mathbf{f}}_\parallel = \left(\tilde{\mathbf{f}}\cdot\hat{\mathbf{k}}\right)\hat{\mathbf{k}}, \qquad \tilde{\mathbf{f}}_\perp = \left(\tilde{\mathbf{f}}\cdot\hat{\mathbf{k}}'\right)\hat{\mathbf{k}}' \qquad (23)$$

of its Fourier transform $\tilde{\mathbf{f}}$ along the unit vectors

$$\hat{\mathbf{k}} = \frac{(k_x, k_z)}{\sqrt{k_x^2 + k_z^2}}, \qquad \hat{\mathbf{k}}' = \frac{(-k_z, k_x)}{\sqrt{k_x^2 + k_z^2}}. \qquad (24)$$

However, on a finite domain, the surface terms cannot be ignored, since they are essential for $\mathbf{f}_\perp$ to have the correct BCs, which by Eq. (22) are the same as those for $\mathbf{u}$ in Eq. (11). Using Eq. (24), the field $\tilde{\mathbf{f}}_\perp$ in Eq. (23) can be seen to be a *particular* solution (the free–space solution) of the Poisson equation

$$\nabla^2 \mathbf{f}_\perp = \mathcal{F}^{-1}\{(-k_z, k_x)(-k_z \tilde{f}_x + k_x \tilde{f}_z)\}, \quad (25)$$

where $\mathcal{F}^{-1}$ stands for the inverse Fourier transform. To be able to impose the required BCs, we need the *general* solution of this equation, which we can get by adding the general solution of the corresponding *homogeneous* equation $\nabla^2 \mathbf{f}_\perp = 0$.

The required transverse component of $\mathbf{f}$ can then be redefined as

$$\mathbf{f}_\perp = \mathbf{v} + \mathbf{w}, \quad (26)$$

with

$$\tilde{\mathbf{v}} = (-k_z, k_x)\frac{-k_z \tilde{f}_x + k_x \tilde{f}_z}{k_x^2 + k_z^2} \quad (27)$$

and

$$\nabla^2 \mathbf{w} = 0, \qquad \nabla \cdot \mathbf{w} = 0, \quad (28)$$

where the last equation is needed to insure the transversality of $\mathbf{w}$ and hence of $\mathbf{f}_\perp$. This requirement can be automatically fulfilled by writing $\mathbf{w}$ explicitly as

$$\tilde{\mathbf{w}} = (-k_z, k_x)\tilde{w} \quad (29)$$

with the *scalar* field w satisfying

$$\nabla^2 w = c \quad (30)$$

where c is a constant. Noting that Eq. (27) can also be rewritten as

$$\tilde{\mathbf{v}} = (-k_z, k_x)\tilde{v}, \qquad \tilde{v} = \frac{-k_z \tilde{f}_x + k_x \tilde{f}_z}{k_x^2 + k_z^2}, \quad (31)$$

we can also rewrite Eq. (26) as

$$\tilde{\mathbf{f}}_\perp = (-k_z, k_x)(\tilde{v} + \tilde{w}), \quad (32)$$

which shows explicitly the transversality of $\mathbf{f}_\perp$. The determination of the value of c, and the treatment of possible divergences in $\tilde{v}$ at $\mathbf{k} = 0$, are closely related and will be dealt with in the next section. The BCs for $\mathbf{w}$ at the lower and upper plates can be obtained from those for $\mathbf{f}_\perp$, and are $\mathbf{w} = -\mathbf{v}$; the BCs on the horizontal direction are periodic but otherwise free, and will be automatically fulfilled by the constructive procedure for $\mathbf{w}$ given in the next section.

IV.   Ultra-fast Laplace solver

We start by solving Eq. (30) with $c = 0$, that is Laplace's equation, on the rectangular domain $0 \leq x \leq L$, $0 \leq z \leq H$, which is an elementary problem in harmonic analysis. Over an *unbounded* domain, the solutions have the form $e^{i\lambda x}e^{\lambda z}$, where λ is an arbitrary separation constant; the particular solutions for the case $\lambda = 0$ are 1, x, z, and xz. Periodicity in x on $[0, L]$ imposes $\lambda = 2\pi p/L$ with $p \in \mathbb{Z}$; the general solution is then

$$w = \sum_p c_p e^{-i2\pi px/L} e^{-2\pi pz/L} + a + bz, \quad (33)$$

with c_p, a, and b (possibly complex) constants. For convenience, and without loss of generality, we will rewrite it in the form

$$w = \sum_p e^{-i2\pi px/L}\left[a_p \cosh\frac{2\pi pz'}{L}\right.$$
$$\left. +(1 - \delta_{p,0})b_p \sinh\frac{2\pi pz'}{L} + \delta_{p,0}b_0 z'\right], \quad (34)$$

where a_p and b_p are constants, and $z' = z - H/2$. Here, we have used that $\cosh(0) = 1$ to absorb all constant terms in a_0, and used that $\sinh(0) = 0$ to absorb the linear term in z' as the particular term of the sum for $p = 0$, leaving Eq. (34) explicitly in the form of a (complex) Fourier series in x. The hyperbolic and linear functions in z' can, in turn, be rather trivially expanded as complex Fourier series in z on the interval $[0, H]$, leaving w in the form of the double Fourier series

$$w = \sum_{p,q} e^{-i2\pi px/L} e^{-i2\pi qz/H} \left[a_p \tilde{C}_{pq} + b_p \tilde{S}_{pq} \right], \quad (35)$$

where $q \in \mathbb{Z}$ and $\tilde{C}_{pq}, \tilde{S}_{pq}$ are the expansion coefficients. Discretizing w on a coordinate grid $(x_n, z_m) = (n\Delta x/N, m\Delta z/M)$ restricts the range of p and q respectively to $[0, N-1]$ and $[0, M-1]$ (so the horizontal and vertical wavenumbers are below the respective Nyquist frequencies), reducing Eq. (35) to a (double) discrete Fourier transform (DFT), whose coefficients

$$\tilde{w}_{pq} = a_p \tilde{C}_{pq} + b_p \tilde{S}_{pq} \qquad (36)$$

give the discretization of w on the wavenumber grid. The matrices $\tilde{C}_{pq}$ and $\tilde{S}_{pq}$ are given explicitly by

$$\tilde{C}_{pq} = \qquad\qquad\qquad\qquad\qquad (37)$$
$$\begin{cases} \dfrac{1}{M} \dfrac{\sinh\left(k_{x,p}\frac{H}{M}\right)\sinh\left(k_{x,p}\frac{H}{2}\right)}{\cosh\left(k_{x,p}\frac{H}{M}\right) - \cos\left(k_{z,q}\frac{H}{M}\right)}, & (p,q) \neq (0,0) \\ 1, & (p,q) = (0,0) \end{cases}$$

$$\tilde{S}_{pq} = \qquad\qquad\qquad\qquad\qquad (38)$$
$$\begin{cases} -\dfrac{1}{M} \dfrac{i\sin\left(k_{z,q}\frac{H}{M}\right)\sinh\left(k_{x,p}\frac{H}{2}\right)}{\cosh\left(k_{x,p}\frac{H}{M}\right) - \cos\left(k_{z,q}\frac{H}{M}\right)}, & p \neq 0 \\ -\dfrac{1}{M} \dfrac{i\sin\left(k_{z,q}\frac{H}{M}\right)\frac{H}{2}}{1 - \cos\left(k_{z,q}\frac{H}{M}\right)}, & p = 0, q \neq 0 \\ 0, & (p,q) = (0,0), \end{cases}$$

where $k_{x,p} = 2\pi p/L$ and $k_{z,q} = 2\pi q/H$. Here we have preferred, for simplicity when writing the numerical code, to replace the intervals $0 \leq p \leq N-1$ and $0 \leq q \leq M-1$ by the equivalent intervals $-N/2 \leq p \leq N/2$ and $-M/2 \leq q \leq M/2$, with their respective extreme points identified.

Equation (36) then provides the general solution of $\nabla^2 w = 0$ on $[0, L] \times [0, H]$, discretized on the wavenumber grid. The corresponding general solution of Eq. (30) can then be expressed formally as

$$\tilde{w}_{pq} = a_p \tilde{C}_{pq} + b_p \tilde{S}_{pq} - \frac{c\delta_{p,0}\delta_{q,0}}{k_{x,p}^2 + k_{z,q}^2}. \qquad (39)$$

Then from Eqs. (31) and (32), the discretization of $\mathbf{f}_\perp$ on the wavenumber grid will be expressed as

$$\tilde{\mathbf{f}}_{\perp,pq} = (-k_{z,q}, k_{x,p}) \Big[a_p \tilde{C}_{pq} + b_p \tilde{S}_{pq}$$
$$+ \frac{-c\delta_{p,0}\delta_{q,0} - k_{z,q}\tilde{f}_{x,p} + k_{x,p}\tilde{f}_{z,q}}{k_{x,p}^2 + k_{z,q}^2} \Big]. \qquad (40)$$

The constant c can now be formally chosen to cancel the possible divergence at $\mathbf{k} = 0$; in practice, we set $c = 0$ and redefine $\tilde{v}_{00}$ to be an arbitrary but finite constant, which without loss of generality can also be taken to vanish.

Imposing the BCs $\mathbf{w} = -\mathbf{v}$, or equivalently $\mathbf{f}_\perp = 0$, at $z = 0$ and $z = H$, is achieved as follows: First we note that any field discretized on the wavenumber grid through a DFT of its discretization on the coordinate grid is automatically periodic in *both* the horizontal and vertical directions, so the BCs at $z = 0$ and $z = H$ will give identical sets of equations. Next we use the fact that

$$\mathbf{f}_{\perp,n0} = \sum_{p,q} e^{-ik_{x,p}x_n} \tilde{\mathbf{f}}_{\perp,pq}, \qquad (41)$$

together with the completitude of the DFT, to rewrite the BC at $z = 0$ as

$$\sum_q \tilde{\mathbf{f}}_{\perp,pq} = 0 \quad \forall p. \qquad (42)$$

We then use Eq. (40) and the parity of $\tilde{C}_{pq}$ (even in q) and $\tilde{S}_{pq}$ (odd in q) to obtain the conditions

$$a_p \sum_q \tilde{C}_{pq} = -\sum_q \tilde{v}_{pq},$$
$$b_p \sum_q k_{z,q}\tilde{S}_{pq} = -\sum_q k_{z,q}\tilde{v}_{pq}, \qquad (43)$$

from which the coefficients a_p and b_p can be immediately retrieved. Finally, substituting into Eq. (40) leads to

$$\tilde{\mathbf{f}}_{\perp,pq} = (-k_{z,q}, k_{x,p}) \Big[\tilde{v}_{pq} - c_{pq} \sum_q \tilde{v}_{pq}$$
$$- s_{pq} \sum_q k_{z,q}\tilde{v}_{pq} \Big], \qquad (44)$$

where it is understood that we take $\tilde{v}_{00} = 0$, and we have introduced the normalized matrices

$$c_{pq} = \frac{\tilde{C}_{pq}}{\sum_{q'} \tilde{C}_{pq'}}, \quad s_{pq} = \frac{\tilde{S}_{pq}}{\sum_{q'} k_{z,q'} \tilde{S}_{pq'}}. \quad (45)$$

Equations (44) and (31) show that the transverse field $\tilde{\mathbf{f}}_{\perp,pq}$ on the wavenumber grid can be obtained directly in terms of the non-transverse field $\tilde{\mathbf{f}}_{pq}$ without needing, at any point, to return to the coordinate grid. Moreover, Eq. (45) shows that c_{pq} and s_{pq} are given matrices that can be computed just once at the start of the simulation, as is the denominator $k_{x,p}^2 + k_{z,q}^2$ in Eq. (31). The only tasks to be performed at each time-step are then: first, computing the scalars $\tilde{v}_{pq}$ from $\tilde{\mathbf{f}}_{pq}$, requiring three multiplications per grid point; second, computing the sums in Eq. (44), which requires one multiplication per grid point; third, multiplying them by c_{pq} and s_{pq}, costing two multiplications per grid point; and fourth, multiplying by $k_{z,q}$ and $k_{x,p}$ at each grid point. The total cost of obtaining $\tilde{\mathbf{f}}_{\perp,pq}$ is then $8\mathcal{N}$ multiplications on a $\mathcal{N} = N \times M$ grid, thus outperforming even the best FPS by a significant factor on large grids.

It must be noted that the method introduced here has some similarity to the streamfunction–vorticity formulation [6], in the sense that the scalar fields $\tilde{v}$ and $\tilde{w}$ play a role similar to these potentials. However, in our method they are not taken as *dynamical* variables, and the evolution equations are not formulated in terms of them but of primitive variables. The method presented here shows also a strong similarity with projection methods [8, 9], but differently from them, pressure is not computed along the time evolution and, in fact, does no longer appear in the evolution equations.

V. Algorithm outline

We present now an outline of the numeric algorithm as we implemented it in the simulations.
Initialization:

- Take zero velocity and Gaussian white noise for the temperature (with amplitude of the order of thermal noise), discretized on the coordinate grid, and take their FFT to get $\tilde{u}_{x,pq}$, $\tilde{u}_{z,pq}$, $\tilde{\theta}_{pq}$ on the wavenumber grid.

- Pre-compute (just once) the matrices c_{pq} and s_{pq}, and $k_{x,p}^2 + k_{z,q}^2$.

Time-stepping:

- Compute convolutions of $\tilde{u}_{x,pq}$, $\tilde{u}_{z,pq}$, $\tilde{\theta}_{pq}$ by FFT with 2/3 rule.

- Compute r.h.s. of evolution equations as

$$
\begin{aligned}
(\partial_t \tilde{u}_x)_{pq} &\leftarrow \quad ik_{x,p}(\tilde{u}_x * \tilde{u}_x)_{pq} + ik_{z,q}(\tilde{u}_z * \tilde{u}_x)_{pq} \\
&\qquad -\sigma k_{pq}^2 \tilde{u}_{x,pq} \\
(\partial_t \tilde{u}_z)_{pq} &\leftarrow \quad ik_{x,p}(\tilde{u}_x * \tilde{u}_z)_{pq} + ik_{z,q}(\tilde{u}_z * \tilde{u}_z)_{pq} \\
&\qquad -\sigma k_{pq}^2 \tilde{u}_{z,pq} + \sigma\tilde{\theta}_{pq} \\
(\partial_t \tilde{\theta})_{pq} &\leftarrow \quad ik_{x,p}(\tilde{u}_x * \tilde{\theta})_{pq} + ik_{z,q}(\tilde{u}_z * \tilde{\theta})_{pq} \\
&\qquad -k_{pq}^2 \tilde{\theta}_{pq} + R\tilde{u}_{z,pq}.
\end{aligned}
$$

- Compute $(\partial_t \tilde{\mathbf{u}})$ as

$$
\begin{aligned}
(\partial_t \tilde{u})_{pq} &\leftarrow \quad \frac{-k_{z,q}(\partial_t \tilde{u}_x)_{pq} + k_{x,p}(\partial_t \tilde{u}_z)_{pq}}{k_{pq}^2} \\
(\partial_t \tilde{\mathbf{u}})_{pq} &\leftarrow \quad (-k_{z,q}, k_{x,p}) \times \\
&\quad [(\partial_t \tilde{u})_{pq} - \tilde{c}_{pq} \Sigma_{q'} (\partial_t \tilde{u})_{pq'} \\
&\quad -\tilde{s}_{pq} \Sigma_{q'} k_{z,q'} (\partial_t \tilde{u})_{pq'}].
\end{aligned}
$$

Here, we have denoted by $(\tilde{f} * \tilde{g})_{pq}$ the convolution product of fields $\tilde{f}_{pq}$ and $\tilde{g}_{pq}$ discretized on the wavenumber grid, which is performed by FFT.

It must be noted that the spatial discretization of the evolution equations has been performed in a closed form, independent of the time-stepping algorithm to be employed to solve the resulting set of ordinary differential equations, outlined above. We must also note that this system does not involve multiplication by matrices with dimension $\mathcal{N}$, the only nonlocal parts being the convolution products (handled through FFT) and the elimination of the longitudinal component of the velocity.

The time-stepping can then be performed by any algorithm designed to solve systems of ordinary differential equations. In our case, we opted for an adaptive-stepsize fifth order Runge–Kutta–Cash–Karp algorithm [19], which in our previous experience we have found efficient, stable, and flexible.

VI. Test runs

Even for a simple system like the one we are studying here, the phenomenology found is rich; we

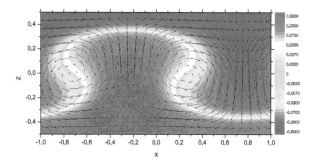

Figure 1: Temperature and velocity fields for $R = 5R_c$ at $t = 2t_c$ on a 32×16 grid.

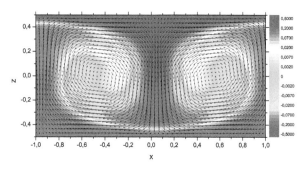

Figure 2: Temperature and velocity fields for $R = 50R_c$ at $t = t_c$ on a 64×32 grid.

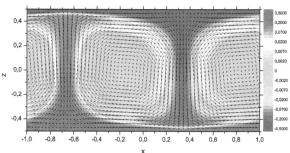

Figure 3: Temperature and velocity fields for $R = 500R_c$ at $t = 0.25t_c$ on a 128×64 grid (velocity decimated to a 64×32 grid).

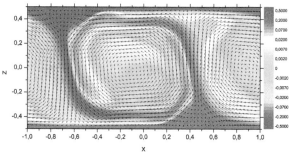

Figure 4: Temperature and velocity fields for $R = 5000R_c$ at $t = 0.05t_c$ on a 192×96 grid (velocity decimated to a 64×32 grid).

present only a brief outline. All results are given in terms of the laterally-infinite cell crytical Rayleigh number $R_c \sim 1701$, and the characteristic vertical diffusion time t_c which for our cell and medium is ~ 11797s. Coordinates and fields are in the dimensionless variables of Eqs. (11)–(13).

At low R, a stationary regime state, consisting in two counter-rotating rolls, is reached in times $\sim t_c$ or less. This time falls rapidly with increasing R, to $\sim 0.1t_c$ at $R \sim 1000R_c$ (see Figs. 1, 2 and 3).

Around $R \sim 5000R_c$, these rolls develop lateral oscillations, and the first "secondary structures" (small whirlpools) appear near the base of the ascending and descending plumes (see Fig. 4).

Above $R \sim 5000R_c$, the regime state becomes disordered and aperiodic, consisting of intermittent plumes and whirlpools in a wide size range (see Fig. 5).

At $R \sim 5 \times 10^5 R_c$, the temperature difference is $\sim$ 65K; the smallest whirlpools are $\sim$ 1cm wide, and the typical wind speeds are $\sim$ 1m/s, in agreement

with experimental observations [1] (see Fig. 6). The time to reach this regime state is rather short, $\sim$ $0.001t_c$.

Note that in all figures, except for Figs. 1 and 2, the velocity grid has been decimated to enhance clarity.

VII. Code performance

Over the full range of R tested here (more than five orders of magnitude), the code maintained the typical velocity divergence and the field values at the bottom and top boundaries at essentially machine-precision zero, showing that the implemented method is sound.

Also over this range of R, the grid spacing needed to achieve "smooth" fields (*i.e.*, to capture all the physical detail down to the smallest present scales)

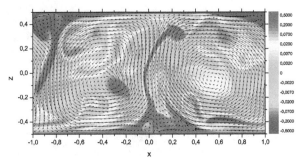

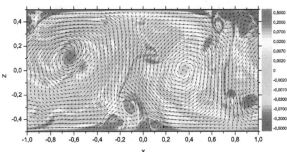

Figure 5: Temperature and velocity fields for $R = 5 \times 10^4 R_c$ at $t = 0.01 t_c$ on a 384×192 grid (velocity decimated to a 64×32 grid).

Figure 6: Temperature and velocity fields for $R = 5 \times 10^5 R_c$ at $t = 0.002 t_c$ on a 512×256 grid (velocity decimated to a 64×32 grid).

is consistent with the width of the (thermal) boundary layer. However, for coarser grids the code still gives qualitatively sound results; typically a checkerboard-like instability develops, but the algorithm keeps it quenched, showing very good stability even in presence of a severe accuracy loss.

The algorithm is also fast: the simulation for $R \sim 10^9$ (see Fig. 6), on a 512×256 grid, took less than one day per simulated minute on a single core of the 3GHz PentiumD CPU on which all our runs were performed, with no code optimization.

It is difficult to find in the literature a directly comparable simulation for more accurate benchmarking. However, from Ref. [6] we can see that, for example, the relaxation to a (steady) regime state in Rayleigh–Bénard convection at $R \lesssim 30000 \sim 20 R_c$ on a ~ 20000-node grid takes 46 minutes on a similar processor at similar speed (3.2GHz Pentium 4), by the method implemented there. In the case of our code, at $R = 20 R_c$ on a 200×100 grid, and with the sole optimization of grouping the real FFTs in pairs (see the 2FFT algorithm in Ref. [19]), the equivalent relaxation took 8 minutes, which is shorter by a factor of ~ 6. But it must be taken into account that with our initial conditions (zero velocities and thermal noise in temperatures) the convection onset is slow, and is followed by a transient stage with strong and disordered convective patterns that decay to the regime state very slowly.

VIII.   Conclusions and outlook

We have been able to show that a Fourier-based pseudospectral method can be adapted to a (admittedly simple) non-free BC setting, at the cost of moderate analytical work on the solutions of Poisson's and Laplace's equations. The method is formulated in primitive variables, but the pressure is not explicitly computed nor referenced, like in a streamfunction-vorticity formulation. It also shares some properties with projection methods, but it decouples the implementation of the incompressibility condition from the time-stepping scheme, allowing great flexibility in the selection of the last. The resulting code is fast and stable, and implements the BCs and the incompressibility condition essentially to machine-precision.

Work on the extension of this scheme to a fully closed Rayleigh-Bénard cell (*i.e.*, with non-free BCs also on the lateral walls) is currently under course.

[1] I De Paul, *Evidence of chaotic heat enhancement in a solar still*, Appl. Thermal Eng. **29**, 1840 (2009).

[2] O Pauluis, *Thermodynamic consistency of the anelastic approximation for a moist atmosphere*, J. Atmos. Sci. **65**, 2719 (2008).

[3] O Pauluis, J Schumacher, *Idealized moist Rayleigh–Bénard convection with piecewise linear equation of state*, Commun. Math. Sci. **8**, 295 (2010).

[4] T Weidauer, J Schumacher, *Moist turbulent Rayleigh–Bénard convection with Neumann and Dirichlet boundary conditions*, Phys. Fluids **24**, 076604 (2012)

[5] B Fornberg, *A practical guide to pseudospectral methods*, Cambridge monographs on applied and computational mathematics, Cambridge University Press, Cambridge, UK (1999).

[6] I Mercader, O Batiste, A Alonso, *An efficient spectral code for incompressible flows in cylindrical geometries*, Computers & Fluids **39**, 215 (2010).

[7] I Mercader, O Batiste, A Alonso, *Continuation of travelling-wave solutions of the Navier–Stokes equations*, Int. J. Numer. Meth. Fluids **52**, 707 (2006).

[8] A J Chorin, *Numerical solution of the Navier–Stokes equations*, Math. Comput. **22**, 745 (1968).

[9] P M Gresho, R L Sani, *On the pressure boundary conditions for the incompressible Navier–Stokes equations*, Int. J. Numer. Meth. Fluids **7**, 1111 (1987).

[10] R L Sani, J Shen, O Pironneau, P M Gresho, *Pressure boundary condition for the time-dependent incompressible Navier–Stokes equations*, Int. J. Numer. Meth. Fluids **50**, 673 (2006).

[11] V Fuka, *PoisFFT - A free parallel fast Poisson solver*, Appl. Math. Comput., **267**, 356 (2015).

[12] E Braverman, M Israeli, A Averbuch, L Vozovoiy, *A fast 3D Poisson Solver of arbitrary order accuracy*, J. Comput. Phys. **144**, 109 (1988).

[13] P M Gresho, D F Griffiths, D J Silvester, *Adaptive time-stepping for incompressible flow. Part I: Scalar advection-diffusion*, SIAM J. Sci. Comput. **30**, 2018 (2008).

[14] P M Gresho, D F Griffiths, D J Silvester, *Adaptive time-stepping for incompressible flow. Part II: Navier–Stokes equations*, SIAM J. Sci. Comput. **32**, 111 (2010).

[15] Y Saad, M H Schultz, *GMRES: A generalized minimal residual algorithm for solving nonsymmetric linear systems*, SIAM J. Sci. Comput. **7**, 856 (1986).

[16] L D Landau, E M Lifshitz, *Fluid mechanics*, Pergamon Press, Oxford (1959).

[17] M C Cross, P C Hohenberg, *Pattern formation outside of equilibrium*, Rev. Mod. Phys. **65**, 851 (1993).

[18] H von Helmholtz, *Über integrale der hydrodynamischen gleichungen, welche den wirbelbewegungen entsprechen*, Celles J. **55**, 25 (1858).

[19] W H Press, S A Teukolsky, W T Vetterling, B P Flannery, *Numerical recipes*, Cambridge University Press, Cambridge, UK (1996).

9

How we move is universal: Scaling in the average shape of human activity

Dante R. Chialvo,[1] Ana María Gonzalez Torrado,[2] Ewa Gudowska-Nowak,[3]
Jeremi K. Ochab,[4] Pedro Montoya,[2] Maciej A. Nowak,[3,4] Enzo Tagliazucchi[5]

Human motor activity is constrained by the rhythmicity of the 24 hours circadian cycle, including the usual 12-15 hours sleep-wake cycle. However, activity fluctuations also appear over a wide range of temporal scales, from days to a few seconds, resulting from the concatenation of a myriad of individual smaller motor events. Furthermore, individuals present different propensity to wakefulness and thus to motor activity throughout the circadian cycle. Are activity fluctuations across temporal scales intrinsically different, or is there a universal description encompassing them? Is this description also universal across individuals, considering the aforementioned variability? Here we establish the presence of universality in motor activity fluctuations based on the empirical study of a month of continuous wristwatch accelerometer recordings. We study the scaling of average fluctuations across temporal scales and determine a universal law characterized by critical exponents α, τ and $1/\mu$. Results are highly reminiscent of the universality described for the average shape of avalanches in systems exhibiting crackling noise. Beyond its theoretical relevance, the present results can be important for developing objective markers of healthy as well as pathological human motor behavior.

[1] Consejo Nacional de Investigaciones Científicas y Tecnológicas (CONICET), Rivadavia 1917, Buenos Aires, Argentina.

[2] Institut Universitari d'Investigacions en Ciències de la Salut (IUNICS) & Universitat de les Illes Balears (UIB), Palma de Mallorca, Spain.

[3] M. Kac Complex Systems Research Center and M. Smoluchowski Institute of Physics, Jagiellonian University, Kraków, Poland.

[4] Biocomplexity Department, Małopolska Center of Biotechnology, Jagiellonian University, Kraków, Poland.

[5] Institute for Medical Psychology, Christian Albrechts University, Kiel, Germany.

I. Introduction

The most obvious periodicity of human (as well as animal) motor activity is the circadian twenty four hours modulation. However, smaller fluctuations are evident on a wide range of temporal scales, from days to a few seconds. Data shows that the activity evolves in bursts of all sizes and durations which are known to be scale-invariant [1–8] regardless of the origins and intended consequences of such activity. Despite the variety of results, the mechanisms underlying the scale-invariant behavior of motor activity remain to be elucidated. Considering the intermittent nature of human motor activity - comprising brief activity excursions separated by periods of quiescence - a natural approach would be to study the average shape of the events, following recent results [9–12] which show that for a large class of processes, the average shape is a scaling function determined mostly by the temporal correlations of the process and its nonlinearities [13].

In the present work, long time series of human motor activity are analyzed, recorded via wristwatch accelerometer, lasting approximately one month. We establish first the presence of truncated scale-invariance in the distribution of the durations of the events as well as in its power spectral density, as described previously in similar type of data. Afterwards, we uncover the average shape of the bursts of activity and derive the scaling function and its associated exponents. Finally, we discuss the origins of such scaling and some possible applications.

II. Materials and methods

The recordings analyzed were part of a larger study and included six healthy, non-smokers, drug-free volunteers (mean age 50.1 years, S.D. = 6.8). The study was approved by the Bioethics Commission of the University of Isles Baleares (Spain). Participants were informed about the procedures and goals of the study, and provided their written consent. After determining their handedness, each subject was provided with a wristwatch-sized activity recorder (Actiwatch from Mini- Mitter Co., OR, USA) measuring acceleration changes in the forearm in any plane. Each data point of activity corresponded to the number of zero crossings in acceleration larger than 0.01 G (sampled at 32 Hz and integrated over a 30-second window length). Records of several thousands of data points were kept in the device's internal memory until being downloaded to a personal computer every week. Subjects wore the device in their non-dominant arm continuously for up to several weeks (mean 28.1 days, S.D.= 4.). After careful visual inspection of the data to exclude sets with gaps (due to subject non-compliance), a combined total of 280 days of data was available for further analysis.

III. Results

For ease of presentation, we will use recordings from a single subject to describe the main results. Nevertheless, results are robust as well as similar for the entire group of subjects in the study. A typical recording is presented in Fig. 1. Panel A shows

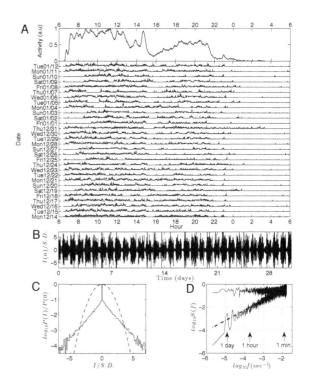

Figure 1: Example data set, distribution of successive increments and their spectral power. Panel A: Time series of activity $x(n)$ recorded continuously from a subject during a month. Individual traces correspond to consecutive days. The top subpanel depicts daily activity averaged over the entire month. Panel B: Time series of successive increments $I(n) = x(n+1) - x(n)$ (normalized by its SD) for the same data. Panel C: Probability density distribution of the time series of successive increments $I(n)$ (continuous line), exhibiting exponential tails (compare with the dotted line, a Gaussian of the same variance). Panel D: Power spectral density (black line) of the time series of successive increments $I(n)$ of panel B. This is scale invariant $S(f) \sim f^\gamma$ with $\gamma = 0.9$ (dashed line). In contrast, for the randomly shuffled increments, the serial correlations vanish and a flat spectral density is obtained (red).

a full month of continuously recorded activity from this subject, who is particularly regular in her daily routines. The subject wakes up with the alarm clock at 6:45 a.m. on week days and has lunch followed by a short nap each day (between 2:00 p.m. and 4:00 p.m. Panel B displays the time series of

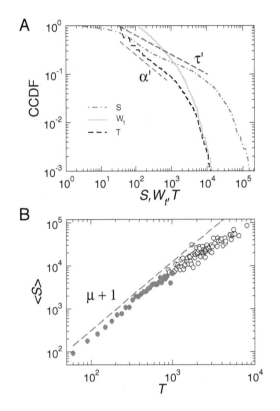

Figure 2: Scaling of activity events in a single subject (same dataset as in Fig. 1). Panel A: The complementary cumulative distribution function (CCDF) for event durations (T) and sizes (S) obeys power-laws with exponents $\alpha' = 0.70$ and $\tau' = 0.44$, respectively (dashed lines). Note that here the densities are cumulative, thus the exponents of the respective PDFs are $\alpha = \alpha' + 1$ and $\tau = \tau' + 1$. The waiting time between events falls exponentially. Panel B: The average size of a given duration is well described (for small T) by $\langle S \rangle(T) \sim T^{\mu+1}$ with $\mu + 1 = 1.59$ (blue dashed line) comparable with results obtained from fitting within the scaling region (red filled symbols) giving $\mu + 1 = 1.61$.

the successive increments of the signal $x(n)$, defined as $I(n) = x(n+1) - x(n)$.

The large-scale statistical features of the time series presented in Fig. 1 are already well known. The density distribution of the successive increments $i(n)$ is non-Gaussian, as can be appreciated by a joint plot with a Gaussian distribution of the same variance (Fig. 1, Panel C). It is known

that the power spectrum of the activity decays as $S(f) \sim f^\beta$ [1, 2]. Because this type of processes are likely to be non-stationary, it is best to estimate the exponents of the spectral density by doing the calculations over the time series of successive increments, whose density distribution is stationary. For instance, for Brownian motion (which is summed white noise), the power spectrum decays $S(f) \sim f^\beta$ with $\beta = -2$ and for white noise $\beta = 0$; the summed time series has an exponent $+2$ larger than the non-summed time series. As discussed in [14], this can be generalized for all self-affine processes: summing a self-affine time series shifts the theoretical power-spectral density exponent by $+2$, and the reverse process is also true: the differences in consecutive values (the "first differences") of a Brownian motion result in white noise, thus taking the first differences shifts the theoretical power-spectral density exponent β by -2. In our case, the exponent obtained for the time series of successive increments $I(n)$ was $\gamma = 0.9$. Thus, the exponent of the raw data is $\beta = \gamma - 2 = -1.1$ [14]. For comparison, the spectral densities of the actual signal and of a surrogate obtained after randomly shuffling the increments are jointly displayed in Panel D of Fig. 1.

To further study the time series from the perspective of individual bursts of activity, we introduce the definition of an event. We consider the time series of activity $x(n)$ and select a threshold value U to be vanishingly small. An event is defined by the consecutive points starting when $x(n) > U$ and ending when $x(n) < U$. This is equivalent to the definition of avalanches in other contexts [9, 15]. In the following part, we will be concerned primarily with the statistics of event lifetimes T, as well as of their average size S and shape. In all subjects, we found that the distributions of event durations and sizes (defined by the area, i.e., the integral of the signal corresponding to the individual events) can be well described, for relatively small values, by a power-law (Fig. 2, Panel A). In contrast, the distribution of waiting times between events demonstrated an exponential decay. In addition to the scale invariance, we found that the longer an event lasted, the stronger the motor activity executed by the subject. The plot of average event size $\langle S \rangle$ as a function of duration T follows a power-law (for small values of T) described by $\langle S \rangle(T) = T^{\mu+1}$ with $\mu + 1 = 1.59$. The exponents in this power-law are

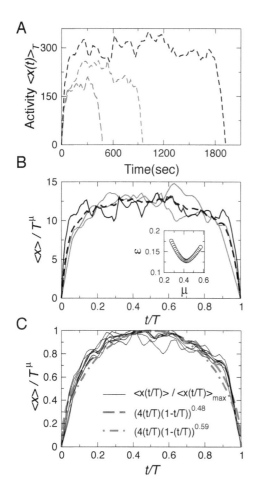

Figure 3: Collapse of events of different duration into a single functional form. Panel A: Three examples of typical events of duration T=480, 960 and 1920 sec.. Panel B: The heterogeneous events shown in Panel A can be collapsed onto the average shape (dashed black line) by normalizing t to t/T and $\langle x(t)\rangle$ to $\langle x(t)\rangle/T^\mu$. The inset shows the cumulative variance for a range of μ. Panel C: The average event shape, i.e., $f_{shape}(t/T)$, recovered from six data sets (thin lines). The best fit using an inverted parabola is shown as a red dashed line ($\mu = 0.49$) as well as the one expected from the critical exponent $\mu = 0.59$ as a dot-dashed blue line.

robust across subjects and to changes of threshold over a reasonable range of values.

This type of scaling is well known in the statistical mechanics of critical phenomena [15]. Examples range from earthquakes [16] to active transport processes in cells [17], crackling noise [11], the statistics of Barkhausen noise in permalloy thin films [10] and plastic deformation of metals [18]. In all these cases, the distributions obey universal functional forms:

$$f(S) \sim S^{-\tau}, \tag{1}$$

$$f(T) \sim T^{-\alpha}, \tag{2}$$

$$\langle S\rangle(T) \sim T^{1/\sigma\nu z}, \tag{3}$$

where f denotes the probability density functions of the size of the event S and its duration T, and $\langle S\rangle(T)$ is the expected size for a given duration. The parameters τ, α and $1/\sigma\nu z$ are the critical exponents of the system and are expected to be independent of the details, being related to each other by the scaling relation:

$$\frac{\alpha - 1}{\tau - 1} = \frac{1}{\sigma\nu z}. \tag{4}$$

We found that the empirical exponents very closely fulfill the expression above. Using the fitting approach introduced by Clauset [19] in the scaling regions depicted in Panel A of Fig. 2, we found $\tau = 1.44$ and $\alpha = 1.70$. Thus, from Eq. (4) a value of $1/\sigma\nu z = \mu + 1 = 1.59$ is expected. The experimental data points are very close to this theoretical expectation (dashed line), especially for the relatively small T values within the scaling region of Panel A (where a linear fit estimates $\mu + 1 = 1.61$), while those for relatively larger T values (corresponding to the cutoff of the distributions) are a bit apart, probably due to undersampling. After repeating this analysis for all subjects in our sample, the average exponents were all within 5% of the reported values.

From scaling arguments, it is expected that the average shape of an event of duration T $\langle x(T,t)\rangle$ scales as :

$$\langle x(T,t)\rangle = T^\mu f_{shape}(t/T). \tag{5}$$

Thus, the shapes of events of different durations T rescaled by μ should collapse on a single scaling function given by $f_{shape}(t/T)$. Note that μ corresponds in this context to the wandering exponent (i.e., the mean squared displacement) of the activity [13, 20].

Examples of this collapse are presented in Panels A and B of Fig. 3. Considering the number of

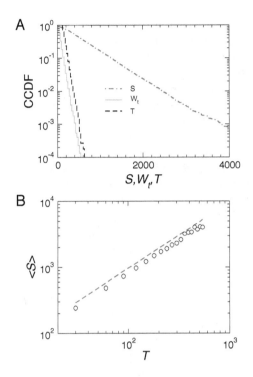

A

B

Figure 4: Scaling is absent in a null model resulting from defining events after randomly reordering the time series $x(n)$. Panel A: Density distributions (CCDF) for event duration, size, and waiting time. All the distributions are exponential (note the logarithmic-linear scale). Panel B: The expected average size for a given duration in the null model is a linear function of T (the dashed line represents the fit with slope 1), therefore, $\mu = 0$ and there is no collapse.

events here averaged (in the order of $N \sim 10^2$), the data collapse is quite satisfactory, while the value of the exponent ($\mu = 0.48$) does not exactly match the one predicted in Eq. (4), $\mu = 0.59$ (likely a consequence of insufficient sampling). To determine the generality of our results, we extended this analysis to six other data sets. For each data set, the value of μ was first determined. Subsequently, the $x(T,t)$ obtained from the events were rescaled with T^μ and their average computed. To account for individual differences in mean activity, shape functions were normalized by their mean value. The results for the six datasets are presented in Panel C of Fig. 3. They can be accurately described by

an inverted parabola, as in other systems previously studied using this method. The best fit disagrees with the empirical functions near their peak, the latter being flatter, likely an effect related to saturation observed in long events.

Finally, we turn to discuss simple null models. We consider two extreme cases, in both of them the raw time series are randomly shuffled to remove serial correlations. In the first case, we remove all temporal correlations by randomly reordering $x(n)$, thus attaining a flat power spectral density. After repeating the above analysis in this surrogate data set, it becomes clear (as shown in Fig. 4) that the scale invariance is absent in all the statistics under study: size S, waiting time W_t and duration T of events (note that the distributions are here plotted using a logarithmic-linear scale). Results in Panel B show that $\mu + 1 = 1$, thus $\mu = 0$, implying that there is not collapse, because with $T^\mu = 1$ in Eq. (5), the amplitude of the individual events remains invariable. To consider the second case, we need first to reorder randomly the time series of increments $I(n)$ and then proceed to integrate the increments. Since each increment is now a random variable, the power spectral density for this surrogate process obeys f^β with $\beta = -2$, and as shown analytically by Baldassari et al. [13], for this case $\mu = 1/2$ and the scaling function is a semicircle. Please note that the fluctuations of human activity described here differ from a simple auto-regressive process: indeed successive increments $I(n)$ are anti-correlated and the power spectral density corresponds to non-trivial power law correlations (i.e., $\beta \neq -2$).

IV.   Discussion

The present findings can be summarized by six stylized facts describing bursts of human activity: I) the spectral density of the time series of activity $x(t)$ obeys a power law, with exponent $\beta \sim 1$; II) successive increments $I(n)$ are anti-correlated with a spectral density obeying a power law with exponent $\gamma \sim 1$, which corresponds to a spectral density for the raw data f^β with $\beta \sim -1$; III) the PDF of the increments $I(n)$ is definitely non-gaussian; IV) the PDF of duration and sizes of events obeys truncated power laws with exponents $1 < \tau < 2$ and $1 < \alpha < 2$; V) the average size of the events scales with its lifetime T as $\langle S \rangle(T) \sim T^\mu$, where $\mu + 1 = (\alpha - 1)/(\tau - 1)$; VI) the time series of individual events can be appropriately rescaled via a transformation of its duration T and amplitude $x(t)$ onto a unique functional shape: $\langle x(T,t) \rangle = T^\mu f_{shape}(t/T)$.

We are aware that these observations are novel only for human activity, because similar statistical regularities of avalanching activity are well known for a large variety of inanimate systems [9–12]. The rescaling of the average shape is not surprising because, placed in the appropriate context, it can be traced back to Mandelbrot's study of the fractal properties of self-affine functions [21]. A curve or a time series are said to be self-affine if a transformation can be found, such that rescaling their x, y coordinates by k and k^μ, respectively, and the variance in y is preserved (with $\mu = 1$ corresponding to self-similarity). In that sense, the successful collapse of the events shape is a trivial consequence of the overall self-affinity of the $x(t)$ time series.

Thus, it is clear that the existence of the scaling uncovered here is not informative per se of the type of mechanism behind: scale-invariance can be constructed via different processes, ranging from critical phenomena [15] to simple stochastic autoregressive dynamics [13, 20]. What is then the mechanism by which the above six facts are generated?

It seems that this question cannot be easily answered by the type of experiments reported here. Fluctuations of this type could have either an intrinsic (i.e., brain-born) origin but also could be the reflection of a collective phenomena (including humans and its environment). In either case, the correlations observed seem to reject the case of independent random events starting and stopping human actions, because neither the distribution of the increments $I(n)$, nor the exponents match the case of a random walk. In terms of brain-born process, it is hard to accept some of the implications of the scaling function in the activity shape. The average parabolic shape means that the very beginning of the motion activity contains information about how long the activity will last, in the same sense that the initial trajectory of a projectile predicts when and where it will land. This proposal is hardly realistic, because there is hardly a reasonable physiological argument in support of any motor planning for the length of time we are observing ($\sim 10^3$ secs). In

terms of collective processes, the results here suggest that the interaction with other humans could determine when and where, on the average, we start and stop moving.

Despite our current relative ignorance, a possibility that sounds interesting is to determine in children, as they grow, if their behavioral product of parental (and otherwise) education are reflected in the shape of their individual scaling function. This seems reasonable given the fact that "tireless running around" is almost a definition of early age well-being, which gives way to less hectic activity as children mature. In the same line of thoughts, if changes in the scaling function can be quantitatively traced to behavioral changes, one could also consider to explore applications of these techniques to monitor eventual progress in the treatment of hyperactivity disturbances such as in the subjects affected by the Attention Deficit Hyperactivity Disorder syndrome. The converse, i.e., cases in which the average activity diminish, as in elderly subjects shall be also explored. Further experiments and analysis should shed light on these possibilities. In, the meantime, the present results provide a guide and six important constraints for the models that should best capture the physics (and biology) of the process.

Acknowledgements - Work supported by National Science Center of Poland (ncn.gov.pl, grant DEC-2011/02/A/ST1/00119); State Secretary for Research and Development (grants PSI2010-19372 and PSI2013-48260) from Spain and by CONICET from Argentina.)

[1] T Nakamura, K Kiyono, K Yoshiuchi, R Nakahara, Z R Struzik, Y Yamamoto, *Universal scaling law in human behavioral organization*, Phys. Rev. Lett. **99**, 138103 (2007).

[2] T Nakamura, et al., *Of mice and men - universality and breakdown of behavioral organization*, PLoS ONE **3**, e2050 (2008).

[3] K Hu, P C Ivanov, Z Chen, M F Hilton, H E Stanley, S A Shea, *Non-random fluctuations and multi-scale dynamics regulation of human activity*, Physica A **337**, 307 (2004).

[4] L A N Amaral, D J B Soares, L R da Silva, L S Lucena, M Saito, H Kumano, N Aoyagi, Y Yamamoto, *Power law temporal auto-correlations in day-long records of human physical activity and their alteration with disease*, Europhys. Lett. **66**, 448 (2004).

[5] C Anteneodo, D R Chialvo, *Unravelling the fluctuations of animal motor activity*, Chaos **19**, 033123 (2009).

[6] K Christensen, D Papavassiliou, A de Figueiredo, N R Franks, A B Sendova-Franks, *Universality in ant behaviour*, J. R. Soc. Interface **12**, 20140985 (2014).

[7] A Proekt, J Banavar, A Maritan, D Pfaff, *Scale invariance in the dynamics of spontaneous behavior*, Proc Natl Acad Sci USA **109**, 10564 (2012).

[8] J K Ochab, et al., *Scale-free fluctuations in behavioral performance: Delineating changes in spontaneous behavior of humans with induced sleep deficiency*, PLoS ONE **9**, e107542 (2014).

[9] L Laurson, X Illa, S Santucci, K T Tallakstad, K J Maloy, M J Alava, *Evolution of the average avalanche shape with the universality class*, Nature Comm. **4**, 2927 (2013).

[10] S Papanikolaou, F Bohn, R L Sommer, G Durin, S Zapperi, J P Sethna, *Universality beyond power laws and the average avalanche shape*, Nature Phys. **7**, 316 (2011).

[11] J P Sethna, K A Dahmen, C R Myers, *Crackling noise*, Nature **410**, 242 (2001).

[12] N Friedman, S Ito, B A W Brinkman, M Shimono, R E L DeVille, K A Dahmen, J M Beggs, T C Butler, *Universal critical dynamics in high resolution neuronal avalanche data*, Phys. Rev. Lett. **108**, 208102 (2012).

[13] A Baldassarri, F Colaiori, C Castellano, *Average shape of a fluctuation: Universality in excursions of stochastic processes*, Phys. Rev. Lett. **90**, 060601 (2003).

[14] B D Malamud, D L Turcotte, *Self-affine time series: I. Generation and analyses*, Adv. Geophys. **40**, 1 (1999).

[15] P Bak. *How nature works. The science of self-organized criticality*, Copernicus, New York (1996).

[16] B Gutenberg, C F Richter, *Magnitude and energy of earthquakes*, Ann. Geofis. **9**, 1 (1956).

[17] B Wang, J Kuo, S Granick, *Burst of active transport in living cells*, Phys. Rev. Lett. **111**, 208102 (2013).

[18] L Laurson, M J Alava, *1/f noise and avalanche scaling in plastic deformation*, Phys. Rev. E **74**, 066106 (2006).

[19] A Clauset, C R Shalizi, M E J. Newman, *Power-law distributions in empirical data*, SIAM Rev. **51**, 661 (2009).

[20] F Colaiori, A Baldassarri, C Castellano, *Average trajectory of returning walks*, Phys. Rev. E **69**, 041105 (2004).

[21] B B Mandelbrot, *Self-affine fractals and fractal dimension*, Physica Scripta **32**, 257 (1985).

10

Fluctuation-induced transport: From the very small to the very large scales

G. P. Suárez,[1] M. Hoyuelos,[1*] D. R. Chialvo[2]

The study of fluctuation-induced transport is concerned with the directed motion of particles on a substrate when subjected to a fluctuating external field. Work over the last two decades provides now precise clues on how the average transport depends on three fundamental aspects: the shape of the substrate, the correlations of the fluctuations and the mass, geometry, interaction and density of the particles. These three aspects, reviewed here, acquire additional relevance because the same notions apply to a bewildering variety of problems at very different scales, from the small nano or micro-scale, where thermal fluctuations effects dominate, up to very large scales including ubiquitous cooperative phenomena in granular materials.

I. Introduction

Much of the efforts devoted to particle transport were triggered by the famous challenge at very small scales presented by Feynman in 1959: "A biological system can be exceedingly small. Many of the cells are very tiny, but they are very active. (...) Consider the possibility that we too can make a thing very small, which does what we want — that we can manufacture an object that maneuvers at that level!" [1].

At the scales discussed by Feynman, our most usual notions of work, energy and transport seem to break down, including some counterintuitive obser-

vations. As discussed in these notes, these findings are not restricted to small scales since work in the last decades shows similar dynamics arising anytime there is a peculiar interplay of fluctuations, nonlinearity and correlations resulting in various classes of fluctuation-induced transport.

To visualize the problem, consider a *gedankenexperiment* involving, for the sake of discussion, our desk. Elementary physics explains how all the objects at the desk stay in place and/or which forces are needed to displace them. Now, consider the imaginary case in which we progressively shrink all the objects up to a size of a few nanometers. It will be noticed that while at the natural scale objects remain steady without any energy expenditure, at the nanometers scale things move around, our "nano cell phone" which was quiet at the natural scale desk, moves and falls off the "nano desk. This exercise reminds us that at the *Brownian domain*, energy would be required *even to stay quiet* since the basic macroscopic methods of controlling energy flow no longer remain valid. This nonintuitive phenomenon in the function of molecular machines was described by Astumian as follows [2]:

---
*E-mail: hoyuelos@mdp.edu.ar

[1] Instituto de Investigaciones Físicas de Mar del Plata (IFIMAR - CONICET) and Departamento de Física, Facultad de Ciencias Exactas y Naturales, Universidad Nacional de Mar del Plata, Deán Funes 3350, 7600 Mar del Plata, Argentina.

[2] Consejo Nacional de Investigaciones Científicas y Técnicas (CONICET), Godoy Cruz 2290, Buenos Aires, Argentina.

"any microscopic machine must either work with Brownian motion or fight against it, and the former seems to be the preferable choice". Analogous observations, with some additional caveats due to inertial forces, can be made if instead of shrinking the mass, we apply an increasingly large external fluctuating field, making now our real size desk to shake around.

This brief review is dedicated to discuss the essence of three elementary results of fluctuation-induced transport including the potential shape, the correlations of the fluctuations and the particle interactions and how they work, calling attention to some common lessons that can be borrowed from problems in apparently far apart scales and fields, from cellular biology to technological applications and applied physics. It should be noted that it is not our intention to cover the extent of the field, this is neither a fair, nor historically correct, exhaustive or updated review of the relevant literature; it only encompasses some interesting results which, in our opinion, warrant further exploration. The reader will find comprehensive reviews covering specific topics, including those on Brownian motors in [3–7], on the more general subject of molecular motors in [2, 8–12], on a more biological perspective of molecular springs and ratchets in [13], or on a systematic analysis of the space-time symmetries of the equations in [14].

The paper is organized as follows. The next section revisits pioneer works on these types of problems, carried on a hundred years ago. Next, we discuss the three fundamental aspects of the problem, including the substrate, the correlations of the fluctuations and the particle interactions. We start by briefly introducing the different realizations of fluctuation-induced transport as popularized two decades ago, i.e., in the so-called correlation ratchets. After that, the two elementary ways to break the symmetry are reviewed, either in the temporal or in the spatial aspects of the system, to conclude introducing yet another way to affect transport, the correlations born out of many particle interactions. The review closes with a discussion of some applications and new directions.

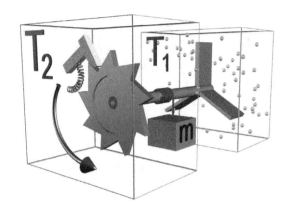

Figure 1: Feynman's imaginary microscopic ratchet, comprised by vanes, a pawl with a spring, two thermal baths at temperatures $T_1 > T_2$, an axle and wheel, and a load m.

II. Smoluchowski-Feynman's ratchet as a heat engine

Feynman famous lectures [15] include an imaginary microscopic ratchet device to illustrate the second law of thermodynamics. The basic idea belongs to Smoluchowski who discussed it during a conference talk in Münster in 1912 (published as proceedings-article in Ref. [16]). As seen in Fig. 1, it consists of a ratchet, a paw and a spring, vanes, two thermal baths at temperatures $T_1 > T_2$, an axle and wheel, and a load. The ratchet is free to rotate in one direction, but rotation in the opposite direction is prevented by the pawl. The system is assumed small so that molecules of the gas at temperature T_1 that collide with the vanes produce large fluctuations in the rotation of the axle. Fluctuations are rectified by the pawl. The net effect is a continuous rotation of the axle that can be used to produce work by, for example, lifting a weight against gravity. The pawl becomes a materialization of Maxwell's demon, a small agent able to manipulate fluctuations at a microscopic level in order to violate the second law of thermodynamics, since in this case a given amount of heat is completely transformed into work. A closer inspection shows that such violation does not really take place. Feynman demonstrated that, if $T_1 = T_2$, no net rotation of the axle is produced. The reason is that the pawl has its own thermal fluctuations that, from time to

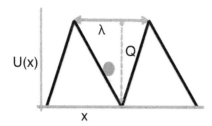

Figure 2: Typical ratchet potential $U(x)$.

time, allow a tooth of the ratchet to slip in the opposite direction. Not even demons are free from thermal fluctuations. In order for the machine to work as intended, the pawl should be colder than the vanes, $T_2 < T_1$. But in this case, there is a heat flux between thermal baths. The mechanical link between vanes and ratchet through the axle implies that the baths are not thermally isolated [17], even when the materials are perfect insulators. The system performs as a heat engine: some work is generated while some heat is transferred from a cold reservoir to a hot reservoir.

In summary, Feynmann's ratchet —and Brownian motors— actually work, but without violating the laws of thermodynamics.

III. Breaking the symmetry: time, space and interactions

Feynman's deep thinking motivated an entire generation of models around the same idea. The model ratchet is a fluctuations-driven overdamped nonlinear dynamical system described by

$$\dot{x} = -U'(x) + f(t),$$
$$U(x) = U(x + \lambda), \quad \langle f(t) \rangle = 0, \tag{1}$$

where $U(x)$ is a periodic potential, such as the one illustrated in Fig. 2, and $f(t)$ is zero-mean fluctuation of some type. In general, the initial theoretical problem is to find the stationary current density $j = \langle \dot{x}(t) \rangle$ in the ratchet given the statistical properties of the fluctuation $f(t)$ and the shape of $U(x)$, and to be able to determine the most efficient conditions for the transformation of fluctuations into a net current.

Multiple variations and extensions of the same problem were studied in the 90s resulting on a jargon of names such as on-off ratchets [8], fluctuating potential ratchets [18,19], temperature ratchets [20,21], chiral ratchets [22–24], and so on. In any case, three elements are always present: a particle which eventually will execute some motion and two forces, one coming from the external applied field and another given by the particular shape of the potential (i.e., the substrate where the particle resides). Thus, an isolated particle "feels" two forces, but while such information is available to an observer, it is important to realize that the particle has no way to distinguish or separate these sources. Thus, the break of symmetry resulting in average directed motion of a particle could come from either spatial or temporal sources. Yet, a third force needs to be considered in the cases in which the concentration of particles becomes relevant and then particles mutual interactions are not negligible anymore, an aspect crucial to understand flow in channels. We will consider all these cases in the following sections.

i. Asymmetries in the substrate

Figure 3 summarizes the two basic ways in which asymmetries in space (or some other degree of freedom of the system, such as phase [25]) contribute to noise-induced transport. The common situation involves an asymmetric periodic potential that breaks the spatial inversion symmetry combined with a temporal, zero mean, forcing periodicity. In Panel A, the case of turning on and off the asymmetric potential is depicted and Panel B shows the case in which a tilting force is added.

The first important result was due to Magnasco in [19] who considered the case of the piecewise linear potential $U(x)$ shown in Fig. 4A, which is exactly solvable [26] for slow fluctuation $f(t)$; it has a characteristic time much larger than the ratchet's relaxation time. The potential is periodic and extends to infinity in both directions. λ measures the spacing of the wells, λ_1 and λ_2 the inverse steepnesses of the potential in opposite directions out of the wells, and Q the well depths. The particle undergoes overdamped Brownian motion due to its coupling with a thermal bath of temperature T, and an external driving $F(t)$ which represents the forces. These two ingredients compose what

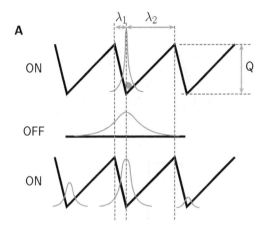

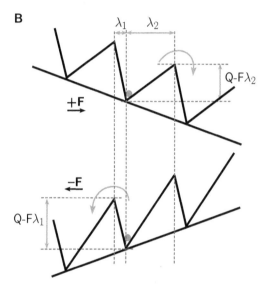

Figure 3: Spatial asymmetry. Panel A: The so-called flashing ratchet is a type of ratchet in which an asymmetric potential is periodically switched off and on. Particles (green circles) diffuse evenly during the off period while the asymmetric potential favors the drift in one direction, producing a net transport to the left. Panel B: In a rocked ratchet, an asymmetric potential is tilted periodically determining a (right directed) transport of the particle trough the relative lowest $Q-F\lambda_2$ value of the right potential barrier.

we called the fluctuation $f(t)$. The expression for the current in the adiabatic limit, which measures the work done by the ratchet was shown to be

$$J(F) = \frac{P_2^2 \sinh(\lambda F/2kT)}{kT \left(\frac{\lambda}{Q}\right)^2 P_3 - \left(\frac{\lambda}{Q}\right) P_1 P_2 \sinh\left(\frac{\lambda F}{2kT}\right)}, \quad (2)$$

$$P_1 = \delta + \frac{\lambda^2 - \delta^2}{4}\frac{F}{Q}, \quad P_2 = \left[1 - \frac{\delta F}{2Q}\right]^2 - \left(\frac{\lambda F}{2Q}\right)^2,$$

$$P_3 = \cosh[(Q - \delta F/2)/kT] - \cosh(\lambda F/2kT),$$

where $\lambda = \lambda_1 + \lambda_2$ and $\delta = \lambda_1 - \lambda_2$. The average current, the quantity of primary interest, is given by

$$j = \langle J \rangle = \frac{1}{\tau} \int_0^\tau J(F(t))\, dt, \quad (3)$$

where τ is the period of the driving force $F(t)$, which is assumed longer than any other time scale of the system in this adiabatic limit. The current is maximized for a given value of the periodic forcing amplitude. Interestingly, numerical computations showed robustness in the results when the forcing is not periodic. The key feature is that it should have a long time correlation.

According to Magnasco, "all that is needed to generate motion and forces in the Brownian domain is loss of symmetry and substantially long time correlations" [19]. Indeed, if the forcing is white noise, the system is at thermal equilibrium and $j = 0$. However, if the fluctuation auto-correlations are non-vanishing, i.e., for colored noise, the system is no longer in thermal equilibrium, and in general $j \neq 0$. Since onset of a current means breaking the "right-left" symmetry, currents may only arise, in the case of additive noise, if the potential $U(x)$ is asymmetric with respect to its extrema. It could be argued [27] that the emergence of current can be viewed as an example of "temporal order coming out of disorder", since the current is apparently time-irreversible, whereas stationary noise does not distinguish "future" from the "past"; we notice, however, that Eq. (1) implies relaxation and is thus time-irreversible itself.

The flashing or pulsating ratchet depicted in Panel A of Fig. 3 was introduced in [28] and re-introduced in a more general theoretical context in [29]. Despite the huge structural complexity of biological Brownian motors, the majority of the models are compatible with a simplified description based on the flashing ratchet. The description is in terms of only one variable x that may represent, for example, the position of a molecule or the coordinate of a complex reaction with many intermediate steps. The environment, composed by some aqueous solution acts, on one hand, as a heat bath and, on the other hand, as a source or sink of ATP, ADP and P_i molecules of the chemical reaction cycle that provides energy to the motor. In this simplified model, a periodic asymmetric potential is periodically turned on and off, as shown in Fig. 3A. The situation is generalized to stochastic variations of the potential with a characteristic correlation time. As happens for the rocked ratchet, the current vanishes for zero correlation time, or fast pulsating limit (white noise). It also disappears in the slow pulsating limit, i.e., when the potential is left on or off for a diverging time. There is an optimum value of the correlation time that maximizes the current.

A recent example of an experimental realization of a rocked ratchet in a mesoscopic scale can be found in [30], where dielectric particles suspended in water are affected by a ratchet potential given by a periodic and asymmetric light pattern.

ii.   Temporal asymmetries

Figure 4B summarizes one type of ratchet in which the higher order statistics of the driving force can be responsible for the transport. Indeed, the work of Millonas [25, 31, 32] and others [33] showed that directed motion can be induced with an unbiased driving force, deterministic or stochastic, as long as it has asymmetric correlations: non zero odd correlation of order higher than one [27].

The case analyzed in the seminal work of Magnasco [19] only considered $F(t)$ symmetric in time $F(t) = F(n\tau - t)$. Instead, the work of Millonas [31] considered the same setting but studying a more general case in which the driving force still is non biased *zero mean*, $\langle F(t) \rangle = 0$, but which is asymmetric in time,

$$F(t) = \begin{cases} \left(\frac{1+\epsilon}{1-\epsilon}\right) A & 0 \le t < \frac{\tau(1-\epsilon)}{2}, \quad \mathrm{mod}\ \tau \\ -A & \frac{\tau(1-\epsilon)}{2} < t \le \tau, \quad \mathrm{mod}\ \tau \end{cases}$$

(4)

as shown in Fig. 4B. In this case, the time averaged

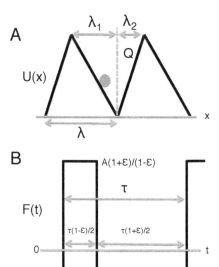

Figure 4: Panel A: The simplest piecewise ratchet potential, where the spatial degree of asymmetry is given by the parameter $\delta = \lambda_1 - \lambda_2$. Panel B: Fluctuation's temporal asymmetry. The driving force $F(t)$ preserves the zero mean $\langle F(t) \rangle = 0$. The temporal asymmetry is given by the parameter ϵ.

current can be easily calculated,

$$\langle J \rangle = \frac{1}{2}(1+\epsilon)J(-A)$$
$$+ \frac{1}{2}(1-\epsilon)J((1+\epsilon)A/(1-\epsilon))$$

(5)

Solving for different values of parameters, it was shown that the current is a peaked function both of kT (see Fig. 5A) and of the amplitude A of the driving. As expected, the driving, the potential, and the thermal noise in fact play cooperative roles. For low temperatures, any transport depends on very large A values, while for large noise the features of the potential and of the driving are washed out.

The most striking results are concerning the competition between the temporal asymmetry and the spatial asymmetry, as pictured in Fig. 5B, resulting on the switching of the direction of the current as the asymmetry factor ϵ is varied. This reversal represents the competition of the spatial asymmetry, which dominates for small ϵ an the temporal asymmetry, which dominates for large ϵ.

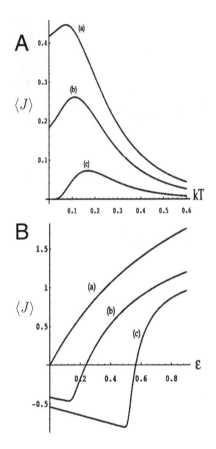

Figure 5: Temporally asymmetric fluctuations with mean zero can optimize or reverse the current in the ratchet of Fig. 4 (with $Q = 1, \lambda = 1$). Panel A corresponds to the case of a symmetric potential (i.e., $\delta = 0$) which shows a peaked function of the net current $\langle J \rangle$ as a function of temperature kT for $\epsilon = 1$ and three values of the driving amplitude $A = 1, A = 0.8, A = 0.5$ labeled (a), (b) and (c), respectively. Panel B shows $\langle J \rangle$ versus the temporal asymmetry parameter ϵ for three asymmetries in the shapes of the potential. Curve (a) is for a symmetric potential, (i.e., $\delta = 0$) and those labeled (b) and (c) for two cases of asymmetric potentials $\delta = -0.3$ and $\delta = -0.7$, respectively (with $kT = 0.01$, and $A = 2.1$)

Temporal asymmetry and spatial asymmetry relate to the problem of nonequilibrium transport in precisely the same way. In both cases, a net effect arises due to an interplay between the strength of a fluctuation, the time it acts, and the underlying dynamics. In the case of a spatial asymmetry, a fluctuation to the right with a given strength which lasts a given time will tend to take the system over the right-hand barrier while the same fluctuation with sign reversed does not lift it over the left-hand barrier. In the case of temporal asymmetry, the probabilities of the fluctuations to the right or to left are different, so the net effect arises in the absence of spatial asymmetries. What both of them show is that even a subtle asymmetry in the *shape* of the potential or in the *shape* of the spectral properties of the noise will give rise to an effect even when the net force due to each vanishes.

The time asymmetry of the mean zero fluctuations discussed above can be cast in several different ways. Dichotomic noise (a type of "Kubo-Anderson" process) was used to demonstrate phase transport in a pair of Josephson junctions [25]. There are also types of continuous noise exhibiting similar asymmetry, including shot noise (common in quantum electronics) which are of this type. Mean zero shot noise, which is temporally asymmetric, can be produced if the frequency and amplitude distribution are slightly different for positive and negative fundamental pulses. Another trivial example of temporally asymmetric driving force is a simple bi-harmonic signal which constitutes a curiosity since it results from adding two (zero mean symmetric) periodic process of harmonic frequencies.

iii.   Particle interactions

The previous discussions were limited to the cases in which an isolated or a few particles were present in the potential. As the concentration is increased, interaction among particles becomes relevant, and it can be the cause of a reduction, and even of a reversal [35–38] of the current. We present two examples.

a.   Vortex current in a 2D array of Josephson junctions

Current reversal has been experimentally observed in a two dimensional array of Josephson junctions [39]. It was numerically analyzed in [40]. A ratchet potential for vortices is generated by modulating the gap between superconducting islands. The density of vortices is controlled by an external magnetic

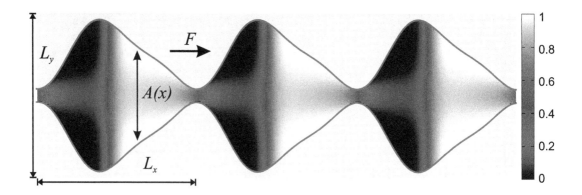

Figure 6: Diffusion in a periodic channel with asymmetric cavities when a force to the right is applied; $A(x)$ is the witdth of the channel. Parameters: total average concentration $c = 0.5$; force $\beta\,a\,F = 0.5$, where a is the lattice spacing; $L_x = L_y = 100\,a$.

field. There is a repulsive vortex-vortex interaction. The results show a preferred direction of the vortex motion, parallel to the ratchet modulation, when an alternating force is applied. But as the vortex concentration is increased, this direction is reversed for appropriate values of the periodic forcing intensity. (Vortex current reversal is also observed for a fixed value of the concentration when the periodic forcing, or AC current amplitude, is varied, see Fig. 4 in [39] or Fig. 2 in [40]).

The vortex current reversal produced by the increase of concentration is a consequence of the following symmetry [39]. Let us consider that the external magnetic field is such that positive vortices are produced. For small concentration (frustration parameter between 0 and 1/2), we have a small discrete number of positive vortices. For large concentration (frustration between 1/2 and 1), we can consider that there is a background of positive vortices in which some negative vortices move. But the movement of this negative vortices is in the opposite direction. For them, the ratchet potential is inverted so the rectification effect of the ratchet is inverted too.

b.   Particle diffusion in a channel with asymmetric cavities

The same effect is observed in a different context. In the next paragraphs, we refer to the hard core interaction between particles that diffuse in a channel with a transverse section $A(x)$ that has

a ratchet shape, see Fig. 6. An external periodic forcing is applied in the direction of the channel. There is a particle-hole symmetry. But before going into the interaction effects, let us consider the low concentration regime, where interactions can be neglected. Several interesting experiments have been performed with particles suspended in a liquid and contained in a channel qualitatively as the one shown in Fig. 6. There are basically two ways to apply the periodic external forcing. In one case, a periodic variation of the pressure is used: particles are drifted back and forth by the movement of the liquid; see [41, 42] and the critical report [43] (cavities of order 5 μm). In the other case, the liquid remains still and the force is directly applied on the particles by an external field as, for example, an electric field on charged particles [44] (cavities of order 50 μm).

Such a system has been proposed for separation of particles of different size [45]. The idea is based on the difference between rectification effects for different size particles. When a periodic —unbiased— forcing is applied, particles move in the forward direction because of the ratchet; but, in general, larger particles move faster than smaller ones. Now we apply a bias, a constant force in the backward direction that reduces the velocity of the larger particles and *reverses* the velocity of the smaller ones. Then we have that larger particles end up in one extreme of the channel and smaller particles in the opposite one, with an estimated purity of 99.997 % according to the authors of [45].

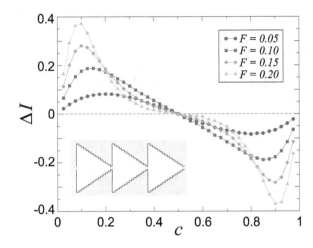

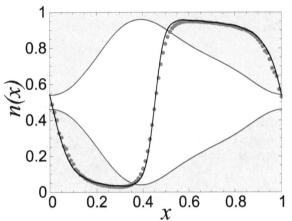

Figure 7: Rectified particle current ΔI against average particle concentration c for different force amplitudes. Units of ΔI: D/a^2, where D is the diffusion coefficient and a is the lattice spacing; units of F: $(\beta a)^{-1}$. Lower left inset: scheme of the channel composed by an array of triangular cavities.

Figure 8: Particle concentration against longitudinal position x for one cavity of the channel depicted in Fig. 6, in a stationary state (cavity length normalized to 1). Dots correspond to Monte Carlo simulations. The curve is obtained from numerical integration of (7). Concetration $c = 0.5$, more details in [49].

The Fick-Jacobs equation [46] gives an appropriate description of the particle density in the channel as long as its dependence on the transverse direction, y, can be neglected, i.e., $n(x, y, t) \simeq n(x, t)$. Let us consider the transverse integral of the concentration: $\rho(x, t) = \int dy\, n(x, y, t) \simeq A(x) n(x, t)$ (the two-dimensional channel can be easily extended to a three-dimensional tube). If D is the diffusion coefficient and $F(t)$ is the total applied force, the Fick-Jacobs equation is

$$\frac{\partial \rho(x, t)}{\partial t} = \frac{\partial}{\partial x}\left[D\left(\frac{\partial \rho}{\partial x} + \beta \frac{\partial H}{\partial x}\rho\right)\right], \quad (6)$$

where $\frac{\partial H}{\partial x} = -F(t) - \beta^{-1}\frac{d}{dx}\ln A(x)$ and $\beta^{-1} = kT$. The expression $\beta^{-1}\ln A(x)$ is called entropic potential due to the similarity with the thermodynamic relation among energy, free energy, temperature and entropy: $H = U - TS$, with $\frac{\partial U}{\partial x} = -F(t)$. In a first approximation, the diffusion coefficient is constant; a further refinement considers a dependence on $A'(x)$, see [47].

Now, let us consider the hard core interaction between particles, of the same size, diffusing in a lattice. A jump of a particle to the right is equivalent to a jump of a hole to the left. A concentration c of

particles subjected to a force F is equivalent to a concentration $1-c$ of holes subjected to a force $-F$. This symmetry is the cause of the shape of Fig. 7, where Monte Carlo results of the rectified current ΔI against the average concentration c for different values of the forcing amplitude is plotted [48]. A square wave in the limit of low frequency was used for the applied force; in this limit, the rectified current is equal to the difference between the current for the force in the positive phase and the current for the force in the negative phase. The particle-hole symmetry is evident in the figure: changing $c \to 1 - c$ and $\Delta I \to -\Delta I$ (a consequence of the change $F \to -F$), we recover the same curves. Le us note the current reversal for large concentration. It is the same effect that was mentioned in the previous section for vortex current in 2D arrays of Josephson junctions. A description based on the Fick-Jacobs equation is also possible for particles with hard-core [49]. Its derivation starts from the non-linear Fokker-Planck equation for fermions [50], where Pauli exclusion principle plays the role of the hard core interaction. Following the same steps used for derivation of the linear Fick-Jacobs equation [47], we can arrive at the following non linear version:

$$\frac{\partial n(x,t)}{\partial t} = \frac{1}{A}\frac{\partial}{\partial x}DA\left(\frac{\partial n}{\partial x} - \beta Fn\left(1-n\right)\right). \quad (7)$$

The non linear term, $n\left(1-n\right)$, is the responsible of the interaction. Fig. 8 shows numerical integration of (7) in good agreement with Monte Carlo simulation results.

IV.   Engineers knew it...

The principles discussed above ruling the correlation ratchets have, in a macro scale, important technological applications. Of course, some of them were applied even before a detailed statistical understanding was available. Vibratory conveyors, or vibratory bowl feeders, are regularly used in many branches of industry such as food processing, synthetic materials or small-parts assembly mechanics, to mention just a few [51]. The conveying speed of these devices was theoretically and experimentally studied, for example, in [52] and references cited therein.

There are many parameters involved in the operation of a vibratory conveyor: amplitude, frequency and mode of the vibrations, inclination angle and friction coefficient are only some of them. A classification in terms of the vibratory modes is as follows: sliding (linear horizontal vibration), ratcheting (linear vertical vibration) or throwing (circular, elliptical or linear tilted vibration).

For throw conveyors, the material being transported loses contact with the through during part of the cycle, see Fig. 9A. Appropriate for granular materials o small objects, particles are forced to perform repeated short flights with a preferred direction, combined with rest and slide phases.

The other two types: sliding and ratcheting involve temporal and spatial asymmetries, respectively.

The sliding type of vibratory conveyors allows transport over a deck that vibrates back and forth with asymmetric motion (see Fig. 5A) in the horizontal direction. The particle or object moves relative to the deck due to alternate stick and slip steps driven by the asymmetric oscillations, as shown in Fig. 9B.

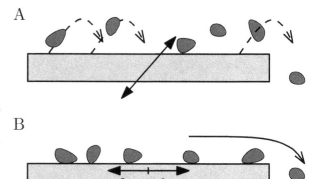

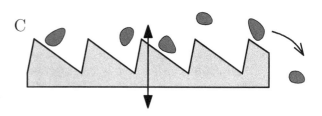

Figure 9: Three types of vibratory conveyors which share some of the principles of small scale ratchets. Panel A: throwing conveyor with linear tilted vibration. Panel B: sliding conveyor; transport is induced by asymmetric horizontal oscillations with zero mean, of the kind shown in Fig. 4B. Panel C: ratchet conveyor with vertical oscillations; similar to the flashing ratchet of Fig. 3A.

The ratchet conveyor achieves transport of granular material using vertical vibrations [38, 53]. Directed motion is caused by the broken space symmetry of the deck's surface, given by a sawtooth-shaped profile, see Fig. 9C. The ratchet conveyor shares qualitative features with the flash or pulsating ratchet depicted in Fig. 3A. One difference is that it includes a ballistic flight phase.

V.   Conclusions

Spatial or temporal asymmetries, or both, are able to generate directed motion in the presence of fluctuations. In addition to a thermal bath, fluctuations with large correlation time, compared to the characteristic relaxation time of the system, should be included. During the last decades, simple models based on these ideas provided a deeper understanding of the complex biological machinery at the nano scale. This success stimulated the study of ratchets in a wide variety of contexts, and in larger scales. Interactions among transported particles are relevant for high concentration; most noticeable, they may produce an inversion of the purported motion direction.

Vibrations inducing directed motion are used in industry for the transport of small —macroscopic— objects since, at least, around 1950. Vibratory conveyors applied the qualitative features of ratchets, with space or time asymmetries, before a detailed theoretical understanding was available. Half a century ago, Feynman called attention to the fact that in his view, "there's plenty of room at the bottom" [1]. We can safely conclude that, even today, there is plenty of room at the top as well.

Acknowledgements - This work was partially supported by Consejo Nacional de Investigaciones Científicas y Técnicas (CONICET, Argentina)

---

[1] R P Feynman, *There's plenty of room at the bottom*, Caltech's Engineering & Science Magazine, Pasadena (1960).

[2] R D Astumian, *Making molecules into motors*, Sci. Am. **285**, 57 (2001).

[3] P Reimann, *Brownian motors: Noisy transport far from equilibrium*, Phys. Rep. **361**, 57 (2002).

[4] P Reimann, P Hänggi, *Introduction to the physics of Brownian motors*, Appl. Phys. A: Mater. Sci. Process. **75**, 169 (2002).

[5] R D Astumian, P Hänggi, *Brownian motors*, Phys. Today **55**, 33 (2002).

[6] J M R Parrondo, B J de Cisneros, *Energetics of Brownian motors: A review*, Appl. Phys. A **75**, 179 (2002).

[7] P Hänggi, F Marchesoni, *Artificial Brownian motors: Controlling transport on the nanoscale*, Rev. Mod. Phys. **81**, 387 (2009).

[8] F Jülicher, A Ajdari, J Prost, *Modeling Molecular Motors*, Rev. Mod. Phys. **69**, 1269 (1997).

[9] M Schliwa, G Woehlke, *Molecular motors*, Nature **422**, 759 (2003).

[10] W R Browne, B L Feringa, *Making molecular machines work*, Nature Nanotech. **1**, 25 (2006).

[11] M von Delius, D A Leigh, *Walking molecules*, Chem. Soc. Rev. **40**, 3656 (2011)

[12] D Chowdhury, *Stochastic mechano-chemical kinetics of molecular motors: A multidisciplinary enterprise from a physicist's perspective*, Phys. Rep. **529**, 1 (2013).

[13] L Mahadevan, P Matsudaira, *Motility powered by supramolecular springs and ratchets*, Science **288**, 95 (2000).

[14] S Denisov, S Flach, P Hänggi, *Tunable transport with broken space-time symmetries*, Phys. Rep. **538**, 77 (2014).

[15] R P Feynman, R B Leighton, M Sands, *The Feynman Lectures on Physics*, Addison-Wesley, MA (1966).

[16] M R von Smoluchowski, *Experimentell nachweisbare derüblichen Thermodynamik widersprechende Molekularphänomene*, Physik. Zeitschr. **13**, 1069 (1912).

[17] J M R Parrondo, P Español, *Criticism of Feynman's analysis of the ratchet as an engine*, Am. J. Phys. **64**, 1125 (1996).

[18] R D Astumian, M Bier, *Fluctuation driven ratchets: Molecular motors*, Phys. Rev. Lett. **72**, 1766 (1994).

[19] M O Magnasco, *Forced thermal ratchets*, Phys. Rev. Lett. **71**, 1477 (1993).

[20] P Reimann, R Bartussek, R Häussler, P Hänggi, *Brownian Motors Driven by Temperature Oscillations*, Phys. Lett. A **215**, 26 (1996).

[21] J D Bao, *Directed current of Brownian ratchet randomly circulating between two thermal sources*, Physica A **273**, 286 (1999).

[22] Z C Tu, Z C Ou-Yang, *A molecular motor constructed from a double-walled carbon nanotube driven by temperature variation*, J. Phys.: Condens. Matter **16**, 1287 (2004).

[23] Z C Tu, X Hu, *Molecular motor constructed from a double-walled carbon nanotube driven by axially varying voltage*, Phys. Rev. B **72**, 033404 (2005).

[24] M van den Broeck, C van den Broeck, *Chiral brownian heat pump*, Phys. Rev. Lett. **100**, 130601 (2008).

[25] M M Millonas, D R Chialvo, *Nonequilibrium fluctuation-induced phase transport in Josephson junctions*, Phys. Rev. E **53**, 2239 (1996).

[26] The straightforward techniques can be found earlier in H. Risken, *The Fokker-Planck Equation*, Springer-Verlag (2nd Ed.) (1984).

[27] M M Millonas, *Self-consistent microscopic theory of fluctuation-induced transport*, Phys. Rev. Lett. **74**, 10 (1995).

[28] A L R Bug, B J Berne, *Shaking-induced transition to a nonequilibrium state*, Phys. Rev. Lett. **59**, 948 (1987).

[29] A Ajdari, J Prost, *Mouvement induit par un potentiel periodique de basse symmetrie: dielectrophorese pulsee*, C. R. Acad. Sci. Paris Sér. II **315**, 1635 (1992).

[30] A V Arzola, K Volke-Sepúlveda, J L Mateos, *Experimental control of transport and current reversals in a deterministic optical rocking ratchet*, Phys. Rev. Lett. **106**, 168104 (2011).

[31] D R Chialvo, M M Millonas, *Asymmetric unbiased fluctuations are sufficient for the operation of a correlation ratchet*, Phys. Lett. A **209**, 26 (1995).

[32] M M Millonas, D R Chialvo, *Control of voltage-dependent biomolecules via nonequilibrium kinetic focusing*, Phys. Rev. Lett. **76**, 550 (1996).

[33] M C Mahato, A M Jayannavar, *Synchronized first-passages in a double-well system driven by an asymmetric periodic field*, Phys. Letters A **209**, 21 (1995).

[34] M M Millonas, D A Hanck, *Nonequilibrium response spectroscopy and the molecular kinetics of proteins*, Phys. Rev. Lett. **80**, 401 (1998).

[35] M Kostur, J Łuczka, *Multiple current reversal in Brownian ratchets*, Phys. Rev. E **63**, 021101 (2001).

[36] S Savel'ev, F Marchesoni, F Nori, *Stochastic transport of interacting particles in periodically driven ratchets*, Phys. Rev. E **70**, 061107 (2004).

[37] Baoquan Ai, Liqiu Wang, Lianggang Liu, *Transport reversal in a thermal ratchet*, Phys. Rev. E 72, 031101 (2005).

[38] I Derényi, P Tegzes, T Vicsek, *Collective transport in locally asymmetric periodic structures*, Chaos **8**, 657 (1998).

[39] D E Shalóm, H Pastoriza, *Vortex motion rectification in Josephson junction arrays with a ratchet potential*, Phys. Rev. Lett. **94**, 177001 (2005).

[40] V I Marconi, *Rocking ratchets in two-dimensional Josephson networks: Collective effects and current reversal*, Phys. Rev. Lett. **98**, 047006 (2007).

[41] C Kettner, P Reimann, P Hänggi, F Müller, *Drift ratchet*, Phys. Rev. E **61**, 312 (2000).

[42] S Matthias, F Müller, *Asymmetric pores in a silicon membrane acting as massively parallel brownian ratchets*, Nature (London) **424**, 53 (2003).

[43] K Mathwig, F Müller, U Gosele, *Particle transport in asymmetrically modulated pores*, New J. of Phys. **13**, 033038 (2011).

[44] C Marquet, A Buguin, L Talini, P Silberzan, *Rectified motion of colloids in asymmetrically structured channels*, Phys. Rev. Lett. **88**, 168301 (2002).

[45] D Reguera, A Luque, P S Burada, G Schmid, J M Rubí, P Hänggi, *Entropic splitter for particle separation*, Phys. Rev. Lett. **108**, 020604 (2012).

[46] J H Jacobs, *Diffusion processes*, Pag. 68, Springer, New York (1967).

[47] R Zwanzig, *Diffusion past an entropy barrier*, J. Phys. Chem. **96**, 3926 (1992).

[48] G P Suárez, M Hoyuelos, H Martin, *Transport in a chain of asymmetric cavities: Effects of the concentration with hard-core interaction*, Phys. Rev. E **88**, 052136 (2013).

[49] G P Suárez, M Hoyuelos, H Martin, *Transport of interacting particles in a chain of cavities: Description through a modified Fick-Jacobs equation*, Phys. Rev. E **91**, 012135 (2015).

[50] T D Frank, *Nonlinear Fokker-Planck equations*, Pag. 280, Springer, Berlin (2005).

[51] C A Kruelle, A Gotzendorfer, R Grochowski, I Rehberg, M Rouijaa, P Walzel, *Granular flow and pattern formation on a vibratory conveyor*, In Traffic and Granular Flow '05, Eds. A Schadschneider, T Poschel, R Kuhne, M Schreckenberg, D. E. Wolf, Pag. 111, Springer-Verlag, Berlin, Heidelberg (2007).

[52] E M Sloot, N P Kruyt, *Theoretical and experimental study of the transport of granular materials by inclined vibratory conveyors*, Powder Technology **87**, 203 (1996).

[53] Z Farkas, P Tegzes, A Vukics, T Vicsek, *Transitions in the horizontal transport of vertically vibrated granular layers*, Phys. Rev. E **60**, 7022 (1999).

11

Autonomous open-source hardware apparatus for quantum key distribution

Ignacio H. López Grande,[1] Christian T. Schmiegelow,[2] Miguel A. Larotonda[1*]

We describe an autonomous, fully functional implementation of the BB84 quantum key distribution protocol using open source hardware microcontrollers for the synchronization, communication, key sifting and real-time key generation diagnostics. The quantum bits are prepared in the polarization of weak optical pulses generated with light emitting diodes, and detected using a sole single-photon counter and a temporally multiplexed scheme. The system generates a shared cryptographic key at a rate of 365 bps, with a raw quantum bit error rate of 2.7%. A detailed description of the peripheral electronics for control, driving and communication between stages is released as supplementary material. The device can be built using simple and reliable hardware and it is presented as an alternative for a practical realization of sophisticated, yet accessible quantum key distribution systems.

I. Introduction

The main goal of cryptography is to obtain a secure method to share information. This is usually achieved by the encryption of the data, using a shared cryptographic key. The security of the protocol then relies on the secrecy of this key. The distribution of a secret key is therefore a crucial task for any symmetric-key cryptographic algorithm. Classically, this can be achieved using the Diffie-Hellman method, or some variation based on it [1].

Quantum Key Distribution (QKD) protocols exploit the quantum no-cloning theorem [2] and the indistinguishability upon measurement of quantum states belonging to non-orthogonal, conjugate bases to accomplish secure distribution of cryptographic keys [3]. These features, combined with the fact that a measurement performed on a quantum system disturbs its original state in some manner, are the fundamental principles in which every QKD protocol is based on, since they allow for the detection of an eventual eavesdropper by monitoring errors on the exchanged key: the attacker cannot completely determine the measured quantum state, nor can she/he copy it; therefore she/he must resend some imperfect copy to the receiver, which may introduce errors in the key. However, a practical real-world QKD implementation is still a technical challenge that combines concepts and technologies from different areas, such as classical and quantum information theory, quantum optics, electronics and optoelectronics [4]. In this work, we describe a functional autonomous apparatus that implements the BB84 quantum key distribution protocol [5] where we implement several solutions that contribute to the affordability of a naturally costly

*E-mail: mlarotonda@citedef.gob.ar

[1] DEILAP-UNIDEF (CITEDEF-CONICET), J. B. de La Salle 4397, B1603ALO Villa Martelli, Buenos Aires, Argentina.

[2] Laboratorio de Iones y Átomos Fríos, Departamento de Física, Facultad de Ciencias Exactas y Naturales, Universidad de Buenos Aires & IFIBA-CONICET, Pabellón 1, Ciudad Universitaria, 1428 C.A.B.A., Argentina.

piece of equipment.

A critical parameter for the security of any quantum cryptography protocol is the Quantum Bit Error Rate (QBER), which is obtained after an error estimation from the sifted keys S_A and S_B —which in theory should be identical— and in the absence of an eavesdropper they are similar up to experimental errors: a small part of the key is randomly selected and used to obtain the QBER, which gives an estimation of the error rate in the whole length of the key. Once the protocol is running, the QBER is routinely monitored by resigning part of the key. It is assumed that any increase of the QBER may be generated by the presence of an eavesdropper; in such case the whole key is discarded. Theoretical upper limits have been found for the QBER rate that if preserved, unconditional security of the key can be granted [6] by applying classical error correction and privacy amplification protocols to the sifted key [7].

The first implementation of a quantum cryptographic protocol dates from 1992 [8]. Since then, the field has rapidly advanced towards sophisticated systems that provide high speed key generation [9], long distance key distribution [10, 11], transmitting photons either over optical fiber or open air, using polarization or time bin [12], or both [13], for qubit-encoding. Such protocols can be based on single photon pulses [14, 15] or on entangled photon states [19]. The use of advanced optoelectronics and high performance detectors is intensive on any QKD implementation. In this work we show that the technologies used in such quantum information algorithms are mature enough to attempt a low cost, yet functional and robust implementation of a quantum key distribution protocol. We give a detailed explanation of the communication scheme and we release the firmware code and the circuit schematics to build the control units as Supplementary Material. The following section is devoted to the description of the optical arrangements used on Alice and Bob stages. Section III. discusses the initial setup, synchronization, transmission and processing routines needed in order to generate a sifted key. The overall performance of the apparatus and its response to different perturbations are discussed thereafter.

II.　Device layout

The developed system comprises an emission stage and a reception stage for the quantum channel, and an *ad-hoc* classical communication system. Quantum bits are encoded in the polarization of weak coherent pulses. These pulses are used as an approximation of a single photon pulsed source. We identify the canonical polarization states $\{|H\rangle, |V\rangle\}$ with the computational basis $B_C = \{|0\rangle, |1\rangle\}$ and the diagonal polarization states $\{|D\rangle, |A\rangle\}$ with the diagonal basis $B_D = \{|+\rangle, |-\rangle\}$. The complete scheme of the apparatus is shown in Fig. 1.

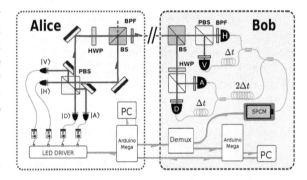

Figure 1: Setup of the QKD system: Polarization selection and spatial overlap between states is obtained with a combination of Polarizing (PBS) and non-polarizing (BS) Beam Splitters. Bob uses a BS to randomly choose the measurement basis. Polarization projections are obtained with a PBS and a half waveplate (HWP). Projected light is coupled into optical fibers and temporally multiplexed with selected delays. A single photon counting module (SPCM) is used for detection and bandpass filters (BPF) are used to reject unwanted light. Δt: 250 ns delay.

Polarized weak light pulses are generated by fast pulsing four infrared LEDs and combining them with Polarizing (PBS) and Non-Polarizing (BS) Beamsplitters: each of the LEDs is used to encode one of the four possible polarization states. The LEDs outputs are coupled and later decoupled to multimode optical fibers to define a propagation direction and divergence, and also to equalize the intensities of the four outputs. This setup is based on off-the-shelf economic infrared LEDs and avoids the use of expensive Pockel's cells and high performance HV drivers for polarization state preparation. The mean photon number per pulse was set to approximately 0.1, measured between the emission and detection stages. Assuming Poissonian photon statistics, this means that in average nearly 90% of the clock pulses carry no photons at all, while less than 0.5% of the pulses are multiphoton pulses. Both empty and multiple detection runs are considered null. It is worth to note that this particular choice of photon number per pulse does not guarantee the generation of a secure key by itself; rather, the conditions for distillation of a secure key from a raw key and the optimum photon rate depend on specific conditions of the setup, such as the length of the quantum channel –that implies distance-dependent losses–, the loss on Bob's receiver stage, and the efficiency and dark count rate of the detectors. Security conditions under different kind of attacks on non-ideal QKD systems have been reported for example in [16,17] and reviewed in [18].

The light paths from the sources entering a polarization beam splitter (PBS) at different inputs were combined by pairs: the reflected beams exit the PBS vertically polarized, while the transmitted outputs are left horizontally polarized. A half-waveplate retarder placed in one of the outputs rotates the polarization of these two paths 45 degrees. A beam splitter cube further combines the paired sources into one common path.

Basis selection at the receiver stage is obtained using a 50% beam splitter cube to randomly obtain either a transmitted photon or a reflected photon. Projection onto the states of the canonical basis is achieved by means of a PBS, while the diagonal basis projections are obtained adding a half-wave plate retarder between the beam splitter and the PBS in one of the paths. A straightforward implementation of the detection stage demands four single photon counting modules (SPCMs), which are expensive devices. With the purpose of obtaining a practical, cost-effective setup we implemented a time multiplexed detection, adding 250 ns delays between the projection paths. The four possible measurement outcomes are encoded into temporal bins: photons are detected using only one commercial single photon counting module and labeled by the time of arrival with respect to a clock reference. Temporal demultiplexing and state determination are obtained measuring coincidences between the single photon detector output and temporal gates

with selected delays. The use of a sole detector also avoids the unbalance of detection efficiencies that is present in multiple detector setups. As a drawback, this scheme presents 4 dB insertion loss per coupler, which attenuates the input signal and lowers the extractable secure key rate, due to the reduced optimal photon rate. This issue can be circumvented by implementing a decoy-state strategy together with the BB84 protocol [20–22]. Such application is currently under development at our laboratory.

The following section deals with the synchronization and control tasks performed by the open source hardware microcontrollers that allow the system to operate in an autonomous manner.

III. Control, driving and synchronization

i. Control and temporal synchronization

Open-source hardware was chosen for the processing of the cryptographic key and controlling units of the system, in order to obtain a practical, small-scale photonic implementation of the quantum protocol: all the synchronization, communication and processing operations, as well as system diagnosis were programmed on Arduino Mega 2560 microcontrollers. A diagram of the key generation protocol is sketched in Fig. 2. The communication scheme is divided in stages where classical information is exchanged (C COM) and a quantum communication stage (Q COM). An initial calibration of the system can be performed, where both parties measure the photon rate per pulse, the total temporal delay of the link and the delay between temporal bins. The communication begins with an exchange of the protocol parameters such as data structure and target key length. Then, after a synchronization sequence, they exchange the quantum bits and the sifting procedure follows: both parties exchange information on basis emission and detection and coincidences between them, keeping only the bits that come from coincident bases. The routine is repeated until the target key length is reached. The shared key is locally transferred to personal computers on each stage via USB ports.

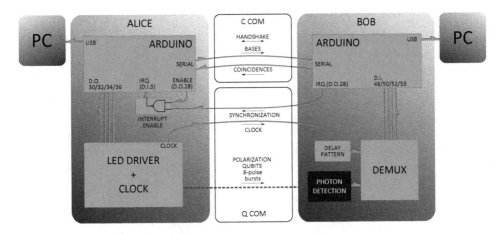

Figure 2: Communication and control setup of the BB84 QKD apparatus. The protocol is controlled by two Arduino Mega microprocessors. The synchronization start byte is generated at Bob's side and sent through an interrupt channel. After the quantum bits are sent and detected, bases are exchanged and the key is sifted. Specific input and output pins of the Arduino controllers are detailed in the figure.

ii. Electronic driving and peripherals

The communication routines described above are implemented directly by the microcontrollers. Specific tasks such as driving the pulsed LEDs, synchronizing the temporal mask and demultiplexing the temporal signals at the receiver side are performed with dedicated electronic peripherals.

Based on a random 2-bit sequence, the Arduino microcontroller sets a logic high on one of the four possible outputs. A monostable multivibrator uses this logic transition to generate a 20 ns pulse that is used as the input for a high speed LED driver. The shunt driving circuit that pulses the current on each LED is constructed using the high-current, low impedance pull-up and pull-down MOSFET transistors at the output of NAND gates and a passive network to provide a prebias current and current overshoot to increase the performance of pulsed LED drivers [23]. The optical pulse duration of 25 ns is limited by the LED response.

At Bob's side, single photon pulses are routed through different delay paths according to their polarization, and the delayed photon clicks are identified as polarization state projections by temporal demultiplexing the digital detections. Pulses from the single photon detector are addressed to the corresponding state channel by comparison with a pulse pattern that repeats the temporal delays added by the optical fibers.

IV. System performance and self-diagnostics

The main cause of bit errors are the non-ideal polarization splitting contrast of the PBSs and low quality waveplates that produce incomplete rotations and distort the ideal linear polarization states at the input and output. Also, off-the plane misalignment of the light paths within the preparation and measuring states can induce undesired rotations of the polarization axes. These are well-known problems for an open air optical setup, and workarounds to minimize them are common to any polarization-sensitive arrangement. Detector dark counts and stray light that leaks through the optical setup are also a source of error. The gated detection helps to minimize these errors. The contribution of this effect to the overall error rate depends linearly on the gate pulse duration.

The other main source of error is the temporal jitter of the signals, which can produce erroneous bit assignment of the temporally multiplexed pulses. The signal jitter is limited by the duration of the light pulse, which is approximately half the Arduino clock period. Larger pulse timing fluctuations can be produced at the generation and detection stages due to missed or added clock pulses at the microprocessors, specifically when handling interrupt signals. These temporal fluctuations can shift states from earlier to later temporal bins, inducing errors on the key. The temporal order of the multiplexed states can be arranged to minimize such errors. A natural choice is to order the detections in the sequence H (first), V, D, A (last). Such choice has an increased probability that temporal jitter can produce an error: assuming delayed detections that deterministically shift the states; in this configuration the probability of producing a bit error is 0.3125. If the delays are arranged to output the temporal sequence H (first), D, V, A (last), consecutive states at the detection pattern do not belong to the same basis. The probability of producing an error provided the states are identified in an adjacent temporal bin in this arrangement is 0.1875, and it is therefore chosen to minimize the error rate. An estimation of the bit error rate produced by this artifact in the actual protocol execution can be obtained as the product of this probability and the state-shift rate due to the overall timing jitter (0.6%), and gives approximately 1.1%. The system was tested using a mean photon rate of μ=0.09. A typical light distribution at the outputs for each polarization state generated by Alice is shown in Fig. 3a).

The apparatus autonomously generates a cryptographic key until the target key length is reached. During the tests, light pulses were emitted in bursts of 19200 pulses per second, while a constant background light of 3000 counts/s at the detector was present in the actual experimental conditions. We obtained a raw key generation rate of 363 bits/s, with a quantum bit error rate (QBER) of 2.7 %. Approximately one third of this rate (0.9 %) corresponds to errors produced by stray light and detector dark counts, while the rest of the errors are due to the electronic jitter as discussed above, and to an imperfect preparation and selection of the polarization states at the optical setup. The measured key generation rate is limited by (and it can be also estimated from) the photon-per-pulse rate, the 50% data that is discarded in average due to non-coincident bases, and the dead times on the communication stage that allow for data processing, which represents roughly two thirds of the total execution time.

During a key generation session, some parameters can be monitored for eavesdropping, inconsistencies or anomalous behavior. The sifted key can be periodically sampled and analyzed for error rate, key generation rate and bias rate (the rela-

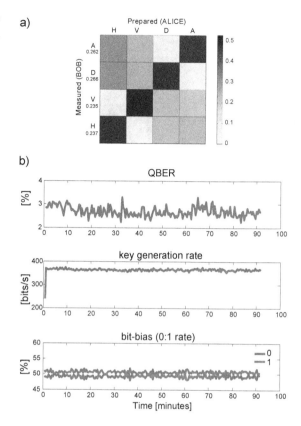

Figure 3: a) Light distribution at the detection channels, for each generated polarization state. Percentages on each row of the graph are the relative amounts of light obtained by adding the counts at each detection channel, for all the emitted states. b) Temporal evolution of different system parameters during normal operation.

tive abundance of "1"s to "0"s in the key, 0.98 in our setup), leading to charts like the one presented on Fig. 3b). Under normal operation conditions, the three parameters are constant through a typical one hour and a half experiment, with a relative dispersion on their average values below 2×10^{-2} for key rate, 7×10^{-3} for bit bias and 2×10^{-3} for QBER (statistics obtained over 20 kbit partitions from a total 1.9 Mbit key).

The response of the system under anomalous conditions was tested disturbing the quantum channel in different manners, while the above parameters were being monitored. Figure 4 shows a sequence of such perturbations: first, in a), the detector was blocked, which caused the key rate to vanish with a characteristic time given by the integration time of the monitoring process. If one of the detection channels (V) is blocked [Fig. 4 b)], the effect is a diminished key rate and a key bias of 2/3. In c), both channels of a basis are blocked. If two channels that encode the same bit are blocked, the key rate remains at half the original rate, but now the series is completely biased, since only one logic bit is produced. More interestingly, during e), a PBS was inserted in the quantum channel, which has the following effect on the transmitted quantum states: $|H\rangle$ are left unchanged —since they are transmitted through the PBS— $|V\rangle$ states are reflected out of the path at the PBS, while $|D\rangle$ and $|A\rangle$ are transmitted as $|H\rangle$ with a 50% chance. This last feature resembles the action of an eavesdropper (Eve) using an intercept-resend strategy, where the bases in which Eve resends bits to Bob are randomly chosen. In this situation, states sent as $|V\rangle$, and (in average) half of the states originally sent on the diagonal basis, are lost at the PBS reflection, leading to a reduction of the key generation rate by a factor of two. More importantly, half of the states originally sent on the diagonal basis are transmitted through the PBS and transformed to the $|H\rangle$ state. If these states are measured on the diagonal basis, they can be detected as either $|D\rangle$ or $|A\rangle$, regardless of the original state. The result of these successive projections is that a $|D\rangle$ ($|A\rangle$) state has a non-negligible probability to be detected as a $|A\rangle$ ($|D\rangle$) state. The quantum bit error rate now raises to 25% for this particular perturbation, signaling a possible eavesdropper. The bit bias of Bob's key is 0.75: the action of the PBS that prevents all the emitted $|V\rangle$ states to be detected generates a ratio of "1"s to "0"s of 3:1. Periodically sampling and an analysis of the generated key thus provides a means for detecting intercept-resend attacks, at the cost of reducing the final key length. With the setup placed on an optical table, QBER variations as low as 0.2% can be detected.

V.   Concluding remarks

We have implemented an open source hardware based autonomous QKD apparatus. Its stability and performance have been tested on megabit-length key distribution sessions, during which some key parameters were monitored. The device was

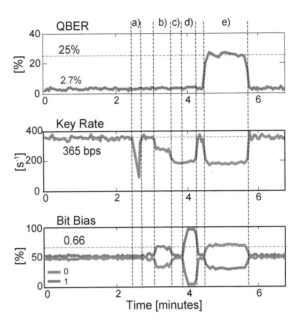

Figure 4: Behavior of the system under different perturbations on the detection stage and the quantum channel, labeled a) to e), consisting in blocking one or more detection channels and inserting a polarizing beamsplitter in the quantum channel. See the text for a detailed explanation.

designed with a cost-effectiveness approach which includes a LED-based single photon probabilistic source, a time multiplexed detection scheme that employs only one SPCM and Arduino-based controlling and processing units for Alice and Bob.

The actual bit error rate can be lowered if the polarization dependent elements (PBS) on Alice and Bob sides are replaced with high-extinction ratio polarizers (at present around 1%). Another way in which the error rate can be improved is by minimizing the incidence of errors originated by detector's dark counts. This can be accomplished with a reduction on the light pulse width that leads to narrower temporal gates. Also, an increase of the mean photon number per pulse can reduce the QBER without compromising security, provided a decoy state protocol is implemented instead.

The overall protocol speed can be raised by replacing the Arduino microcontrollers with faster FPGA-based boards, where the communication and the processing blocks may be parallelized. Also, as mentioned above, the temporal demultiplexing can be done directly on the board. Faster clock boards allow for an additional reduction of the temporal delays between channels on the time multiplexed detection scheme. These can be set to be as short as 50 ns, depending on pulse width and temporal jitter.

The developed apparatus is able to autonomously generate a cryptographic key with limited yet simply improvable performance. The whole system can be used to establish a small-scale secure information channel between eye of sight distance sites, for academic purposes, or it can serve as a testbed for different quantum information-related resources, such as original protocols, detectors, light sources, or the development of alternative physical quantum channels. We understand that a cryptographic system based on well-known, simple and available technology that can be fully mastered and controlled by the end user may turn out more useful and secure than a sophisticated, "black box" type system that has many parts that are beyond the user's control, and which may depend on third party services to be operated or maintained.

Acknowledgements - This work was supported by the ANPCyT PICT 2010-2483 and MINDEF PIDDEF 012/11 grants. M.A.L. is a CONICET fellow, C.T.S. and I.H.L.G. were funded by CONICET scholarships.

––––––––––

[1] W Diffie, M Hellman, *New directions in cryptography,* IEEE T Inform. Theory **22**, 644 (1976).

[2] W K Wootters, W H Zurek, *A single quantum cannot be cloned,* Nature **299**, 802 (1982).

[3] M Planat, H C Rosu, S Perrine, *A survey of finite algebraic geometrical structures underlying mutually unbiased quantum measurements,* Found. Phys. **36**, 1662 (2006).

[4] N Gisin, G Ribordy, W Tittel, H Zbinden, *Quantum cryptography,* Rev. Mod. Phys. **74**, 145 (2002).

[5] C H Bennett, G Brassard, *Quantum cryptography: Public key distribution and coin tossing,* Theor. Comput. Sci. **560**, 7 (2014).

[6] N J Cerf, M Bourennane, A Karlsson, N Gisin, *Security of quantum key distribution using d-level systems,* Phys. Rev. Lett. **88**, 127902 (2002).

[7] C H Bennett, G Brassard, C Crépeau, U M Maurer, *Generalized privacy amplification,* IEEE T Inform. Theory **41**, 1915 (1995).

[8] C H Bennett *et al.,* *Experimental quantum cryptography,* J. Cryptol. **5**, 3 (1992).

[9] A R Dixon *et al.,* *Gigahertz decoy quantum key distribution with 1 mbit/s secure key rate,* Opt. Express **16**, 18790 (2008).

[10] P A Hiskett *et al.,* *Long-distance quantum key distribution in optical fibre,* New J. Phys. **8**, 193 (2006).

[11] R Ursin *et al.,* *Entanglement-based quantum communication over 144 km,* Nat. Phys. **3** 481 (2007).

[12] I Marcikic *et al.,* *Distribution of time-bin entangled qubits over 50 km of optical fiber,* Phys. Rev. Lett. **93**, 180502 (2004).

[13] W T Buttler *et al.,* *Practical four-dimensional quantum key distribution without entanglement,* Quantum Inf. Comput. **12**, 1 (2012).

[14] C H Bennett, *Quantum cryptography using any two nonorthogonal states,* Phys. Rev. Lett. **68**, 3121 (1992).

[15] H Bechmann-Pasquinucci, W Tittel, *Quantum cryptography using larger alphabets,* Phys. Rev. A **61**, 062308 (2000).

[16] N Lütkenhaus, *Security against individual attacks for realistic quantum key distribution,* Phys. Rev. A **61**, 052304 (2000).

[17] A Acin *et al.,* *Device-independent security of quantum cryptography against collective attacks,* Phys. Rev. Lett. **98**, 230501 (2007).

[18] V Scarani *et al.,* *The security of practical quantum key distribution,* Rev. Mod. Phys. **81**, 1301 (2009).

[19] A K Ekert, *Quantum cryptography based on bell's theorem,* Phys. Rev. Lett. **67**, 661 (1991).

[20] W Y Hwang, *Quantum key distribution with high loss: Toward global secure communication,* Phys. Rev. Lett. **91**, 057901 (2003).

[21] Y Zhao *et al., Experimental quantum key distribution with decoy states,* Phys. Rev. Lett. **96**, 070502 (2006).

[22] Z L Yuan *et al., Unconditionally secure one-way quantum key distribution using decoy pulses,* Appl. Phys. Lett. **90**, 011118 (2007).

[23] Agilent Application Bulletin 78, *Low cost fiber-optic links for digital applications up to 155 MBd,* Agilent Technologies Inc. (1999).

12

Increasing granular flow rate with obstructions

Alan Murray,[1] Fernando Alonso-Marroquin[2]*

We describe a simple experiment involving spheres rolling down an inclined plane towards a bottleneck and through a gap. Results of the experiment indicate that flow rate can be increased by placing an obstruction at optimal positions near the bottleneck. We use the experiment to develop a computer simulation using the PhysX physics engine. Simulations confirm the experimental results and we state several considerations necessary to obtain a model that agrees well with experiment. We demonstrate that the model exhibits clogging, intermittent and continuous flow, and that it can be used as a tool for further investigations in granular flow.

I. Introduction

When does an obstruction placed near a bottleneck increase the flow rate of discrete objects moving through the bottleneck? Answering this question is of great utility on many scales. For example, the safe evacuation of pedestrians moving through confined environments (train stations, stadiums, concert halls, etc) in emergency situations can be of vital importance. And grain falling through a silo, where efficient flow is necessary for production and clogging is undesirable, is another example.

As stated by Magalhaes et al. in Ref. [1], 'Granular materials are ubiquitous either in nature or in industrial processes' and so a fundamental understanding of their motions is of intrinsic interest, both from physics and engineering perspectives. The quantitative study of the flow of granular materials has been performed for many decades [2], yet there is enormous scope for many interesting and creative techniques to be developed: exper-

imental, theoretical and computational. Granular flow around an obstruction placed near a bottleneck is of particular interest as it has many potential applications and this has been investigated by many authors [1, 3–7].

Increases in flow rate achieved through placement of an obstruction near a bottleneck have been reported by several investigators [8–10]. Another commonly reported phenomenon is that of 'clogging', also described as the formation of arches [4]. There are also many novel investigations of granular flow. For example, Wilson et al. [3] consider granular flow of particles that are completely submerged under water and Lumay et al. [7] investigate the flow of charged particles that repel each other. Clearly, granular flow is a very wide and active area of research.

Several authors have investigated the use of physics engines used in games, an example of which can be found in the work by Carlevaro and Pugnaloni [15]. In this paper we present a real time, 3D computer simulation using the PhysX physics engine. The model can be used in the study of granular flow as it exhibits continuous flow, intermittent flow, clogging, and has several parameters that an investigator can control as the simulation

*E-mail: fernando.alonso@sydney.edu.au

[1] SAE Creative Media Institute, Sydney, Australia.

[2] School of Civil Engineering, The University of Sydney, NSW 2006, Australia.

is occurring. The computer model is based on an experiment that uses spheres for the particles and three different obstruction shapes. We use experimental data to calibrate and validate our model. The model can then be used to determine how the highest flow rates may be achieved both with regard to obstruction position and to find optimal obstruction shapes.

The paper is organized as follows: we present the model in section II. Then we describe the experiments in two parts. Section III deals with flow rate without obstructions, and section IV investigates flow rate with obstructions. We discuss the limitations in section V, and conclude in section VI.

II.   The Model

We use PhysX [12], a freely available real time, 3D physics engine for computations. PhysX is widely used in the games industry as, when necessary, it favors speed and stability of computation over accuracy. Integration of the equations of motion is done using an unconditionally stable, semi implicit Euler scheme yielding algebraic equations that are solved using a progressive Gauss-Seidel algorithm, whilst enforcing the Signorini conditions [16]. The time step can be adjusted for greater accuracy at the cost of speed and hence, real time performance of the engine. For this study, we left the time step at its default value of 0.02 seconds.

PhysX uses the Coulomb model for friction and restitution is a measure of the loss of kinetic energy between colliding particles. Details and further references about the rigid body system and friction model used in PhysX can be found in the work by Tonge [16].

The model incorporates the following parameters which are freely specifiable and chosen to agree with experimental results: friction, restitution, and sphere diameter. In addition, our model allows us to vary the angle of inclination of the plane, the gap width and the distance from the obstruction to the gap. We incorporate imperfections in the sphere diameters by allowing them to vary by up to 5%. The model uses 1551 particles whilst ≈ 1000 particles were used in the experiment. Each simulation is performed 20 times.

Unity [13], a freely available game engine, is used as our programming interface to PhysX. We also use Blender [14], a freely available 3D modeling package for modeling the obstructions.

III.   Flow rate without obstructions

The basic set up of the experiment with obstructions is shown in Fig. 1(a), where particles are released from above the obstruction. The particles themselves are spheres with diameters: 6.15 mm, 7.75 mm and 11.9 mm, referred to as 'small', 'medium', and 'large' respectively, and are shown in Fig. 1 (b). Note that without obstructions, the particles are released at the gap as shown in Fig. 1(c).

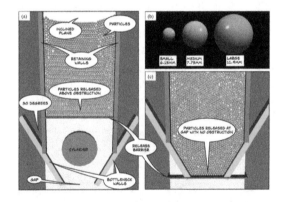

Figure 1: (a) Basic set up of the experiment with an obstruction. (b) Particles are spheres of three different diameters. (c) Experiment of flow rate without obstructions, the particles are released at the gap.

We use an inclined plane that makes an angle of 8° with the horizontal. The plane has retaining walls, walls that form a bottleneck and release barriers as shown in Fig. 1(a). We pack a known number of spheres of a specified size, shown without texture in Fig. 1(b), onto the plane and release them from rest so that they collide with the bottleneck walls, and flow through the gap. The walls of the bottleneck make an angle of 30° to the retaining walls.

Under gravity, the spheres move down the inclined plane, come into contact with the bottleneck walls and flow through the gap. The number of particles is known and the time taken for all particles to flow through the gap is measured, from which the flow rate, J, is determined. The flow rate is thus defined as the number of particles passing through the gap per second.

The experiment is repeated for various gap widths, with the corresponding flow rates measured. The entire experiment is performed individually for small, medium, and large spheres. All dimensions (particles, plane, bottleneck angles, etc) are known, as are the coefficients of friction and restitution of the particles. As per the experiment, in our model we release the particles at rest from the gap as shown in Fig. 1 (c).

It was found that there is a relatively sharp rise in flow rate as the gap is increased. It is precisely this range of distances where the flow rate transitions from clogging to intermittent to continuous flow rates. The model also exhibits clogging as shown in Fig. 2 (a), and intermittent flow. Therefore, in order to get reliable and reproducible flow rates, we decided to perturb the particles just sufficiently to prevent extended periods of clogging which thereby affect flow rates. As described by Garcimartin et al. [11], the problem of clogging can effectively be e.

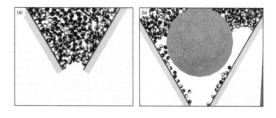

Figure 2: The model exhibits clogging (a) at the gap and between the obstruction and the bottleneck walls (b), consistent with the experiment.

In a similar way, we use the technique of 'shaking'. In the case of flow rate without obstructions, the bottleneck walls were shaken as shown in Fig. 3(a) from a maximum of 0.5 mm diminishing to zero over one second, every five seconds. The parameters that define a shake are: shake amplitude, direction of shake, shake duration and time between shakes. These parameters can be set arbitrarily in the model and, whilst other times and distances were tested, the above were found to be just suffi-

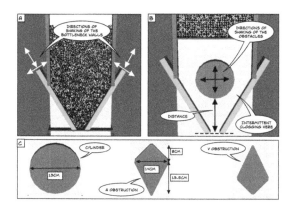

Figure 3: (a) directions of shaking of the bottleneck walls in the experiment of flow rate without obstructions. (b) directions of shaking in the experiment of flow rate with obstructions. Shaking is necessary to avoid clogging between the obstruction and the bottleneck walls. (c) obstructions used in experiments. The V obstruction is the A obstruction rotated through 180°.

cient to yield reproducible results.

We varied the values of coefficient of friction and coefficient of restitution in the model until we got good agreement with experimental results. It was found that varying the coefficient of restitution had little to no effect on flow rate.

We found a coefficient of friction of 0.24 yielded model results that gave good agreement with experimental results. The experimentally measured value of the spheres was found to be 0.28±0.4. The results for each diameter sphere are shown in Fig. 4. The model results agree with the experimental results quite well.

IV.   Flow rate with obstructions

The experiment is repeated with medium spheres (of diameter 7.75 mm). However, this time, obstructions (shown in Fig. 3 (c)) are placed at varying distances away from the gap, which is now fixed at a width of 3.3 cm.

It was found that there was a relatively sharp rise in flow rate as an obstruction is moved away

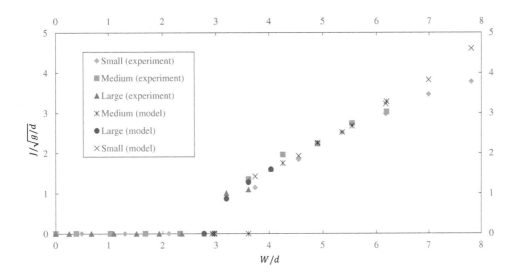

Figure 4: Flow rate without obstruction versus gap in dimensionless form. Comparison of experiment with simulation using spheres of three different diameters. J is the flow rate, g is the acceleration due to gravity, d is the particle diameter and W is the gap width.

from the gap. It is precisely this range of distances where the flow rate transitions from clogging to intermittent to continuous flow rates. The model also exhibited clogging as shown in Fig. 2(b), and intermittent flow. Therefore, in order to get reliable and reproducible flow rates, we decided to perturb the particles just sufficiently to prevent extended periods of clogging which thereby affect flow rates.

With the gap set at 3.3 cm, no clogging was observed for spheres of 7.75 mm at the gap. However, with the presence of an obstruction, clogging was observed in areas between the obstruction and the bottleneck walls as shown in Fig. 2(b) and Fig. 3(b). To get consistently measurable flow rates in the model, once again, we decided to perturb the system using the technique of 'shaking', described above. This time, we decided to shake the obstruction also shown in Fig. 3(b). The obstruction was shaken by a maximum of 1 mm diminishing to zero over one second, every five seconds. The freely specifiable parameters that define a 'shake' are the same as those for shaking the bottleneck walls. Figure 5 shows the maximum flow rates for the three obstructions, where the 'waiting room' effect, discussed in Ref. [8], occurs for the cylinder and the V obstruction but not the A obstruction.

Figure 5: Peak flow rates with obstructions, where the 'waiting room' effect is present for the cylinder and the V obstruction but not for the A obstruction.

Figure 6 shows flow rates as measured experimentally and with the model. Both sets of results indicate increases in flow rates for certain ranges of distance of obstruction to the gap. For the model, each graph shows error bars and data that were more than two standard deviations away from the mean were not included in the statistical analysis. This was necessary as, even with the shaking technique described above, the particles did occasionally clog for extended periods of time, thereby affecting the flow rate measurements.

Clearly, the model shows all obstructions im-

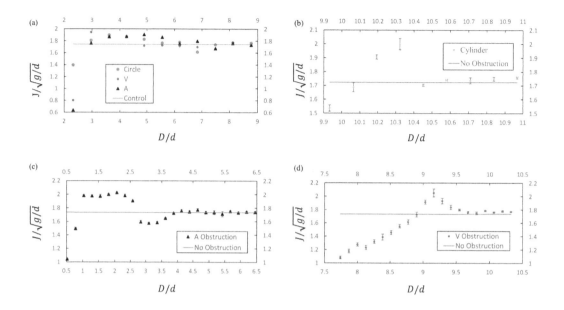

Figure 6: Results of flow rate with obstructions versus distance in dimensionless form. J is the flow rate, g is the acceleration due to gravity, d is the particle diameter, and D is the distance from the obstruction to the gap. The experimental results in (a) show that each obstruction has a range of distances where flow rate is greater than with no obstruction. The numerical results (b-d) show that the model also predicts that each obstruction has a range of distances from the gap where the flow rate is greater than flow rate with no obstruction.

prove flow rate for a range of distances, and reassuringly, the further the obstruction is moved away from the gap, the closer we get to flow rates without obstruction.

In the case of the cylinder, due to its shape and size relative to the bottleneck, we could only start at distances ~ 7 cm from the gap. We see that there is a steady rise in flow rate as the cylinder is moved further away from the gap. A maximum flow rate is reached and then a slightly sharper decrease is reached until we get flow rates the same as if no obstruction is present.

In the case of the A obstruction, due to its shape and size relative to the bottleneck, we were able to place the obstruction very close to the gap, which is why distances start at ~ 0.4 cm from the gap. We see a sharp rise in flow rate as the obstruction is moved away from the gap, and then we see a leveling out of the flow rate over a range of distances from 0.8 cm to 2.8 cm. There is a small peak in this range, but we consider this a statistical fluctuation.

Moving further away, from 2.8 cm to 3.2 cm, we

see a sharp decrease in flow rate to values that are actually below flow rates with no obstacle. This is an intriguing phenomenon, showing what we might expect: obstructions decrease flow rates.

The flow rate remains constant to a distance of 3.6 cm and then starts rising. At a distance of 4 cm, the flow rate is approximately equal to that with no obstructions.

In the case of the V obstruction, due to its shape and size relative to the bottleneck, we could only start at distances ~ 6 cm from the gap. We see that there is a steady rise in flow rate that is slower than that of the A obstruction, as the V obstruction is moved further away from the gap. A maximum flow rate is reached, higher than that with no obstruction, and then a decrease occurs until we get flow rates the same as if no obstruction were present.

V. Limitations of PhysX engine

We observed that the model exhibits continuous flow, intermittent flow, and clogging, all phenomena that have been identified and observed experimentally. The model was quite sensitive to the absolute sizes of the geometric structures used. This is true for all physics engines, as they have to be 'tuned' to the range we would like them to work most accurately in. For example, an engine might be tuned to work best with objects whose sizes are of the order of meters but it will not work for objects at the scale of millimeters. To overcome this limitation, we ran the simulations in the dimensions where the performance of the engine was optimal, and then used dimensional analysis to extrapolate the result to the real experimental conditions.

We found that the particles in the model needed to be perturbed more often than in actual experiments, in order to get consistent flow rates. This can be attributed to a real time optimization employed by PhysX, known as 'sleeping'. When a particle's angular and/or linear velocity falls below certain threshold values for more than a few frames, these velocities are set to zero and the particle goes into a 'sleep' state, in which no collision detection occurs and hence the particle's velocity remains at zero. The particle 'wakes up' when it is subjected to net forces. The reason for this optimization is that it relieves the processor of having to perform needless computations especially with regard to collision detection, thereby allowing much larger numbers of particles to be present and only performing the necessary computations as required. For this study, we left the threshold 'sleep velocities' at their default values.

In spite of these limitations, we can confirm that, as reported in various experiments, it is possible to increase flow rate of discrete particles by placing an obstruction at a suitable location near a gap.

VI. Conclusions

We have developed a model that exhibited reasonable quantitative accuracy in the case of no obstructions, and good qualitative agreement when obstructions were introduced. As in experiments, we showed that the flow exhibits clogging, intermittent flow rates, and continuous flow rates. The model also confirms experimental results that placement of an obstruction near a gap can actually increase flow rate through the gap. Better experimental agreement with obstructions will be possible if the

roundness of the obstructions is more accurately modeled.

The model is quite flexible in that many parameters can be changed via user input, and without modifying the actual code. In this way, the model can be used as a convenient tool to suggest further experiments and allows us to investigate different obstruction shapes, both of which may help us gain a greater understanding of granular flow.

We have also described a suite of freely available realtime 3D software, which is mature and has many online resources in the form of documentation and tutorials and is being actively developed, which can be used together to create simulations of interest in the area of granular flow.

Acknowledgements - We thank Ivan Gojkovic and Jaeris Wu for performing the experimental part of this paper.

[1] C Magalhaes, A Atman, G Combe, J Moreira, *Jamming transition in a two-dimensional open granular pile with rolling resistance*, Pap. Phys. **6**, 060007 (2014).

[2] W Beverloo, H Leniger, J van der Velde, *The flow of granular solids through orifices*, Chem. Eng. Sci. **15**, 260 (1961).

[3] T Wilson, C Pfeifer, N Mesyngier, D. Durian, *Granular discharge rate for submerged hoppers*, Pap. Phys. **6**, 060009 (2014).

[4] I Zuriguel, *Clogging of granular materials in bottlenecks*, Pap. Phys. **6**, 060014 (2014).

[5] P Mort, *Characterizing flowability of granular materials by onset of jamming in orifice flows*, Pap. Phys. **7**, 070004 (2015).

[6] O Rodolfo, J Sales, M Gargiulo, A Vidales, *Density distribution of particles upon jamming after an avalanche in a 2D silo*, Pap. Phys. **7**, 070007 (2015).

[7] G Lumay, J Schockmel, D Hernandez-Enriquez, S Dorbolo, N Vandewalle, F Pacheco-Vazquez, *Flow of magnetic repelling grains in a two-dimensional silo*, Pap. Phys. **7**, 070013 (2015).

[8] F Alonso-Marroquin, S Azeezullah, S Gallindo-Torres, L Olsen-Kettle, *Bottlenecks in granular flow: When does an obstacle increase flow rate in an hourglass?*, Phys. Rev. E **85**, 020301 (2012).

[9] S Yang, S Hsiau, *The simulation and experimental study of granular materials discharged from a silo with the placement of inserts*, Powder Tech. **120**, 244 (2001).

[10] I Zuriguel, A Janda, A Garcimartín, C Lozano, R Arévalo, D Maza, *Silo clogging reduction by the presence of an obstacle*, Phys. Rev. Lett. **107**, 278001 (2011).

[11] A Garcimartín, C Lozano, G Lumay, I Zuriguel, *Avoiding clogs: The shape of arches and their stability against vibrations*, AIP Conf. Proc. **1542**, 686 (2013).

[12] K Kumar, *Learning physics modeling with PhysX*, Packt Publishing Ltd, Birmingham (2013).

[13] W Goldstone, *Unity game development essentials*, Packt Publishing Ltd, Birmingham (2009).

[14] J Blain, *The complete guide to Blender graphics*, (Second Edition), A K Peters/CRC Press, Boca Raton (2014).

[15] C Carlevaro, L Pugnaloni, *Steady state of trapped granular polygons*, J. Stat. Mech. P01007 (2011).

[16] R Tonge, *Iterative rigid body solvers*, http://www.gdcvault.com/play/1018160/Physics-for-Game.

13

Knife-bladed vortices in non-Newtonian fluids

E. Freyssingeas,[1] D. Frelat,[1] Y. Dossmann,[1] J.-C. Géminard[1]*

A tank is filled with a non-Newtonian fluid. We report on the deformation of the free surface resulting from the presence of an underlying vortex. In a tiny range of the experimental parameters, the flow spontaneously loses its initial axi-symmetry, leading to the formation of a stationary knife-bladed vortex. We report on the series of patterns observed experimentally and summarize the conditions of the existence of the latter by establishing a state diagram.

I. Introduction

Flows in complex fluids exhibit many intriguing phenomena [1]. Among them, one of the most striking is the appearance of a cusp at the trailing end of a bubble rising through a non-Newtonian, viscoelastic fluid [2, 3]. Interestingly, the cusp assumes a knife-edge shape which is the result of an instability leading to spontaneous symmetry breaking of the interface. Such symmetry breaking, which is forbidden in Newtonian fluids and thus only possible due to the non-Newtonian properties of the fluid, has also been revealed observing the deformation of the free surface resulting from the settling of a solid sphere [4,5]. Symmetry breaking has also recently been observed in a jet of a visco-elastic fluid impinging on a wall at right angle [6]. Following the same research line, we were seeking for an experimental situation involving the free surface, making possible to observe a spontaneous loss of axi-symmetry in a configuration similar to that in Ref. [4], but in the steady state. We came to the idea of producing a vortex in a complex fluid and observing that the free surface was an adequate and convenient experimental configuration to achieve.

Conversely to vortices in Newtonian fluids, vortices in complex fluids, and specially the resulting deformation of the free surface, have attracted little or no attention. However, we mention here a recent study of the formation of vortices in non-Newtonian fluids [7]. One can notice that, in many practical cases, complex fluids, such as viscoelastic fluids (e.g., polymers or wormlike micelles solutions) or yield-stress fluids (e.g., gels or foams), are set in rotation, leading to the formation of vortices in the bulk and to the deformation of the free surface. Our aim is thus to provide a first experimental study of the deformation of the free surface of a non-Newtonian fluid and to show that, in a tiny range of the experimental parameters, one can indeed observe a spontaneous loss of the flow symmetry (Fig. 1). To do so, we first characterize the fluids in use and then report a series of experimental observations; among them, the state diagrams and some geometrical characteristics of the observed patterns.

II. Setup and protocol

We aim at characterizing at best, using simple imaging techniques, the deformation of the free surface of a non-Newtonian fluid due to the presence of an underlying vortex. We first describe the experimental device, then the fluids under study and finally, the experimental protocol.

*E-mail: jean-christophe.geminard@ens-lyon.fr

[1] Univ Lyon, Ens de Lyon, Univ Claude Bernard, CNRS, Laboratoire de Physique, F-69342 Lyon, France.

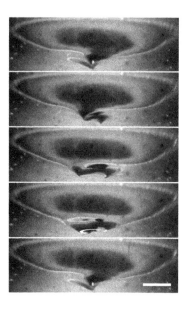

Figure 1: Asymmetric vortex observed at the free surface of a non-Newtonian fluid. The free surface is observed from below. One clearly observes that the vortex lost the initial symmetry and exhibits two tips. The scalebar is worth 1 cm (Camera 3, Ω = 4.5 Hz, h = 4 cm, sample 1). Associated movies Fig.1.avi, Fig.1.c1.avi, Fig.1.c2.avi, and Fig.1.c4.avi can be found as supplementary material [8].

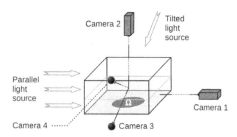

Figure 2: Sketch of the experimental device. The cameras 2, 3 and 4 are in the same vertical plane, perpendicular to the axis of camera 1. Depending on the observation angle, either of the two light sources is used. The horizontal size of the container is 20×20 cm^2, and its depth is 10 cm.

i. Experimental device

The main part of the experimental device (Fig. 2) consists of a square tank (in-plane size 20×20 cm^2, depth 10 cm). The side walls are made of glass plates, whereas the bottom is made of a thick PVC (Polyvinyl chloride) plate. A cylindrical housing, machined in the bottom, receives a PVC disc (diameter 10.1 cm) whose surface is flush with the bottom of the tank. The disk is driven in rotation by a DC motor fed by a power supply, so that the rotation velocity can be tuned in the range from 0 to 6.25 Hz. Flat side walls avoid optical deformations when observing from side. The lateral dimensions of the base have been chosen to ensure that, in our experimental conditions, the fluid flow never spatially extends to the lateral walls so that the observed patterns are not altered by lateral boundary conditions.

Our experiments rely on visual characterization of the deformation of the free surface. We use 2 CCD cameras (Jai, CB-080; COSMICAR TV Zoom Lens 12.5–75mm, 1:18) to quantitatively assess geometrical properties of the free surface. These two cameras (Camera 1 and Camera 2) are set in front and above the tank respectively, to make it possible to visualize the vertical and in-plane profiles of the free surface. Both CCD cameras provide accurate images, easy to analyze. We also use 2 additional cameras (Webcams, Logitech Quickcam Pro 9000) to report qualitative observations. These two webcams (Camera 3 and Camera 4) are placed one above and the second below the free surface plane. They both point toward the center of the free surface, making angles of about 30 deg and 20 deg with the horizontal. Cameras 2, 3 and 4 are all in the same vertical plane, perpendicular to the axis of Camera 1. In order to assess the vertical profile from the front view (Camera 1), the fluid is lit with a parallel, horizontal white light, which casts the shadow of the free surface onto a sheet of tracing paper covering the output side wall. Cameras 2, 3 and 4 are used with light from a second white source which provides parallel light, making a 45-degrees angle with the vertical.

ii. Fluid samples

The yield-stress fluids in use are mixtures of various concentrations of a commercial hair gel (Styling gel extra strong fixing, Auchan production, mainly made up of Carbopol[1]) and distilled water. The rheological properties of such mixtures do not allow us to generate

---

[1]Carbopol is a family of polymers that are used as thickeners, suspending agents and stabilizers. They are utilized in a wide variety of personal care products, pharmaceuticals and household cleaners. Most Carbopol polymers are high-molecular-weight acrylic-acid chains, usually crosslinked. The crosslinked polymers are not actually water soluble, but swell into hydrated spheres that give the product its rheological properties.

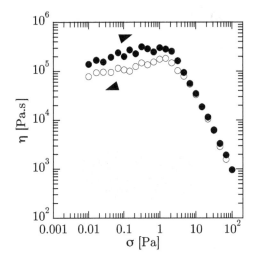

Figure 3: Shear viscosity η vs. shear-stress σ (Sample 3).

vortices in too concentrated mixtures (typically over 50 wt.% in gel). We therefore considered only mixtures with concentrations in hair gel smaller than 50 wt.%.

To prepare these mixtures, chosen masses of gel and distilled water are mixed in a large container (1 to 3 liters of fluid are necessary for the experiments), and then stirred for several minutes. The mixture is then let to rest at room temperature for at least 12 hours for good homogenization. Then, to get rid of air bubbles in the mixture that could prevent correct visualization, the mixtures are centrifuged at 1100 rpm for 5 minutes.

We used two different batches of hair gel which, although of the same brand, have slightly different rheological properties. Batch 1 exhibits a yield stress larger than batch 2. We mainly focused our investigation on four different mixtures: 45 wt.% and 40 wt.% of gel from batch 1 and 45 wt.% and 35 wt.% from batch 2 in water. A fifth mixture, 20 wt.% of gel from batch 1 was studied at a single height (4.3 cm). Samples are numbered and labeled according to Table 1.

We assessed rheological properties of the samples using a rheometer (Bohlin; C-VOR 150) in plane-plane geometry (plate diameters: 60 mm) at room temperature (22°C). In order to avoid sliding at the walls, sandpaper was glued on both surfaces. In order to insure reliable measurements, we considered two different gaps, i.e., 1 mm and 0.5 mm, between the plates and checked that the results were similar for both gaps. We report averages over 3 measurements.

The viscosity η was obtained by measuring the shear-rate at constant imposed shear-stress, σ (Fig. 3). One observes two regimes. For small σ, data are very scattered: the imposed shear stress is below the yield, i.e., the torque applied by the rheometer is not enough to make the fluid flow. In contrast, beyond a threshold shear stress, σ_0, the applied torque induces a continuous flow. The viscous behavior of the flow is dominant. We remark that the measurements do not reveal any significant hysteresis upon increasing and decreasing shear stress (or equivalently shear rate) in this regime. Above the yield, the viscosity decreases when σ is increased, according to $\eta(i) \propto \sigma^{-1.5}$. We have $\eta(1) > \eta(2) > \eta(3) > \eta(4) > \eta(5)$ (Fig. 4a).

In addition, the storage modulus G' was obtained by imposing a periodic shear-stress at 1 Hz and by measuring the resulting shear-strain (Fig. 4b). Again, two regimes are observed. For small σ, the measurements exhibit a plateau while, beyond a threshold shear-stress, the storage modulus G' decreases when the shear stress is increased. Note that the values of the yield stress are very similar to those obtained above, from the measurement of the viscosity. In this regime, $G' \propto \sigma^{-1.5}$ and $G'(1) > G'(2) > G'(3) > G'(4) > G'(5)$.

From these two experiments, we assess that all samples exhibit a well-defined yield stress, σ_0 (Table 1). Samples 1 and 2 have similar rheological behaviors, $\eta(1)$ and $\eta(2)$, $G'(1)$ and $G'(2)$ as well as $\sigma_0(1)$ and $\sigma_0(2)$ being of the same order of magnitude. Samples 3, 4 and 5 are much less viscous and elastic, with smaller values of the yield stress than samples 1 and 2.

iii. Experimental protocol

The chosen fluid is poured in the tank and leveled to a chosen height, h. The central disc is set in rotation. The rotation velocity, Ω, is increased by steps up to the maximum rotation velocity of about 6 Hz and, then, decreased still by steps until rest. At each step, we record movies from the various cameras in the steady-state.

III. Experimental results

i. Control experiment in a Newtonian fluid

We first report on a control experiment performed with a Newtonian fluid (a water/glycerine mixture) to validate the experimental set-up. The free surface exhibits a parabolic shape and deepens when the rotation velocity Ω is increased. The maximum depth of the profile,

Sample# (symbol)	1 (○)	2 (●)	3 (△)	4 (▲)	5 (⊞)
Batch 1 (%.wt)	45	40	–	–	20
Batch 2 (%.wt)	–	–	45	35	–
σ_0 (Pa)	3.5 ± 0.5	2.5 ± 0.3	1.0 ± 0.2	0.8 ± 0.2	0.5 ± 0.1

Table 1: Numbering, labeling, composition and yield stress, σ_0 of the fluids in use.

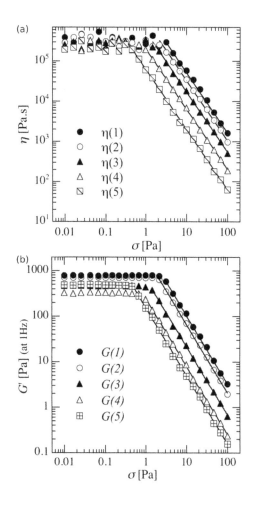

Figure 4: Rheological properties of the samples – (a) Shear viscosity η and (b) storage modulus G' vs. shear-stress σ.

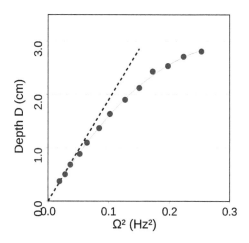

Figure 5: Depth D vs. Ω^2 for a Newtonian fluid. At large rotation velocity Ω, the depth D is limited by the bottom ($h = 3$ cm).

and decreasing Ω.

ii. First experimental observations in weakly non-Newtonian fluids

Let us first consider the behavior of the free surface for samples which exhibit rather small storage modulus G' (Samples 3, 4 and 5, Table 1).

Upon increasing the rotation velocity Ω, above a threshold, which depends on the depth h, a depression appears at the vertical of the rotating disk (Fig. 6). Initially, the deformation is concave. But, upon increasing Ω, one observes the appearance of shapes involving a change in the sign of the curvature of the profile (Fig. 6, $\Omega \geq 3.55$ Hz). One can immediately notice from the images that the geometry of the free surface is mainly governed by geometrical constraints. We report in Fig. 7 a summary of the main geometrical characteristics of the observed deformation of the free surface as function of Ω. We focus on three different characteris-

D, increases linearly with Ω^2, for limited values of Ω (Fig. 5). This behavior is in accordance with the prediction for a Newtonian flow in solid body rotation. When the rotation velocity is further increased, the vertical extent of the central depression is restricted by the bottom of the tank. No hysteresis is observed upon increasing

tics: the depth D, i.e., the distance between the lowest point of the profile and the free surface; the width W, i.e., the typical diameter of the depression at the free surface; the width Wi, i.e., the typical distance between the inflection points in the profile from camera 1. On the one hand, the lateral size of the depression, W or Wi, is of the order of the diameter of the rotating disk. Hence, we observe that the fluid can only be set in motion above the rotating disk. The fluid motion is thus confined in a vertical column having a diameter of about the diameter of the disk. The remaining volume of fluid in the tank remains at rest. On the other hand, the flow extends to the free surface only if Ω is large enough. Upon further increase of Ω, the depth D of the depression at the free surface is rapidly limited by the depth of fluid, h. At large Ω, the bottom of the depression is flat. Note, however, that the disk is never dewetted and a thin layer of fluid remains at center. In this conditions, the profile of the free surface exhibits a significant hysteresis. Indeed, upon decreasing Ω, one observes that the bottom of the depression progressively detaches from the bottom of the container but that the central part, that is in solid body rotation, remains flat and horizontal. The depth D decreases until it vanishes.

We finally mention that the crater can lose axi-symmetry and exhibit polygonal shapes like smoke rings [9, 10], liquid tori levitating due to Leidenfrost effect [11], hydraulic jumps [12, 13] or simple liquids in the same experimental conditions as ours [14, 15]. Vortex rings and their stability have attracted much attention in the past [16] and we will not develop further the case of weakly non-Newtonian fluids, but rather focus on the behavior of the free surface in the case of strongly non-linear fluids. We shall see that the free surface can then loose its initial axi-symmetry due to a qualitatively different physical mechanism.

iii.   Strongly non-Newtonian fluids

The samples 1 and 2 exhibit similar rheological properties and, in accordance, the behavior of the free surface is very similar in both fluids. In addition, in spite of a slight hysteresis, the system exhibits similar behaviors for increasing and decreasing rotation velocities. Exploring a large range of height, h, and increasing rotation velocity Ω, we report for sample 1, the set of 5 typical shapes of the free surface in the steady-state that we managed to observe experimentally.

a.   Bulge

For small rotation velocity Ω, we observe an axisymmetric bulge above the disk at the center (Fig. 8). Around this bulge, the fluid is at rest, as can be directly deduced from the roughness of the surface, which remains static. In the central region, the fluid of the free surface is in solid body rotation, as can be proven by marking the surface with a tool and observing that this mark remains unchanged. The rotation velocity at the surface, ω, which is always smaller than the rotation velocity of the disk, Ω, decreases when the depth h is increased. We observe that, initially, the height of the bulge increases when Ω is increased.

b.   Cap

At a slightly larger rotation velocity Ω, a circular cap pops up at the center of the bulge (Fig. 9). The transition depends slightly on the depth of fluid, the critical rotation velocity increasing almost linearly with h (see state diagram in Fig. 13). In this regime, the fluid around the bulge is still at rest, the cap at the center is in solid-body rotation, but the free surface in the remaining part of the bulge flows. The rotation velocity of the cap is always smaller than that of the bulge in the central region. As Ω is further increased, the bulge flattens while the height of the cap above the free surface decreases (from 7-8 mm at the lowest velocities, down to about 2 mm). Meanwhile, the diameter of the cap is of about 2 cm in diameter, almost independent of Ω, for small h (typically less than 3.5 cm). We note however a slight decrease of its lateral size when Ω is increased for larger h. These behaviors are reversible. We note that a further increase of Ω leads to the appearance of a depression around the cap.

c.   Crater

When Ω is further increased, the cap sinks below the free surface. We observe the formation of a crater (Fig. 10), at the bottom of which the cap remains, at least close to the transition which is discontinuous. Indeed, the center of the free surface suddenly sinks from several millimeters (Fig. 12). The onset increases with h (see state diagram in Fig. 13). The depth of the crater increases with Ω. By contrast, the diameter of the crater at the top (typically 4-6 cm) does not depend significantly on Ω but decreases when h is increased.

Depending on h, we observe two scenarios. For small

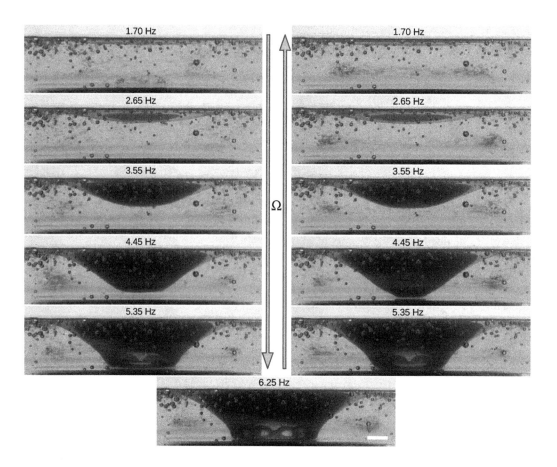

Figure 6: Evolution of the free surface profile upon increasing rotation velocity, Ω. Note that the profile exhibits significant hysteresis when Ω is cycled. The scalebar is worth 1 cm (Camera 1, $h = 2.4$ cm, sample 3).

height h (typically smaller than 3 cm), the center of the crater remains circular at any rotation velocities and its diameter decreases when Ω is increased. By contrast, for larger height h (typically above $3 \sim 3.5$ cm), as the rotation velocity increases, the central part of the crater loses its original circular symmetry for taking an oval shape. The cap becomes increasingly asymmetric when Ω is increased; the size of the major axis always remains constant while the size of the minor axis decreases progressively. In this case, a further increase of Ω leads successively to two more states of the free surface, both asymmetric, that we describe in the next sections.

d. Knife-edge

When the cap at the center of the crater is reduced only to a furrow, the bottom of the depression sinks in and takes the form of a twisted knife-edge that rotates around the vertical axis (see Fig. 1). One thus notice the occurrence of a furrow (with forks at both ends). The transition seems to be continuous; the circular shape of the negative cap turning, gradually, into a furrow that sinks into the gel. The rotation velocity of the knife-edge, ω, is again much smaller than that of the rotating disc, Ω. Its length, W, and depth, D, depend on Ω; in particular, W decreases when Ω is increased. Note that the obtention of this steady knife-edge, equivalent to that observed in Ref. [4], constitutes the main achievement of our experimental study. It is observed in a very narrrow range of the experimental parameters as can be seen in the state diagram reported in Fig. 13.

e. Singular point

Finally, when Ω is further increased, the length of the knife-edge, W, decreases until it vanishes. The lat-

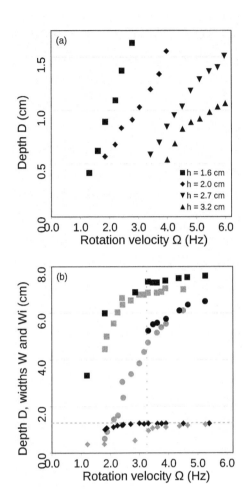

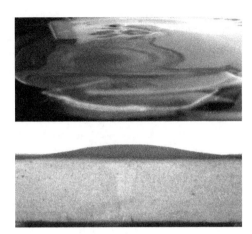

Figure 8: For small rotation velocity Ω, on observes an axisymmetric bulge at the vertical of the rotating disk (Top: camera 4; Bottom: camera 1, Ω = 1.25 Hz, h = 2.7 cm, sample 1). Associated movies Fig.8.c1.avi, and Fig.8.c4.avi can be found as Supplementary material [8].

Figure 7: (a) Typical depth D of the pattern at the free surface vs. rotation velocity Ω for various depth of the fluid bed, h. Note that the formation of the depression at the free surface requires a larger rotation velocity Ω when the depth h is larger (The typical accuracy of the measurements is of about ±0.25 cm). (b) Typical depth, D (◆), and widths W (■) and Wi (●) of the pattern at the free surface vs. rotation velocity Ω (Black symbols: upon increasing Ω; Gray symbols: upon decreasing Ω, h = 1.3 cm, sample 3).

Figure 9: At intermediate rotation velocity Ω, a circular cap pops up at the vertical of the rotation axis (Top: camera 4; Bottom: camera 1, Ω = 2.6 Hz, h = 2.7 cm, sample 1). Associated movies Fig.9.c1.avi, and Fig.9.c4.avi can be found as supplementary material [8].

ter then reduces to a single point (Fig. 11). The transition between the knife edge and the singular point is again continuous. Note that the surface is still not axi-symmetric but takes the shape of an inverted and slightly twisted triangular-based pyramid. The latter ro-

tates around the vertical axis with a rotation velocity smaller than that of the disc. Its typical size does significantly depend on Ω.

Figure 10: At sufficiently large rotation velocity Ω, the free surface sinks below its original level, leading to the formation of a crater. One can still observe the cap at center (Top: camera 4; Bottom: camera 1, Ω = 3.3 Hz, h = 2.4 cm, sample 1). Associated movies Fig.10.c1.avi, and Fig.10.c4.avi can be found as supplementary material [8].

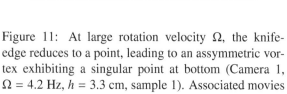
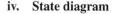

Figure 11: At large rotation velocity Ω, the knife-edge reduces to a point, leading to an assymmetric vortex exhibiting a singular point at bottom (Camera 1, Ω = 4.2 Hz, h = 3.3 cm, sample 1). Associated movies Fig.11.c1.avi, and Fig.11.c2.avi as supplementary material [8].

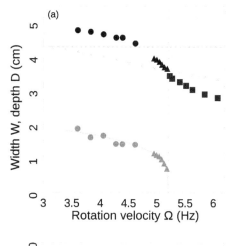
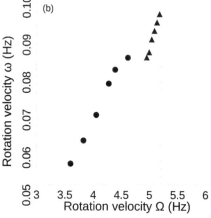

Figure 12: (a) Typical width, W, and depth, D, of the pattern at the free surface vs. rotation velocity, Ω. (b) Rotation velocity of the pattern at the free surface, ω, vs. rotation velocity of the disk Ω (h = 4.3 cm).

iv. State diagram

We have qualitatively distinguished 5 typical shapes of the free surface. We report, in Fig. 12, an example of the evolution of the characteristics of the pattern at the free surface as function of the rotation velocity Ω for a given depth of the fluid bed.

In order to go further, we propose, in Fig. 13, state diagrams where the various patterns are located in the plane (h,Ω). As only little hysteresis is observed between increasing and decreasing Ω, we report data for increasing rotation velocity, only. We note that the knife-edge is seen in a narrow region at intermediate rotation velocity Ω for sufficiently large depth h of the fluid bath. In order to account for the effect of the fluid proporties, we report diagrams for samples 1 and 2 (Table 1). Sample 2 exhibits a slightly smaller storage modulus G' than sample 1. As a consequence, the domains of the knige-edge (▲) and of the singular point (□) are slightly shifted toward larger depth, h, and thus slightly reduced, compared to what was observed for sample 1.

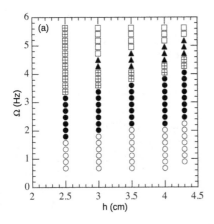

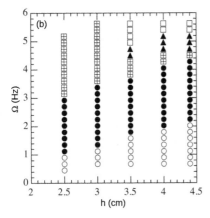

Figure 13: State diagrams – We place in the plane (h,Ω) the domain of existence of the various typical shapes of the free surface observed experimentally. (a) Sample 1. (b) Sample2 ($\circ$: Bulge; $\bullet$: Cap; $\boxplus$: Negative cap; $\blacktriangle$: Knife edge; $\square$: Singular point, Sample 1).

IV. Discussion

In this section, we qualitatively explain the origin of the observed patterns. From images from side, on can have a clue on the structure of the flow in the material (Fig. 14).

First, due to the non-sliding condition at the disk surface, a rotation of the fluid with the orthoradial velocity $r\Omega$ is imposed at the bottom of the fluid bed above the disk (r is the radial coordinate). The fluid enters in motion in the radial direction if the associated stress $\rho(r\Omega)^2$ exceeds the yield stress σ_0. Denoting R the radius of the rotating disk, we can define a minimal rota-

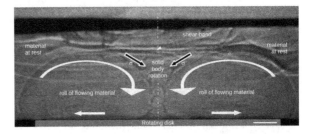

Figure 14: Direct imaging of the convection rolls. When new material is added, because of the difference in temperature, one can observe the convection rolls thanks to the inhomogeneity of the refraction index. One can clearly observe at small velocity, the formation of the convection rolls (Camera 1).

tion velocity $\Omega_m \equiv \sqrt{\sigma_0/\rho}/R$ below which the fluid, in the vicinity of the disk surface, does not flow in the radial direction (ρ denotes the density of the fluid). From the experimental values of σ_0, ρ and R, we estimate $\Omega_m = (0.3 \pm 0.1)$ Hz, depending of the sample. The latter value is compatible with the lower bound of the rotation velocity range in which we observed deformation of the free surface.

Once the radial flow is initiated, it extends away from the disks over a distance again limited by the yield stress, σ_0. Considering the orthoradial velocity $R\Omega$ imposed at the edge of the rotating disk, assuming that the flow extends over the distance L, we can write that the viscous stress, $\eta R\Omega/L$, is constant and equals the yield, the condition being imposed by the fact that at $R + L$ the fluid remains at rest. We get $L = \eta R\Omega/\sigma_0$. Here, η takes the value measure for $\sigma = \sigma_0$ (Fig. 4). For instance, considering the most rigid fluid, with $\sigma_0 \sim 800$ Pa, we estimate $\eta \sim 50$ Pa.s and, thus, $L \sim 2$ cm for $\Omega \sim 1$ Hz. This value is again compatible with the observed radial extension of the roll observed in Fig. 14.

Thus far, we have showed that the rotation of the disk puts the fluid in radial motion in a region of typical diameter $2(R + L)$ in the disk plane. In turn, this outward flow has a pumping effect, i.e., fluid flows back above the disk as can be seen in Fig. 14 and rolls form. Their typical vertical extension is imposed by the geometrical constraints so that it is of the order of a fraction of the radius of the disk, R. We assume here that the depth, h, of the fluid bed does not limit the flow which thus does no extend to the free surface. In this limit of deep fluid beds, we predict that the fluid remains at rest around the disk, at a radial distance larger that $R + L$, and above the disk, at a distance of about αR, with $\alpha \lesssim 1$. The region above the disk, in which the fluid is in motion, is thus separated from the rest of the fluid which remains at rest by a horizontal shear band which clearly visible in Fig. 14. The thickness, T, of the shear band is such that the shear stress equals the yield and we get, following the same reasoning as above, that $T \sim L/2 \sim 1$ cm, in agreement with the observations.

From this analysis of the underlying rolls, we can qualitatively understand the sequence of the patterns we observe experimentally. Note first that, in our experimental conditions, the depth of the fluid bed, h, is always limited, smaller than R so that the rolls can reach the free surface. Upon increasing Ω, one observes the following sequence of patterns:

- Very small Ω, typically smaller than Ω_m – The rotation velocity is so small that it does not induce any radial flow. No pattern is observed.

- Rotation velocity $\Omega \gtrsim \Omega_m$ – The rotation velocity is large enough to induce a radial flow, but only in a limited region of the disk. Indeed, the radial stress induced by the rotation overcomes the yield stress for $r > r_m \simeq \sqrt{\sigma_0/\rho}/\Omega$, only. As a consequence, rolls form but their typical radial size is of the order of $R - r_m$. Accordingly, their typical vertical size is of about $\alpha(R - r_m)$. Thus, because of the solid-like behavior of the material below the yield, a layer seats at rest on top of them. However, the inward flow at the top of the rolls produces an inward creep of the material above, leading to the formation of the bulge.

- Rotation velocity $\Omega \gtrsim \Omega_h$, such that the rolls reach the free surface – Due to the toroidal shape of the rolls, the outer flow lines form a cusp at center. In this region, above the cusp, the material does not flow but, again due to the inward stress, the creep leads to the formation of the cap, i.e., a volume of fluid in solid body rotation surrounded by flowing material. Considering that the top of the rolls reaches the free surface when $\alpha(R - r_m) = h$, we get $\Omega_h = \sqrt{\sigma_0/\rho} \frac{\alpha}{\alpha R - h}$. Typically, we get a value of Ω_h

that is of the order of a few Hz for the most rigid fluid and increases with h. In spite of a quantitative discrepancy, the transition is correctly predicted by this argument.

- For moderate values of Ω above Ω_h, the horizontal stress σ (Fig. 14) exerted by the rolls on the fluid sitting on top still makes the free surface rise. But, conversely, when Ω is further increased, the increase of the vertical component of σ at the center, which is oriented downward, overcomes the inward radial contribution, which leads to a downward displacement of the free surface. We oberve what we called the 'negative' cap. Indeed, even in this case, a volume of the fluid can remain in solid body rotation at the center. This transition is difficult to predict quantitatively without the knowledge of the structure of the rolls, but the transition is qualitatively well understood.

- Finally, further increase of the rotation velocity leads to the erosion of the domain of fluid in solid rotation at the center. When it disappears, we observe either the knife-edge or the singular point. The transition is again difficult to predict precisely but we can estimate the maximum value of the vertical component of the velocity in the roll around, v_M, which is compatible with no flow at center. To do so, we write that, at the external edge of the central column of fluid, the shear stress, $\eta v_M/(2R)$, is of about the yield stress σ_0. Assuming further that v_M is of the order of the orthoradial velocity of the fluid at the outer egde of the disk $R\Omega_M$, we estimate this central region which remains in solid body rotation disappears for $\Omega_M > 6$ Hz. This value is compatible, in order of magnitude, with the velocity measured at the transition between the cap and the knife-edge for the deepest fluid beds (Fig. 13). Note, however, that the dependence on h is not accounted for by our argument, which would require the introduction of a precise description of the rolls to account for the details of the transition.

Finally, we remark that, even if we qualitatively understand why the cap, submitted to the compressive stress σ, can buckle and lose the axi-symmetry, we are not able to describe the knife-edge and the singular point. The description of these specific patterns in detail deserves a special theoretical effort.

V.   Conclusion

We carried out an experimental investigation of the deformation of the free surface of a complex fluid induced by an underlying vortex. We observe, upon increasing rotation velocity Ω, a sequence of patterns that we described and understood, at least qualitatively, by considering the main characteristics of the underlying vortex. We reported a state diagram in which the domains of observation of the various patterns are placed in the plane rotation velocity - depth of the fluid bed, $(\Omega - h)$.

The main achievement of our study is the observation of the knife-edge and of the singular point that are very specific patterns that would not be observed in Newtonian fluids. They are the steady-state version of the previously observed folding of the free surface of a complex fluid during the penetration of a sinking bead [4]. Our experimental configuration is particularly interesting as it makes possible the observation of patterns in the steady state whereas the previous experiments only gave access to transients.

However, the present study remains purely qualitative and thereby raises several theoretical points. First, a better description of the fluid flow, including a precise account of the fluid rheology, would be necessary for a precise description of the state diagrams. But, in addition, the experimental configuration is interesting to answer several questions: are the patterns of the free surface observed in this investigation generic to the non-Newtonian fluids or for yield stress fluids only? Are the same types of patterns observed for a viscoelastic fluid (entangled solutions of polymers or wormlike micelles)? The applications of such a study are numerous, especially in terms of geophysics, where complex fluids play an important role (landslides, lava flow).

---

[1] P Coussot, *Yield stress fluid flows: A review of experimental data,* J. Non-Newton. Fluid **211**, 31 (2014).

[2] D De Kee, R P Chhabra, *A photographic study of shapes of bubbles and coalescence in Non-Newtonian polymer solutions,* Rheol. Acta **27**, 656 (1988).

[3] N Dubash, I A Frigaard, *Propagation and stopping of air bubbles in Carbopol solutions,* J. Non-Newton. Fluid **142**, 123 (2007).

[4] T Podgorski, A Belmonte, *Surface folds during the penetration of a viscoelastic fluid by a sphere,* J. Fluid Mech. **460**, 337 (2002).

[5] T Podgorski, A Belmonte, *Surface folding of viscoelastic fluids: Finite elasticity membrane model,* Eur. J. Appl. Math. **15**, 385 (2004).

[6] H Lhuissier, B Néel, L Limat, *Viscoelasticity breaks the symmetry of impacting jets,* Phys. Rev. Lett. **113**, 194502 (2014).

[7] C Palacios-Morales, R Zenit, *The formation of vortex rings in shear-thinning liquids,* J. Non-Newton. Fluid **194**, 1 (2013).

[8] Supplementary material can be found at http://dx.doi.org/10.4279/PIP.080007

[9] R H Hernández, B Cibert, C Béchet, *Experiments with vortex rings in air,* Europhys. Lett. **75**, 743 (2006).

[10] I S Sullivan, J J Niemela, R E Hershberger, D Bolster, R J Donnell, *Dynamics of thin vortex rings,* J. Fluid Mech. **609**, 319 (2008).

[11] S Perrard, Y Couder, E Fort, L Limat, *Leidenfrost levitated liquid tori,* Europhys. Lett. **100**, 54006 (2012).

[12] T Bohr, C Ellegaard, A Espe Hansen, A Haaning, *Hydraulic jumps, flow separation and wave breaking: An experimental study,* Physica B **228**, 1 (1996).

[13] A Andersen, T Bohr, T Schnipper, *Separation vortices and pattern formation,* Theor. Comp. Fluid Dyn. **24**, 329 (2010).

[14] T R N Jansson, M P Haspang, J H Jensen, P Hersen, T Bohr, *Polygons on a Rotating Fluid Surface,* Phys. Rev. Lett. **96**, 174502 (2006).

[15] H A Abderrahmane, K Siddiqui, G H Vatistas, *Rotating waves within a hollow vortex core,* Exp. Fluids **50**, 677 (2011).

[16] K Shariff, A Leonard, *Vortex Rings,* Annu. Rev. Fluid Mech. **24**, 235 (1992).

14

Observation of the two-way shape memory effect in an atomistic model of martensitic transformation

E. A. Jagla[1]*

We study a system of classical particles in two dimensions interacting through an isotropic pair potential that displays a martensitic phase transition between a triangular and a rhomboidal structure upon the change of a single parameter. It has been previously shown that this potential is able to reproduce the shape memory effect and super-elasticity, among other well-known features of the phenomenology of martensites. Here, we extend those previous studies and describe the development of the more subtle two-way shape memory effect. We show that in a poly-crystalline sample, the effect is mostly due to the existence of retained martensite within the austenite phase. We also study the case of a single crystal sample where the effect is associated to particular orientations of the dislocations, either induced by training or by an ad hoc construction of a starting sample.

I. Introduction

Martensitic transformations are temperature driven non-diffusive phase transitions in which the atoms of a solid perform small individual atomic displacements between a high temperature high symmetry austenite (A) phase, and a low temperature lower symmetry martensite (M) phase [1]. When the symmetry of the martensite phase is a sub-group of that of the austenite phase, the system displays the remarkable shape memory effect (SME) [2]: the sample in the martensite phase can be mechanically deformed significantly, yet, when the temperature is raised and the austenite becomes stable, the sample returns to its original shape. In simple terms, the origin of the shape memory effect is associated to the fact that

in the martensite phase, and even if the sample is mechanically deformed, each atom preserves the atomic neighborhood it had in the austenite, and when temperature is raised and austenite becomes stable again, each single atom returns to its original position.

It is quite remarkable that in many cases, after a sample was submitted to one (or more) of these cycles, it is able to "remember" the kind of deformation it was submitted to, and upon a new cooling step, it deforms spontaneously to the remembered shape, without the application of an external deformation. The shape change under this cooling-heating cycling is termed the two-way (or all-round) shape memory effect (TWSME). The origin of the TWSME is more subtle than the simpler SME. It does not originate in general symmetry considerations but in the remaining, after the initial cooling-deformation-heating "training" cycles, of atomic arrangements within the austenite that "remember" the form in which the sample was stressed; upon a new cooling, they favor the nucleation of martensite in preferential orientations. It

*E-mail: jagla@cab.cnea.gov.ar

[1] Comisión Nacional de Energía Atómica, Instituto Balseiro (UNCu), and CONICET. Centro Atómico Bariloche, Avda. E. Bustillo 9500, 8400 Bariloche, Argentina.

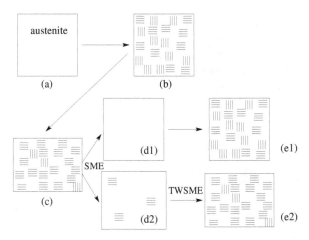

Figure 1: Sketch of the shape changes in a sample displaying a martensitic transformation. Martensite is schematically indicated by the dashed regions, which can be vertical or horizontal, indicating a martensitic variant elongated in the corresponding direction. (a) Starting austenite sample. (b) Isotropic martensite after cooling. (c) Martensite deforms due to interconversion of variants, after an applied stretching along x. (d) Sample returns to the austenite phase after heating, recovering totally (d1) or partially (d2) its original form (the amount or remaining deformation in (d2) may be hardly visible experimentally). This is the SME. In the last case, the retained martensite in (d2) can trigger the spontaneous shape change of the sample after cooling (e2), giving rise to the TWSME.

has been observed that the TWSME can be triggered by small remaining pieces of the martensite phase within the austenite (this is referred to as retained martensite), collections of dislocations with particular orientations or regions of the austenite with anisotropic remnant strains [3–8]. The generic scheme giving rise to the TWSME is qualitatively depicted in Fig. 1.

Although well understood in qualitative terms, many details of the TWSME have not been totally elucidated. Particularly, the thermodynamical or dynamical reasons that favor the formation of retained martensite or other kind of elastic distortions that trigger the effect are only poorly understood. Also, other interesting questions can be asked, as for instance to what extent a starting austenite sample can be engineered, in such a way

to obtain a predefined deformation effect upon cooling, without the necessity of training cycles.

In this paper, we contribute to answer these questions. In section II, we present an atomistic two-dimensional model that displays a martensitic phase transition between a triangular and a rhomboidal phase. In section III, we construct a polycrystalline sample with periodic boundary conditions, and apply a cooling-deformation-heating training cycle. After that, we investigate if the TWSME appears. We observe the appearance of the TWSME for the case in which the deformation during the training cycle is large enough, which is correlated with only a partial recovery of shape after heating in the training cycle, and with the existence of preferential orientations of retained martensite in the austenite phase. In section IV, we show the same effect for the more realistic situation of a long bar upon bending. In section V, we study the TWSME in a single crystal sample with dislocations, showing how preferential orientations of the dislocations can justify the appearance of the TWSME. We also address the potentially promising possibility of designing a sample with an ad hoc distribution of defects that displays the TWSME. We show that a particular spatial distribution of dislocations can be proposed for which a robust and persistent TWSME is obtained. Finally, in section VI, we summarize and conclude.

II.   Model

Martensitic transformations typically occur due to an *entropic stabilization* mechanism of the austenite phase [9]. The bare interaction potential between the particles has the deepest minimum at the structure of the martensite phase, whereas the austenite phase is at most a secondary metastable minimum. Yet, the basin of attraction is wider for the austenite than for the martensite, so that thermal fluctuations generate a decrease of the free energy of the austenite with respect to the martensite as temperature is raised. At a certain temperature, the two structures can exchange stability, and the transition occurs.

A full numerical simulation of a martensitic transformation requires the simulation of the dynamics during a long time, to observe the transition between the structures taking place. In addition, the elementary time step of the simulation must be small enough in order to reproduce the thermal movement of the particles. On the whole, this fully first principle simulation requires a large temporal span that makes the simulation very time-consuming. Simulations along these lines have in fact been done [10–12], however, the subtle effects we want to detect are hard to observe with this technique.

In an alternative approach, we consider an effective interaction potential that already includes the effects of entropy. In this sense, this atomic potential must be actually considered to be a free energy functional in the spirit of Ginzburg-Landau free energies used in general descriptions of phase transitions. We will assume that the effect of temperature is to change the form of this effective free energy potential. The stability range between martensite and austenite structures is directly encoded in the form of the potential, and then the simulation will display the transition without the need to truly simulate the atomic thermal vibrations.

We numerically study the system by solving the time dependence of the particle coordinates according to a standard Verlet scheme in the presence of a frictional term proportional to velocity. This term takes out of the system any kinetic energy generated during the transformation, and effectively keeps the system at a local energy minimum. As we did in [13], we numerically study the system by solving the time dependence of the particle coordinates according to a Verlet scheme [14] of the form

$$
\begin{aligned}
\mathbf{r(t)} &= 2\mathbf{r}(t-dt) - \mathbf{r}(t-2dt) \\
&\quad + dt^2[\mathbf{f}(t-dt) - \mu \mathbf{v}(t-dt)] \quad (1) \\
\mathbf{v(t)} &= [\mathbf{r}(t) - \mathbf{r}(t-dt)]/dt, \quad (2)
\end{aligned}
$$

with $\mathbf{r}(t)$ the position of each particle, $\mathbf{v}(t)$ its velocity and $\mathbf{f}(t)$ the force acting on the particle coming from all neighbor particles at time t. Note that a frictional term proportional to the velocities (and to a friction coefficient μ) has been included. A kind of term like this is neccesary as the conversion between martensite and austenite occurs in general at values of the control parameter that are not precisely the equilibrium ones (see below). This means that there is some conversion from potential to kinetic energy at the transition, and this energy

excess has to be taken out of the sample to avoid generating very large particle velocities. The local friction term efficiently accomplishes this goal, whereas protocols depending on the global energy (the Nose-Hoover thermostat, for instance) would not be effective [14]. In all this paper, we use $\mu = 6$ and $dt = 0.01$.

The potential we use and its basic features were presented in [13]. For completeness, we give the form of the potential here. The interparticle potential $V(r)$ is composed of several parts:

$$V(r) = V_0 + V_1 + V_2 + V_3, \quad (3)$$

where

$$
\begin{aligned}
V_0 &= A_0\left[\frac{1}{r^{12}} - \frac{2}{r^6} + 1\right] \text{ if } r < 1 \quad (4) \\
V_1 &= \left[\frac{(r-1)^2(r+1-2c)^2}{(c-1)^4}\right] - 1 \text{ if } r < c \,(5) \\
V_2 &= -\frac{A_2}{s_2^4}[(r-d_2-s_2)^2(r-d_2+s_2)^2] \\
&\qquad\qquad \text{if } d_2 - s_2 < r < d_2 + s_2 \quad (6) \\
V_3 &= \frac{A_3}{s_3^4}[(r-d_3-s_3)^2(r-d_3+s_3)^2] \\
&\qquad\qquad \text{if } d_3 - s_3 < r < d_3 + s_3, \quad (7)
\end{aligned}
$$

and 0 otherwise in all cases. V_0 is the repulsive part of a LJ potential and its weight in the total potential is measured by the parameter A_0. The quartic term V_1 contributes with an attractive well to the total potential. The last two terms are fine tuning terms that provide a small minimum of amplitude A_2 centered at d_2, and a small maximum of amplitude A_3 centered at d_3. They were adjusted to penalize appropriately the triangular lattice, and/or favoring the martensitically related structure. The potential is fully determined by the set of parameters $P = \{A_0, A_2, A_3, c, d_2, s_2, d_3, s_3\}$. To study the triangular-rhomboidal transition (T-R), we use $P = \{A_0, 0.003, 0.01, 1.722, 0.98, 0.04, 1.74, 0.2\}$ with variable A_0. A_0 is the parameter that models the effect of temperature. There is a critical value of $A_0^c \simeq 0.067$ such that for $A_0 < A_0^c$ ($A_0 > A_0^c$) the martensite (austenite) phase is globally more stable.

The lattice parameter of the austenite phase is approximately 0.995 at the transition point, increasing slightly with the value of A_0. The austenite phase is characterized by two different inter-atomic distances that are approximately 0.93 and 1.04 at the transition. In order to visualize parts of the sample in the austenite or martensite phase, and since a direct examination of every atomic position is impractical, we make the following analysis: given a configuration of the particles, neighbor particles located at a distance compatible with the largest interatomic distance in the martensite phase are identified (for practical reasons, we use a distance window between 1.015 and 1.2), and those links are plotted as segments. This highlights regions of the sample in the martensitic phase. In addition, since our attention is focused on the shape of the sample, the following color code is implemented: when the large link of the martensite is oriented mostly horizontally, the link is plotted in red, whereas when the link is oriented mostly vertically, it is plotted in green. This color coding allows to identify at a glance if the sample (or a part of it) is preferentially stretched in the x or y direction.

III. Shape memory and two-way shape memory in a periodic, polycrystalline sample

The "shape" of a sample has the characteristic of being strongly connected to the existence of free surfaces. Yet, as a first example and since we are looking for tiny effects that must show up in the simulations, a different set up will be more convenient. In this section, we will study a system with periodic boundary conditions within a rectangular box of size l_x, l_y. In this set up, the shape of the sample is characterized by the ratio l_x/l_y. The values of l_x of l_y are allowed to adjust dynamically during the simulation according to the following mechanism. The internal compressions f_x and f_y in the system in both directions x and y are calculated along the simulation and are driven towards externally imposed values f_x^{ext} and f_y^{ext}, by the adjustment of l_x of l_y, according to the equations

$$\frac{dl_{x,y}}{dt} = \lambda(f_{x,y} - f_{x,y}^{ext}). \qquad (8)$$

If $f_{x,y}^{ext}$ are set to zero, the equilibrium values of $l_{x,y}$ determine the shape of a "free" sample. Instead, non-zero values of $f_{x,y}^{ext}$ model the application of external forces along the x or y directions.

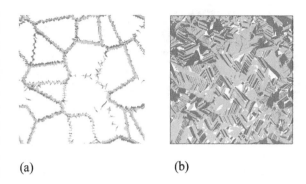

(a)   (b)

Figure 2: (a) The starting polycrystalline sample after initial relaxation in the austenite phase. Plotting details and the meaning of the color code are provided in the text. Note that although the bulk of the grains is in the austenite phase, atoms with interatomic distances compatible with the martensite are widely present at the grain boundaries. (b) The sample after transformation to the martensite phase. There is no appreciable shape change during this transformation.

The starting sample consists of a poly-crystal constructed in the following way: a number of grain centers and orientations are randomly chosen, and they are used to generate a given grain orientation around each of the centers. Each grain extends in the sample according to a Voronoi tessellation criterion, namely each sector of the sample is dominated by the nearest grain center. As a first step, this initial configuration is relaxed with the true interatomic potential and $A_0 = 0.085$, namely well within the stability range of the austenite phase. The values of $l_{x,y}$ are also allowed to relax during the process, under $f_{x,y}^{ext} = 0$. The configuration obtained is shown in Fig. 2(a). We observe that although the grains are in the austenite phase in bulk, in the grain boundaries there are atoms that are at the appropriate relative distance so as to trigger the martensite transformation. As A_0 is reduced and the martensite becomes progressively more stable, martensite crystals grow from the grain boundaries and invade the bulk of the grains, until all the sample is transformed [Fig. 2(b)]. In this transformation, the form of the sample does not change, beyond some fluctuations associated to finite size effects.

At this stage, we apply an external deformation

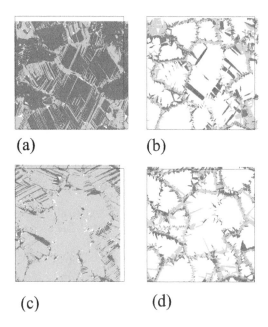

(a)   (b)

(c)   (d)

Figure 3: (a) The sample in Fig. 2(b) after external stretching in the x direction and further relaxation once the stretching is withdrawn. (b) State after re-transformation to the austenite. (c) and (d) The same for an initial compression along x direction. The black square is an outline of the sample in Fig. 2(b), for comparison. Note that the remaining martensite in (b) and (d) corresponds to the deformation produced by the applied stretching.

through the setting of non-zero values of $f_x^{ext} = -f_y^{ext}$, until we get a prescribed change in the shape of the sample, characterized by the ratio l_x/l_y. Figure 3(a) depicts the structure of the sample after a 9% change in l_x/l_y, and ulterior relieve of the external force and relaxation. Interconversion of variants caused by the external deformation (favoring x-elongated red ones) is apparent, and the sample retains a deformation of about 5%. From this configuration, the value of A_0 is taken back to 0.085, and the final configuration is shown in Fig. 3(b). Note that although the sample mostly returns to its initial configuration, displaying SME, there are remains of martensitic variants, mostly red ones, that correspond to pieces of martensite compatible with the previously applied deformation. This produces that the shape of the sample in Fig. 3(b) is not exactly equal to that in Fig. 2(b), but some

elongation along x remains. Figure 3(c) and (d) show the equivalent situation that occurs when the external stretching is performed along the y direction. In this case, the remaining martensite in Fig. 3(d) is mostly of the green type, and the sample retains an elongation along the y direction.

The detailed evolution of the shape of the sample, as measured by l_x/l_y, is presented in Fig. 4(a): the shape does not change in the first $A \rightarrow M$ transformation. The sample elongates upon the application of the external stress (S). When the stress is relieved (R), the deformation is reduced, but the sample remains deformed. When the sample returns to the austenite phase (A_2), it mostly recovers its original shape. This is the standard SME. From the configurations A_2, we perform new austenite-martensite cycling, without applying any external stress. The evolution of the shape during this process is seen in Fig. 4 $(A_2 \rightarrow M_2 \rightarrow A_3...)$. We see that a systematic TWSME is observed and that follows the training imposed by the original external deformation. To better quantify this effect, we first plot in Fig. 4(b) the amount of remnant deformation after the first re-transformation to the austenite (A_2) as a function of the deformation caused by the external loading (R). We see that this dependence is strongly non-linear. It is vanishingly small if the amount of applied deformation is low, but it rapidly increases at larger applied deformation. In (c), we see that the intensity of the TWSME is proportional to the remaining deformation after the training step indicating that preferential orientation of the remaining martensite within the austenite is the main responsible of TWSME. In fact, the remaining deformation in the austenite can be argued to be proportional to the imbalance between the different orientations of the retained martensite. In turn, this imbalance is responsible for the preferential growth of conveniently oriented variants when the sample transforms to the martensite phase.

It is worth to be mentioned that the TWSME we observe is robust with respect to cycling between martensite and austenite. Typically, the effect is maximum in the first cycle, it diminishes about 20% in the second cycle, and then it conserves this value for at least 20 cycles. That was the longest run we performed.

In experimental realizations, it has been observed that the TWSME can be almost perfect: the sam-

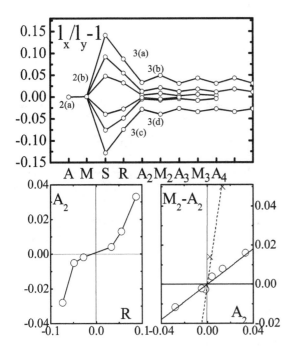

Figure 5: The bar shaped sample. The black lines outline the grains in the starting sample. The actual configuration of the sample after initial relaxation in the austenite phase is shown.

Figure 4: (a) Evolution of the shape (l_x/l_y) of the sample according to the following protocol. A: starting austenite [Fig. 2(a)]. M: transformation to the martensite [2(b)]. S: stretched sample upon application of different external loads. R: relaxed sample after external force is set to zero [examples in 3(a),(c)]. A_2: transformation to the austenite, and observation of the SME. M_2, S_2 and the following: shape change under cyclic austenite-martensite transformation, in the absence of external loading. This is the TWSME. (b) The remnant deformation as a function of the maximum deformation. The form of this curve is clearly non-linear. (c) The amount of shape recovery in the TWSME as a function of the remnant deformation in the austenite phase. This relation is seen to be almost linear in the range of deformations analyzed, with a proportionality factor of about 0.5. The two crosses and the dotted line show a few results in a sample with larger grain sizes, where this factor is much bigger.

ple may recover the deformed shape, even if the remnant deformation is so small that it is undetectable. On the contrary, the plot in Fig. 4(c) shows that the amount of the TWSME in our case

is tiny: it approximately corresponds to a spontaneous deformation upon transition to the martensite that is about one half of the remnant deformation in the austenite after training. We think that this is related to the grain size of our sample. In fact, in additional simulations using smaller grains, the TWSME was almost unobservable. Instead, in a more demanding simulation using grains of linear size about three times those in Fig. 2, the effect was much stronger, as the two crosses in Fig. 4(c) indicate. The dotted line indicates in this case that the shape recovery is about five times larger than the remnant deformations. We think that this trend makes plausible that the TWSME we are observing is compatible with that observed experimentally, where the grain sizes are much larger than those in our samples.

IV.   SME and TWSME in a free bar

The results in the previous section show clearly that our model contains all the necessary ingredients to explain the origin of the TWSME. It would be desirable, however, to show the effect appearing in an experimentally achievable configuration, particularly in a sample with free surfaces.

In order to do this, we consider the case of the bending of a bar shaped sample, with free surfaces. The internal structure of the sample is again polycrystalline but, in this case, the spatial distribution of the grains was chosen by hand. The reason for this choice is that truly random grains produce too strong variations from sample to sample in our small systems. The outline of the grains and the actual initial configuration after relaxation are shown in Fig. 5. The system is submitted to

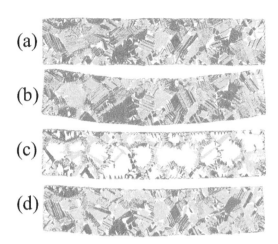

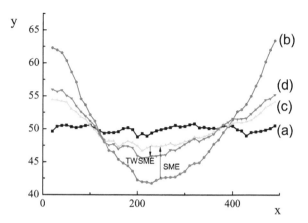

Figure 7: Profiles of the samples in the previous figure, to quantify the degree of bending in each stage. Both axis are in units of the inter-atomic distance.

Figure 6: Evolution of the bar along the treatment. (a) Free sample in the martensite state. (b) Bending caused by an external loading, after the loading has been relieved. (c) Transformation to austenite. The sample straights, showing SME, but an appreciable deformation remains. (d) Re-transformation to the martensite phase. The sample bends (a small amount) spontaneously. This is the TWSME.

a sequence of transformations as displayed in Fig. 6. First: reduce A_0 down to 0.057 to reach the martensitic state (a). In this process, the shape of the sample does not change appreciably. Second: application of an external bending stress followed by the removal of the stress. The final configuration obtained is shown in (b), where it is seen that an important bending deformation remains in the sample. Third: transform back to the austenite, by increasing A_0 to 0.085 (c). In this stage the SME shows up, and the bar tends to recover its original straight shape. Note, however, that the recovery is not complete: a certain amount of deformation remains that is originated in pieces of x-elongated (green) martensite in the lower part of the sample, and y-elongated (red) martensite in the upper part. This imbalance of martensitic variants is the germ of the spontaneous bending of the sample when we transform to the martensite phase ($A_0 = 0.057$) in the absence of external stresses. The final configuration is shown in (d).

To quantify the TWSME, which is tiny for our small samples, we calculated the average vertical coordinate of the particles in slices of the bar at different horizontal coordinates, and these values are plotted in Fig. 7 for the configurations in Fig. 6. We see that in addition to the clear SME between (b) and (c), a small but clearly noticeable TWSME between (c) and (d) exists, in which the sample bends spontaneously without the application of any stress. This effect is persistent upon successive cycling between austenite and martensite, and it is of the same intensity as the one described in the previous section for a sample with periodic boundary conditions. In fact, quantifying the effect by the curvature of the bar, we see that the shape recovery when the sample transforms to the martensite (the "TWSME" arrow in Fig. 7) is about 0.4-0.5 of the remnant deformation (the (c) profile in the same figure), very much as in Fig. 4(c).

V. SME and TWSME in single crystals with dislocations

In the previous sections, the origin of the TWSME was related to the remaining of martensite variants within the austenite structure. For the systems analyzed, these remains were favored by the polycrystal nature of the sample that allows for low energy atomic rearrangements at the grain boundaries.

TWSME occurs also in the case of single crystal

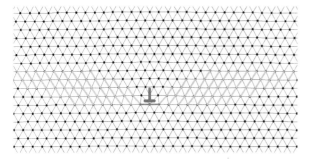

Figure 8: Lattice structure around a single dislocation in a perfect triangular structure. The dislocation core is indicated by the "⊥" symbol. Highlighted interatomic links indicate those that are in the range [1.015, 1.2], well different than the equilibrium lattice parameter of the triangular structure (0.995).

samples. In fact, although a perfect austenite single crystal is not able to encode within its structure a memory of a previous deformation in the martensite phase, the unavoidable existence of defects – particularly dislocations– in the crystal structure provides a mechanism to generate the TWSME. Dislocations are defects that produce anisotropic stresses in the crystal structure, which can favor the growth of martensite variants of particular orientations upon cooling [3, 15, 16].

Experimentally, it is known that the TWSME can be induced in single crystals by training, this means that dislocations acquire a convenient distribution to trigger the shape transformation in some prescribed way, after the removal of the external stress. It is also possible to consider the case of a sample that is engineered with a collection of dislocations distributed in some appropriate manner, such that the sample may have a tendency to spontaneously elongate in one direction and contract in the perpendicular one upon the martensitic transformation. Although at present it is not feasible to produce such an ad hoc distribution of dislocations, this seems an interesting theoretical situation to consider.

The elastic distortions that a single dislocation generates on an isotropic elastic solid are well-known. However, it is not necessary to consider this deformation in full detail here. The following analysis is more appropriate to our purposes. We constructed by hand a single crystal structure

with a dislocation in an otherwise perfect single crystal sample. After preparation, the structure was relaxed using the interatomic potential in the austenite phase ($A_0 = 0.085$). The final structure is shown in Fig. 8. We also plot in that figure the interatomic distances that are in the range appropriate to become the largest lattice parameter of the martensite structure, i.e., they will be the germs on which martensite crystals will grow. It is thus seen in the particular case of Fig. 8 that this dislocation will induce a tendency to elongate the sample along the insertion line of the dislocation (y direction in this case) and a tendency to contract along the glide direction (x direction).

In order to see to what extent dislocations are effective to encode the TWSME, we first consider a simple case. To avoid the existence of free surfaces that can blur the effect of the dislocations, a single crystal with periodic boundary conditions is necessary. However, it is not possible to accommodate a single dislocation with these boundary conditions. Instead, we used a crystal with two dislocations as sketched in Fig. 9(a). Note that both dislocations have the same orientation of the glide plane, thus contributing in the same way to a possible TWSME. The separation between the two glide planes was chosen by hand, however, note that the relative position of the two dislocation along this direction (the x separation of the dislocation cores) is adjusted by the system, as dislocations are very mobile along x. As the A_0 parameter of the potential is reduced, the system remains within the austenite phase up to $A_0 \simeq 0.60$ [Fig. 9(a)], well below the equilibrium transition value which is 0.65. This indicates that there is an appreciable energy barrier to the nucleation of the martensite in the present case. Reducing A_0 further produces an abrupt transition to the martensite phase [Fig. 9(b)-(e)], in which only two of the three possible martensitic variants are present. These variants are precisely those favored by the dislocations in the system, and generate a global contraction along x and expansion along y, as indicated in the last panel of Fig. 9 by the outline of the initial and final shapes. The amount of the TWSME is, in this case, very close to the ideal value that can be expected in this case (about 7 % in l_x/l_y) This is a clear evidence that the sample displays a strong TWSME.

The previous example shows –in a rather ideal

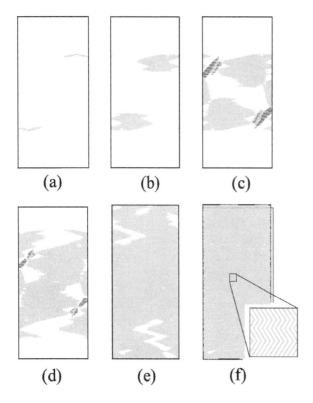

Figure 9: Martensitic transition of a sample with two dislocations and periodic boundary conditions. (a) Stable configuration for $A_0 = 0.060$. (b) to (e) spontaneous transformation for $A_0 = 0.059$. (f) final configuration at $A_0 = 0.057$. The red rectangle is the shape of the original sample in (a), and allows to see that there was an important shape change.

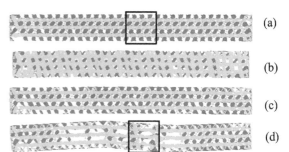

Figure 10: A free bar engineered with a collection of dislocations to induce the TWSME. (a) Structure after a relaxation with a parameter A_0 corresponding to the austenite phase. (b) The sample after the first transition to the martensitic, and (c) back to the austenite. (d) The austenite sample after a few cycles, where some degradation is observed. Details of the configurations in the two outlined boxes are shown in the next figure.

situation– that the TWSME may be induced in our model by dislocations. Looking for the same effect in a more realistic set up, we tried to train a single crystal to display TWSME. Since some amount of disorder in the starting sample is necessary, we started with a sample with a small concentration of vacancies, and performed austenite-martensite cycles in the presence of an external stress. It was observed that the vacancies reaccommodated in the form of dislocations, that were distributed in the sample in a more or less random fashion. However, we did not observe an overabundance of dislocations with the insertion plane parallel to the stretching direction that would favor the TWSME. In fact, this was not observed beyond the limit of sample to sample variations. It is not clear to us

why this attempt was unsuccessful.

We thus decided to insert by hand an ad hoc distribution of dislocations that produce the TWSME in a sample with free boundaries. We concatenated many elemental pieces similar to those in Fig. 9(a), although the separation between dislocations was much shorter. The sample obtained after a relaxation step in the austenite phase is shown in Fig. 10(a). A detail displaying the distribution of dislocations can be seen in Fig. 11(a). Note that all dislocations have the same glide directions, and all contribute to a shrinkage along x and expansion along y upon the martensitic transformation. The sample after the transition to the martensitic phase is shown in Fig. 10(b), where an appreciable decrease of length is apparent. The sample returns to its initial length after back-transforming to the austenite [Fig. 10(c)]. The shape change is persistent, as Fig. 12 shows, although we also see in Fig. 10(d) and 11(b) that the sample degrades after cycling. We suspect that the degradation effect would be lesser in larger samples, and also in a three dimensional set up where the movement of dislocations is more limited due to dislocation entanglement.

The present example is a theoretical verification that dislocations can be appropriately distributed

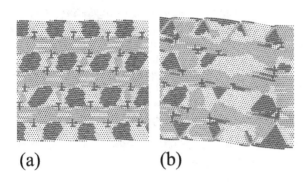

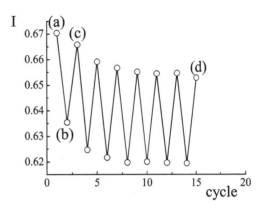

Figure 11: Detail of the configurations in the two square boxes highlighted in the previous figure. In (a), note the almost periodic arrange of dislocations that were introduce by hand. In (b), we show a particular part of the sample that has suffered of an important degradation due to dislocation movement.

Figure 12: The shape of the free bar in Fig. 10 as a function of successive austenite-martensite transformations. The I parameter reported is proportional to the moment of inertia with respect to x, defined as $I = \sum_i \overline{(x_i - \overline{x_i})^2}$.

in a single crystalline sample in order to induce the TWSME.

VI.   Conclusions

In this paper, we have used a tunable, classical, isotropic potential for identical particles in two dimensions to study the origin of the two-way shape memory effect (TWSME) in martensitic transformations. Our model starts with an effective model at intermediate time scales, in which the actual atomic vibrations are not explicitly resolved, but their effect is incorporated in an effective change of the interatomic potential. In this way, subtle effects such as the TWSME can be studied without much numerical effort, something that would not be possible from a full ab initio simulation.

We first studied the case of a polycrystalline sample, and made clear how the TWSME has its origin in the remaining of martensite pieces of the appropriate orientation inside the austenite, after a cooling-deformation-heating protocol. In addition, we were able to show that the same effect is observable in the more natural experimental configuration of a bar, under bending stresses.

In the second part of the work, we have seen how the presence of dislocations may induce the TWSME of single crystal samples, as dislocations produce a strain field on its neighborhood that fa-

vors the martensitic transformation into variants that produce a stretching of the sample in the insertion plane, and a contraction along the glide direction. We have seen that a properly chosen periodic arrange of dislocations is able to induce a bulk effect with an important shape change.

These findings indicate, first of all, that the present model and simulation techniques are appropriate to discuss subtle effects, as the TWSME, that are well understood on a qualitative basis, but only rather superficially in their microscopical details. It also allows to study problems that are not immediately accessible to present experimental capabilities, as for instance to address the effect of a given dislocation distribution in the shape change of a given sample under martensitic transformation. All these results are promising, and give confidence that other even more demanding studies, as for instance the interplay of martensitic transformation and cracking and fracturing of a sample, can also be studied with the kind of model studied here.

Acknowledgements - I acknowldege fruitful discussions with G. Bertolino, J. L. Pelegrina, M. Sade and F. Lovey. This research was financially supported by Consejo Nacional de Investigaciones Científicas y Técnicas (CONICET), Argentina.

[1] K Otsuka, M Wayman (Eds.), *Shape memory materials*, Cambridge University Press, Cambridge (1998).

[2] K Bhattacharya, *Microstructure of martensite: why it forms and how it gives rise to the shape-memory effect*, Oxford University Press, Oxford (2003).

[3] R Stalmans, J Van Humbeeck, K Delaey, *Thermomechanical cycling, two way memory and concomitant effects in Cu-Zn-Al alloys*, Acta Metall. Mater. **40**, 501 (1992).

[4] H Xu, S Tan, *Calorimetric investigation of a Cu-Zn-Al alloy with two way shape memory*, Scripta Metall. Mater. **33**, 749 (1995).

[5] J Perkins, R O Sponholz, *Stress-induced martensitic transfomation cycling and the two-way shape memory training in Cu-Zn-Al alloys*, Metall. Trans. A **15**, 313 (1964).

[6] D Ríos Jara, G Guénin, *On the characterization and origin of the dislocations associated with the two way memory effect in Cu-Zn-Al thermoelastic alloys-I. Quantitative analysis of the dislocations*, Acta Metall. Mater. **35**, 109 (1987).

[7] E Cingolani, M Ahlers, M Sade, *The two way shape memory effect in Cu-Zn-Al single crystals: role of dislocations and stabilization*, Acta Metall. Mater. **43**, 2451 (1995).

[8] E Cingolani, M Ahlers, *On the origin of the two way shape memory effect in Cu-Zn-Al single crystals*, Mater. Sci. Eng. A **273**, 595 (1999).

[9] R S Elliott, D S Karls, *Entropic stabilization of austenite in shape memory alloys*, J. Mech. Phys. Solids **61**, 2522 (2013).

[10] O Kastner, G J Ackland, *Mesoscale kinetics produces martensitic microstructure*, J. Mech. Phys. Solids **57**, 107 (2009).

[11] O Kastner, G Eggeler, W Weiss, G J Ackland, *Molecular dynamics simulation study of microstructure evolution during cyclic martensitic transformations*, J. Mech. Phys. Solids **59**, 1888 (2011).

[12] O Kastner, *First principles modelling of shape memory alloys: Molecular dynamics simulations*, Springer, Berlin (2012).

[13] M F Laguna, E A Jagla, *Classical isotropic two-body potentials generating martensitic transformations*, J. Stat. Mech-Theory E. **2009**, P09002 (2009).

[14] M P Allen and D J Tildesley, *Computer simulation of liquids*, Oxford University Press, Oxford (1987).

[15] M Sade, A Hazarabedian, A Uribarri, F Lovey, *An electron-microscopy study of dislocation structures in fatigued Cu-Zn-Al shape-memory alloys*, Philos. Mag. **55**, 445 (1987).

[16] F Lovey, A Hazarabedian, J Garcés, *The relative stability of dislocations embedded in the β phase matrix and in martensite phases in copper based alloys*, Acta Metall. Mater. **37**, 2321 (1989).

15

An efficient impurity-solver for the dynamical mean field theory algorithm

Y. Núñez Fernández,[1*] K. Hallberg[1]

One of the most reliable and widely used methods to calculate electronic structure of strongly correlated models is the Dynamical Mean Field Theory (DMFT) developed over two decades ago. It is a non-perturbative algorithm which, in its simplest version, takes into account strong local interactions by mapping the original lattice model on to a single impurity model. This model has to be solved using some many-body technique. Several methods have been used, the most reliable and promising of which is the Density Matrix Renormalization technique. In this paper, we present an optimized implementation of this method based on using the star geometry and correction-vector algorithms to solve the related impurity Hamiltonian and obtain dynamical properties on the real frequency axis. We show results for the half-filled and doped one-band Hubbard models on a square lattice.

I. Introduction

Materials with strongly correlated electrons have attracted researchers in the last decades. The fact that most of them show interesting emergent phenomena like superconductivity, ferroelectricity, magnetism, metal-insulator transitions, among other properties, has triggered a great deal of research.

The presence of strongly interacting local orbitals that causes strong interactions among electrons makes these materials very difficult to treat theoretically. Very successful methods to calculate electronic structure of weakly correlated materials, such as the Density Functional Theory (DFT) [1], lead to wrong results when used in some of these systems. The DFT-based local density approxima-

tion (LDA) [2] and its generalizations are unable to describe accurately the strong electron correlations. Also, other analytical methods based on perturbations are no longer valid in this case so other methods had to be envisaged and developed.

More than two decades ago, the Dynamical Mean Field Theory (DMFT) was developed to study these materials. This method and its successive improvements [3–8] have been successful in incorporating the electronic correlations and more reliable calculations were done. The combination of the DMFT with LDA allowed for band structure calculations of a large variety of correlated materials (for reviews, see Refs. [9, 10]), where the DMFT accounts more reliably for the local correlations [11, 12].

The DMFT relies on the mapping of the correlated lattice onto an interacting impurity for which the fermionic environment has to be determined self-consistently until convergence of the local Green's function and the local self-energy is reached. This approach is exact for the infinitely

*E-mail: yurielnf@gmail.com

[1] Centro Atómico Bariloche and Instituto Balseiro, CNEA, CONICET, Avda. E. Bustillo 9500, 8400 San Carlos de Bariloche, Río Negro, Argentina

coordinated system (infinite dimensions), the non-interacting model and in the atomic limit. Therefore, the possibility to obtain reliable DMFT solutions of lattice Hamiltonians relies directly on the ability to solve (complex) quantum impurity models.

Since the development of the DMFT, several quantum impurity solvers were proposed and used successfully; among these, we can mention the iterated perturbation theory (IPT) [13,14], exact diagonalization (ED) [15], the Hirsch-Fye quantum Monte Carlo (HFQMC) [16], the continuous time quantum Monte Carlo (CTQMC) [17–20], non-crossing approximations (NCA) [21], and the numerical renormalization group (NRG) [22,23]. All of these methods imply certain approximations. For a more detailed description, see [24].

Some years ago, we proposed the Density Matrix Renormalization Group (DMRG) as a reliable impurity-solver [25–27] which allows to surmount some of the problems existing in other solvers, giving, for example, the possibility of calculating dynamical properties directly on the real frequency axis. Other related methods followed, such as in [28,29]. This way, more accurate results can be obtained than, for example, using algorithms based on Monte Carlo techniques. The scope of this paper is to detail the implementation of this method and to show recent applications and potential uses.

II. DMFT in the square lattice

We will consider the Hubbard model on a square lattice:

$$H = t \sum_{\langle ij\rangle\sigma} c_{i\sigma}^\dagger c_{j\sigma} + U \sum_i n_{i\uparrow} n_{i\downarrow} - \mu \sum_i n_i, \quad (1)$$

where $c_{i\sigma}$ $\left(c_{i\sigma}^\dagger\right)$ annihilates (creates) an electron with spin $\sigma =\uparrow, \downarrow$ at site i, $n_{i\sigma} = c_{i\sigma}^\dagger c_{i\sigma}$ is the density operator, $n_i = n_{i\downarrow} + n_{i\uparrow}$, U is the Coulomb repulsion, μ is the chemical potential, and $\langle ij\rangle$ represents nearest neighbor sites.

Changing to the Bloch basis $d_{\mathbf{k}}^\dagger$, the non-interacting part becomes:

$$H^0 = \sum_{k,\sigma} t(\mathbf{k}) d_{\mathbf{k}\sigma}^\dagger d_{\mathbf{k}\sigma}, \quad (2)$$

with $t(\mathbf{k}) = 2t\left(\cos k_x + \cos k_y\right) - \mu$. The Green's function for (1) is hence given by:

$$G(\mathbf{k},\omega) = [\omega - t(\mathbf{k}) - \Sigma(\mathbf{k},\omega)]^{-1}, \quad (3)$$

where $\Sigma(\mathbf{k},\omega)$ is the self-energy.

The DMFT makes a local approximation of $\Sigma(\mathbf{k},\omega)$, that is, $\Sigma(\mathbf{k},\omega) \approx \Sigma(\omega)$. This locality of the magnitudes allows us to map the lattice problem onto an auxiliar impurity problem that has the same local magnitudes $G(\omega)$ and $\Sigma(\omega)$. The impurity is coupled to a non-interacting bath, which should be determined iteratively. The Hamiltonian can be written:

$$H_{imp} = H_{loc} + H_b, \quad (4)$$

where H_{loc} is the local part of (1)

$$H_{loc} = -\mu n_0 + U n_{0\uparrow} n_{0\downarrow}, \quad (5)$$

and the non-interacting part H_b representing the bath is:

$$H_b = \sum_{i\sigma} \lambda_i b_{i\sigma}^\dagger b_{i\sigma} + \sum_{i\sigma} v_i \left[b_{i\sigma}^\dagger c_{0\sigma} + H.c. \right], \quad (6)$$

where $b_{i\sigma}^\dagger$ represents the creation operator for the bath-site i and spin σ, label "0" corresponds to the interacting site.

The algorithm is summarized as:

(i) Start with $\Sigma(\omega) = 0$.

(ii) Calculate the Green's function for the local interacting lattice site:

$$\begin{aligned} G(\omega) &= \frac{1}{N} \sum_k G(\mathbf{k},\omega) \quad (7)\\ &= \frac{1}{N} \sum_k [\omega - t(\mathbf{k}) - \Sigma(\omega)]^{-1}. \end{aligned}$$

(iii) Calculate the hybridization

$$\Gamma(\omega) = \omega + \mu - \Sigma(\omega) - [G(\omega)]^{-1}. \quad (8)$$

(iv) Find a Hamiltonian representation H_{imp} with hybridization $\Gamma_d(\omega)$ to approximate $\Gamma(\omega)$. The hybridization $\Gamma_d(z)$ is characterized by the parameters v_i and λ_i of H_{imp} through:

$$\Gamma_d(\omega) = \sum_i \frac{v_i^2}{\omega - \lambda_i}. \qquad (9)$$

(v) Calculate the Green's function $G_{imp}(\omega)$ at the impurity of the Hamiltonian H_{imp} using DMRG.

(vi) Obtain the self-energy

$$\Sigma(\omega) = \omega + \mu - [G_{imp}(\omega)]^{-1} - \Gamma_d(\omega). \qquad (10)$$

Return to **(ii)** until convergence.

At step (iv) we should find the parameters v_i and λ_i by fitting the calculated hybridization $\Gamma(\omega)$ using expression (9). At half-filling, because of the electron-hole symmetry, we have $\Gamma(\omega) = \Gamma(-\omega)$ and hence $\lambda_{-i} = -\lambda_i$, and $v_{-i} = v_i$, where the bath index i goes from $-p$ to p, and it does not include $i = 0$ for an even number of bath sites $2p$.

Almost all of the computational time is spent at step **(v)**, where the dynamics of a single impurity Anderson model (SIAM) (see Fig. 1) is calculated. We use the correction-vector for DMRG following [30]. The one-dimensional representation of the problem (needed for a DMRG calculation) is as showed in Fig. 1, except that for the spin degree of freedom we duplicate the graph, generating two identical chains, one for each spin. Moreover, it should be noticed that this is not a local or short-range 1D Hamiltonian (usually called chain geometry, where the DMRG is supposed to work very well). However, we refer to [31, 32] where strong evidence of better performance of the DMRG for this kind of geometry (star geometry) compared to chain geometry is presented.

The correction-vector for DMRG consists of targeting not only the ground state $|E_0\rangle$ of the system but also the correction-vector $|V_i\rangle$ associated to the frequency ω_i (and its neighborhood), that is:

$$(\omega_i + i\eta - H_{imp} - E_0)|V_i\rangle = c_0^\dagger |E_0\rangle, \qquad (11)$$

where a Lorentzian broadering η is introduced to deal with the poles of a finite-length SIAM. For a better matching between the ω windows (with width approximately η), we target the correction vectors of the extremes of the window. Once the DMRG is converged, the Green's function is evaluated for a finer mesh (around 0.2 of the original

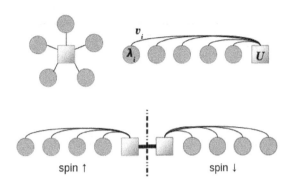

Figure 1: Schematic representation of the impurity problem for the DMFT. The circles (square) represent the non-interacting (interacting) sites, and the lines correspond to the hoppings. Top: star geometry drawn in two ways. Bottom: 1D representation as used for DMRG calculations.

window) [30]. In this way, a suitable renormalized representation of the operators is obtained to calculate the properties of the excitations around ω_i, particularly the Green's function.

In what follows, we present results for a paradigmatic correlated model using the method described above.

III. Results

We have used this method to calculate the density of states (DOS) of the Hamiltonian (Eq. 1) on a square lattice with unit of energy $t = 0.25$, for several dopings, given by the chemical potential. We consider a discarded weight of 10^{-11} in the DMRG procedure for which a maximum of around $m = 128$ states were kept, even for the largest systems (50 sites). For these large systems, the ground state takes around 20 minutes to converge and each frequency window, between 5 and 20 minutes. This is an indication of the good efficiency of the method.

The metal-insulator Mott's transition at half-filling is showed in Fig. 2. The transition occurs between $U = 3$ and $U = 4$. In Fig. 3, we observe that the metallic character of the bands remains robust under doping for a given value of the interaction, showing a weight transfer between the bands due to the correlations. The metallic character is also seen in the variation of the filling with μ.

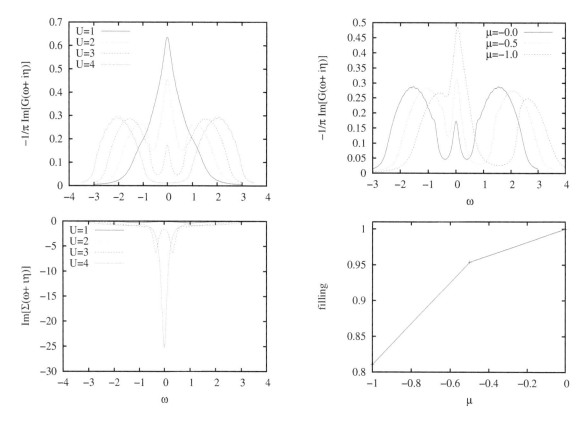

Figure 2: Top: Density of states for $U = 1, 2, 3, 4$ at half-filling. We use a bath with 30-50 sites per spin and a Lorentzian broadening $\eta = 0.12$. The Fermi energy is located at $\omega = 0$. Bottom: Imaginary part of the self-energy.

Figure 3: Top: Density of states for $U = 3$, same parameters as in Fig. 2, and several chemical potentials ($\mu = 0$ corresponds to the half-filled case). Bottom: Filling vs chemical potential showing a metallic behavior.

Figure 4 shows our results for a larger value of the interaction U, for which we find a regime of doping having an insulating character. However, for a large enough doping (obtained for a large negative value of the chemical potential), the systems turn metallic and acquire a large density of states at the Fermi energy. While the system is insulating, changing the chemical potential only results in a rigid shift of the density of states. The small finite values of the DOS at the Fermi energy for the insulating cases are due to the Lorentzian broadening η, see Eq. (11).

IV. Conclusions

We have presented here an efficient algorithm to calculate dynamical properties of correlated sys-

tems such as the electronic structure for any doping. It is based on the Dynamical Mean Field theory method where we use the Density Matrix Renormalization Group (DMRG) as the impurity solver. By using the star geometry for the hybridization function (which reduces the entanglement enhancing the performance of the DMRG for larger bath sizes) together with the correction vector technique(which accurately calculates the dynamical response functions within the DMRG) we were able to obtain reliable real axis response functions, in particular, the density of states, for any doping, for the Hubbard model on a square lattice. This improvement will allow for the calculation of dynamical properties on the real energy axis for complex and more realistic correlated systems.

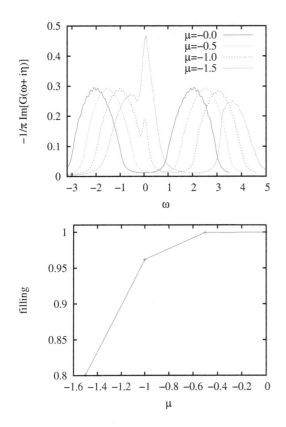

Figure 4: Top: Density of states for $U = 4$, same parameters as in Fig. 2, and several chemical potentials ($\mu = 0$ corresponds to the half-filled case). Bottom: Filling vs chemical potential showing the transition from a metal to an insulator.

Acknowledgements - We thank Daniel García for useful discussions.

[1] P Hohenberg, W Kohn, *Inhomogeneous electron gas*, Phys. Rev. **136**, B864 (1964).

[2] R O Jones, O Gunnarsson, *The density functional formalism, its applications and prospects*, Rev. Mod. Phys. **61**, 689 (1989).

[3] G Kotliar, D Vollhardt, *Strongly correlated materials: Insights from dynamical mean-field theory*, Physics Today **57**, 53 (2004).

[4] A Georges, G Kotliar, W Krauth, M J Rozenberg, *Dynamical mean-field theory of strongly correlated fermion systems and the limit of infinite dimensions*, Rev. Mod. Phys. **68**, 13 (1996).

[5] G Kotliar, S Y Savrasov, G Pálsson, G Biroli, *Cellular dynamical mean field approach to strongly correlated systems*, Phys. Rev. Lett. **87**, 186401 (2001).

[6] T Maier, M Jarrell, T Pruschke, M H Hettler, *Quantum cluster theories*, Rev. Mod. Phys. **77**, 1027 (2005).

[7] M H Hettler, A N Tahvildar-Zadeh, M Jarrell, T Pruschke, H R Krishnamurthy, *Nonlocal dynamical correlations of strongly interacting electron systems*, Phys. Rev. B **58**, R7475 (1998).

[8] D Sénéchal, D Perez, M Pioro-Ladrière, *Spectral weight of the Hubbard model through cluster perturbation theory*, Phys. Rev. Lett. **84**, 522 (2000).

[9] M Imada, T Miyake, *Electronic structure calculation by first principles for strongly correlated electron systems*, J. Phys. Soc. Jpn. **79**, 112001 (2010).

[10] K Held, *Electronic structure calculations using dynamical mean-field theory*, Adv. in Phys. **56**, 829 (2007).

[11] V I Anisimov, A I Poteryaev, M A Korotin, A O Anokhin, G Kotliar, *First-principles calculations of the electronic structure and spectra of strongly correlated systems: dynamical mean-field theory*, J. Phys. Condens. Mat. **9**, 7359 (1997).

[12] A I Lichtenstein, M I Katsnelson, *Ab initio calculations of quasiparticle band structure in correlated systems: LDA++ approach*, Phys. Rev. B **57**, 6884 (1998).

[13] A Georges, G Kotliar, *Hubbard model in infinite dimensions*, Phys. Rev. B **45**, 6479 (1992).

[14] M J Rozenberg, G Kotliar, X Y Zhang, *Mott-Hubbard transition in infinite dimensions. II*, Phys. Rev. B **49**, 10181 (1994).

[15] M Caffarel, W Krauth, *Exact diagonalization approach to correlated fermions in infinite dimensions: Mott transition and superconductivity*, Phys. Rev. Lett. **72**, 1545 (1994).

[16] J E Hirsch, R M Fye, *Monte Carlo method for magnetic impurities in metals*, Phys. Rev. Lett. **56**, 2521 (1986).

[17] A N Rubtsov, V V Savkin, A I Lichtenstein, *Continuous-time quantum Monte Carlo method for fermions*, Phys. Rev. Lett. **72**, 035122 (2005).

[18] P Werner, A Comanac, L de Medici, M Troyer, A J Millis, *Continuous-Time Solver for Quantum Impurity Models*, Phys. Rev. Lett. **97**, 076405 (2006).

[19] H Park, K Haule, G Kotliar, *Cluster Dynamical Mean Field Theory of the Mott Transition*, Phys. Rev. Lett. **101**, 186403 (2008).

[20] E Gull, A J Millis, A I Lichtenstein, A N Rubtsov, M Troyer, P Werner, *Continuous-time Monte Carlo methods for quantum impurity models*, Rev. Mod. Phys. **83**, 349 (2011).

[21] T Pruschke, D L Cox, M Jarrell, *Hubbard model at infinite dimensions: Thermodynamic and transport properties*, Phys. Rev. Lett. **47**, 3553 (1993).

[22] K G Wilson, *The renormalization group: Critical phenomena and the Kondo problem*, Rev. Mod. Phys. **47**, 773 (1975).

[23] R Bulla, *Zero temperature metal-insulator transition in the infinite-dimensional hubbard model*, Phys. Rev. Lett. **83**, 136 (1999);

R Bulla, A C Hewson, T Pruschke, *Numerical renormalization group calculations for the self-energy of the impurity Anderson model*, J. Phys. Condens. Mat. **10**, 8365 (1998).

[24] K Hallberg, D J García, P Cornaglia, J Facio, Y Núñez-Fernández, *State-of-the-art techniques for calculating spectral functions in models for correlated materials*, EPL **112**, 17001 (2015).

[25] D J García, K Hallberg, M J Rozenberg, *Dynamical mean field theory with the density matrix renormalization group*, Phys. Rev. Lett. **93**, 246403 (2004).

[26] D J García, E Miranda, K Hallberg, M J Rozenberg, *Mott transition in the Hubbard model away from particle-hole symmetry*, Phys. Rev. B **75**, 121102 (2007);

E Miranda, D J García, K Hallberg, M J Rozenberg, *The metal-insulator transition in the paramagnetic Hubbard Model*, Physica B: Cond. Mat. **403**, 1465 (2008);

D J García, E Miranda, K Hallberg, M J Rozenberg, *Metal-insulator transition in correlated systems: A new numerical approach*, Physica B: Cond. Mat. **398**, 407 (2007);

S Nishimoto, F Gebhard, E Jeckelmann, *Dynamical density-matrix renormalization group for the Mott-Hubbard insulator in high dimensions*, J. Phys. Condens. Mat. **16**, 7063 (2004);

M Karski, C Raas, G Uhrig, *Electron spectra close to a metal-to-insulator transition*, Phys. Rev. B **72**, 113110 (2005);

M Karski, C Raas, G Uhrig, *Single-particle dynamics in the vicinity of the Mott-Hubbard metal-to-insulator transition*, Phys. Rev. B **75**, 075116 (2008);

C Raas, P Grete, G Uhrig, *Emergent Collective Modes and Kinks in Electronic Dispersions*, Phys. Rev. Lett. **102**, 076406 (2009).

[27] Y Núñez Fernández, D García, K Hallberg, *The two orbital Hubbard model in a square lattice: a DMFT + DMRG approach*, J. Phys.: Conf. Ser. **568**, 042009 (2014).

[28] M Ganahl et al, *Efficient DMFT impurity solver using real-time dynamics with matrix product states*, Phys. Rev. B **92**, 155132 (2015).

[29] F Wolf, J Justiniano, I McCulloch, U Schollwöck, *Spectral functions and time evolution from the Chebyshev recursion*, Phys. Rev. B **91**, 115144 (2015).

[30] T D Kühner, S R White, *Dynamical correlation functions using the density matrix renormalization group*, Phys. Rev. B **60**, 335 (1999).

[31] A Holzner, A Weichselbaum, J von Delft, *Matrix product state approach for a two-lead multilevel Anderson impurity model*, Phys. Rev. B **81**, 125126 (2010).

[32] F Alexander Wolf, I McCulloch, U Schollwöck, *Solving nonequilibrium dynamical mean-field theory using matrix product states*, Phys. Rev. B **90**, 235131 (2014).

16

An improvement to the measurement of Dalitz plot parameters

S. Ghosh,[1*] A. Roy[1†]

Precise measurement of the Dalitz plot parameters of the $\eta' \to \eta\, \pi^+\, \pi^-$ decay gives a better insight of the dynamics of the heavy pseudo-scalar meson. Various measurements of the Dalitz plot parameters of the η' meson have been performed with the detection of all the final state particles including the neutral decay of $\eta \to 2\gamma$. In many experiments, reconstruction of the η meson from the neutral decay modes comes with the disadvantage of poor resolution and low efficiency compared to that of the charged particles. In this article, a study of the Dalitz plot parameters keeping the η meson as a missing particle is presented. The method is found to be advantageous in the case of poor photon resolution. Effect of the charged particle resolution on the Dalitz variable Y is also examined. This work may provide guidance to select a suitable method for the Dalitz plot analysis, depending on the detector resolution.

I. Introduction

Low energy quantum chromodynamics (QCD) is effectively studied in the framework of the chiral perturbation theory (ChPT). The Lagrangians for the hadronic processes are derived from ChPT and the effective degrees of freedom are, usually, the octet of pseudo-scalar mesons (π, K, η) [1]. The observed mass of the η' meson in nature is much higher than the other pseudo-scalar mesons, which gives the axial $U(1)$ anomaly [2,3]. This suggests the possible existence of many unsolved problems that could provide inputs to the Lagrangian. Study of the production and decay of the η' meson is, therefore, important to both theory and experiment [4]. Some possible dynamics that could be studied in the η' decay model are the new sym-

metry and the symmetry breaking during interactions [5], gluonic contributions, gauge field configurations with non-zero winding number, and the quark instantons [6,7]. This has motivated many groups to study both charged and uncharged decays of the η' meson [1,8].

The Dalitz plot plays an important role as a useful tool to study decay dynamics of a meson decaying into three bodies. As the three-body decay has two degrees of freedom, one can define two linearly independent variables to represent the decay in the phase space. In this article, we shall study the decay of $\eta' \to \eta\, \pi^+\, \pi^-$ and for that we define the Dalitz plot variables [9] as

$$X = \frac{\sqrt{3}(T_{\pi^+} - T_{\pi^-})}{Q},$$

$$Y = \frac{(m_\eta + 2m_\pi)}{m_\pi} \cdot \frac{T_\eta}{Q} - 1. \tag{1}$$

Here T and m are, respectively, the kinetic energy (in the rest frame of η') and the mass of the particles indicated by the subscripts. Then, Q = T_{π^+}

---

*E-mail: phd12115113@iiti.ac.in
†E-mail: ankhi@iiti.ac.in

[1] Discipline of Physics, School of Basic Sciences, Indian Institute of Technology Indore, Khandwa Road, Simrol, MP-453552, India.

$+ T_{\pi^+} + T_{\pi^+} = m_{\eta'}$ - m_η - $2m_\pi$, is the available energy of the reaction. The decay dynamics of the η' mesons is studied in the form of Dalitz plot parameters which are obtained by fitting the Dalitz plot with the following general parameterization:

$$f(X,Y) = A \left(1 + aY + bY^2 + cX + dX^2 \right). \quad (2)$$

Here, a, b, c, and d are the Dalitz plot parameters of the decay and A is the overall normalization constant. The measurement of Dalitz plot parameters are specifically important to understand and crosscheck the correct inputs to the theoretical distribution of the Lagrangian [10]. It is, therefore, important to measure the Dalitz plot parameters precisely with good statistics and resolution.

The Dalitz plot parameters of the $\eta' \to \eta\ \pi^+ \pi^-$ decay have been measured in the experiments of the VES [11] and the BESIII [12] collaborations. GAMS-4π [13] and IHEP-IISN-LANL-LAPP collaborations [7] have reported the Dalitz plot parameters for the neutral decay mode ($\eta\ \pi^0\ \pi^0$) of the η' meson. The values of the Dalitz plot parameters for both of these decay modes of the η' meson should be the same under the isospin limit. However, this is not observed in the above experiments. This discrepancy may be reconciled by precise measurements of the small Dalitz plot parameters with high statistics. The previous experiments have identified the final state particles (π^+, π^- and γ) of the $\eta' \to (\eta)\ \pi^+\pi^- \to (2\gamma)\ \pi^+\pi^-$ decay and used the invariant mass method for the Dalitz plot analysis. If the photon resolution is poor, the reconstruction efficiency of the η meson will be significantly low and will also affect the Dalitz variable Y and the parameters related to this variable.

Here, we will be reporting a method that improves the Dalitz plot parameters in the measurements of the η' decay modes and found particularly advantageous in the case of poor photon resolution. We shall consider the production of η' meson through photo-production reaction first and then introduce the detector resolution to all the final state particles π^+, π^-, p and γ of the channel $\gamma + p \to \eta'p \to (\eta)\ \pi^+\pi^-p \to (2\gamma)\ \pi^+\pi^-$ p to simulate the detector environment. The Dalitz plot parameters are then calculated using the following methods: (a) exclusive measurement or invariant mass method and (b) missing measurement method. The exclusive measurement method is

commonly used, where all the final state particles (π^+, π^- and γ) are considered and η is reconstructed as the invariant mass ($\eta \to 2\gamma$). In the missing measurement method, we have reconstructed the η' meson as a missing particle using the information of the recoiled proton, the beam photon and the target proton ($\gamma_{beam} + p_{target} \to \eta'\ p_{recoil}$). The η' information along with π^+ and π^- is used to reconstruct the η meson as a missing particle. The calculations of the Dalitz plot parameters using both these methods are described and compared. In addition to that, a bin width study is also performed with missing analysis to optimize the bin size for the extraction of the Dalitz plot parameters. In the missing measurement method, a dependence of the Dalitz variable Y on the detector resolutions of the charged particles has also been examined. Further, to study the effect of the background, method (b) was subjected to a combinatoric background channel and the Dalitz plot parameters are then systematically studied with varying background component in the mixture of the signal and the background. The background is then subtracted from every Dalitz plot bin and the parameters are reported.

II. Model to fold the detector resolutions

The Pluto simulation framework (version 5.42) developed by the HADES collaboration was used to generate hadronic physics reactions for this analysis [14]. Dalitz plot parameters of the $\eta' \to \eta\ \pi^+ \pi^-$ decay from the BESIII [12] experiment were used as input parameters for the generated events. A total of 10^5 γ + p $\to \eta'$ p events, each with a photon beam energy of 2.5 GeV were generated in the phase space model for this analysis. However, the energy of the η' meson is arbitrary as the Dalitz plot parameters are independent of the initial energy of the η' meson. The detector effects are absent in the events generated by the Pluto event generator. However, in reality, the momentum of the particle passing through the detector is modified according to the detector resolution. Generally, the detector response is distributed with a resolution which is dependent on the magnitude of the particle's momentum. To incorporate the detector response, each component of the momentum

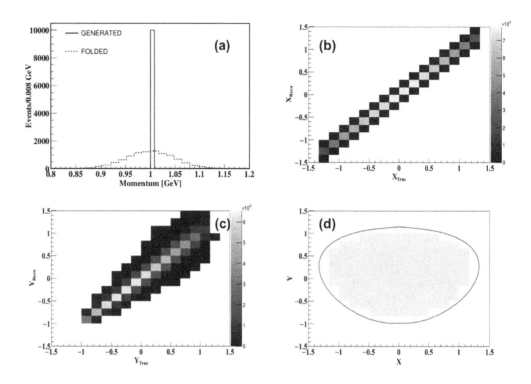

Figure 1: (a) Number of events as a function of the momentum, where the solid histogram is for the events with unit momentum before folding and the dashed histogram is for the events after the folding with a Gaussian distribution whose σ is 3% of the mean. (b) Number of generated vs. reconstructed events in the Dalitz variable bins X. (c) Number of generated vs. reconstructed events in the Dalitz variable bins Y. (d) The Dalitz variable X vs. Y showing the bins completely inside the Dalitz plot boundary.

of the particles are convoluted with a random number sampled from a Gaussian distribution of mean 1 and $\sigma = 3\%$ of the mean, which introduces the momentum dependent resolution to the particle. This effect transforms particles with a single momentum into a particle with a momentum distribution as shown in Fig. 1(a) [15].

Since we are studying a case where the photon resolution is poor compared to the charged particle resolution, the final state charged particles (π^+, π^- and p) are folded with 1% resolution [16,17] in the momentum components to simulate a detector-like effect, whereas 6% resolution is used for the photons. There are experiments that suffer from poor photon resolution for which 6% is rather an underestimation [18]. Dalitz plot variables X and Y, after the addition of resolution to the final state particles in momentum components, are compared

with the generated Dalitz plot variables as shown in Fig. 1(b) and (c). It is observed in Fig. 1(c) that at higher values of Y, the bin migration is more significant. The reason for this is that a higher value of Y corresponds to a high energy η meson, which decays to energetic photons with poor resolution. This leads to a enhanced bin migration at higher value of Y.

Folding has the effect of migrating the events from one bin to the neighboring bins of the Dalitz plot. It distributes the events in the bins away from the diagonal, creating a homogeneous migration of the events and thus allows to calculate the parameters from a properly binned Dalitz plot. Both Dalitz plot variables, shown in Fig. 1, are divided into 18 bins from −1.5 to 1.5. Though the migration is observed for the individual Dalitz variables shown in Fig 1(b) and (c), the maximum number

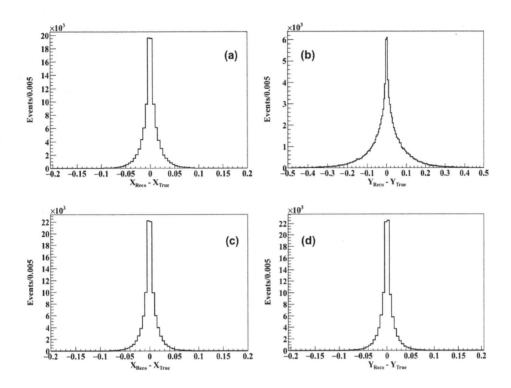

Figure 2: (a) Number of events vs. difference of the reconstructed (after folding) and the true value of the Dalitz variable X and (b) the Dalitz variable Y for the exclusive measurement method. (c) Number of events vs. difference of the reconstructed and the true value of Dalitz variable X and (d) the Dalitz variable Y for the missing measurement method (b).

of events, however, lie on the diagonal and confined well within the resolution of the bins.

The general parameterization of Eq. (2) was used to parametrize the decay. A least square fitting procedure MINUIT [19] was used to minimize the χ^2 given in Eq. (3) to fit each bin of the Dalitz plot.

$$\chi^2 = \sum \left(\frac{N_j - f(X_j, Y_j)}{\Delta N_j} \right)^2. \qquad (3)$$

Where N_j and ΔN_j denotes the number of events and their statistical uncertainties for the Dalitz plot bin j ($j = 1, 2, ..., n$). X_j and Y_j are the central coordinates of each bin, and $f(X_j, Y_j)$ denotes the fitted form of the polynomial. The bins on the boundary of the Dalitz plot are removed and only bins which are completely inside the boundary are considered as shown in Fig. 1(d).

i.   Method (a): Exclusive measurement

In this method, measurement of the Dalitz plot parameters are performed with all the detected final state particles π^+, π^- and γ. To calculate the resolution of Dalitz variables X and Y, first a difference of the reconstructed (after folding) and the true value of the Dalitz variable is considered. Then, the standard deviation of X and Y for a large number of events is calculated using

$$\sigma = \sqrt{\left(\frac{\sum(x - \mu)^2}{N} \right)}, \qquad (4)$$

where x represents each value of the Dalitz variables and μ is the mean, which should be zero in this case, and N is the total number of events. As seen from Fig. 2(a) and (b), the resolutions of the Dalitz variables X and Y are respectively 0.019 and 0.099, after the folding in the momentum compo-

Table 1: Dalitz plot parameters with varying number of bins in the Dalitz variables X and Y for methods (a) and (b).

Par	Method (a)			Method (b)		
	17×17	18×18	19×19	17×17	18×18	19×19
a	-0.067 ± 0.008	-0.062 ± 0.008	-0.066 ± 0.008	-0.045 ± 0.008	-0.044 ± 0.008	-0.044 ± 0.008
b	-0.134 ± 0.015	-0.142 ± 0.015	-0.145 ± 0.014	-0.058 ± 0.016	-0.063 ± 0.015	-0.081 ± 0.015
c	0.014 ± 0.005	0.014 ± 0.005	0.009 ± 0.005	0.014 ± 0.006	0.014 ± 0.005	0.009 ± 0.006
d	-0.098 ± 0.009	-0.096 ± 0.009	-0.093 ± 0.009	-0.085 ± 0.010	-0.081 ± 0.009	-0.092 ± 0.009
$\frac{\chi^2}{ndf}$	1.25	1.04	1.09	1.08	1.14	0.95

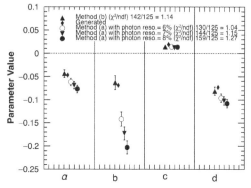
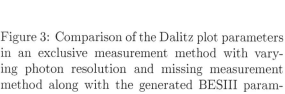

Figure 3: Comparison of the Dalitz plot parameters in an exclusive measurement method with varying photon resolution and missing measurement method along with the generated BESIII parameters.

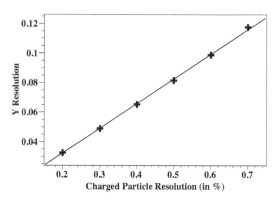

Figure 4: Variation of the Y resolution with the charged particle resolution.

nents. The resolution of the variables X and Y are independent in this method as the former was calculated from π^+ and π^- but the later was calculated from the photons and, therefore, causing the resolution to be poor.

ii. Method (b): Missing measurement

In the missing measurement method, the Dalitz variables X and Y are calculated without using the information of the final photons from the decay of the η meson. Only π^+ and π^- are folded with the resolutions and the produced η' information is used to calculate the Dalitz variables X and Y. The calculated resolutions of the Dalitz variables X and Y are both 0.016 and are shown in Fig. 2(c) and (d). In contrast to the exclusive measurement method, in this method, the resolutions of X and Y are same.

This is due to the fact that both X and Y use the information of the same particles (π^+ and π^-) with added resolution [20].

iii. Comparison of the method (a) and (b)

To compare the methods described above, we have extracted the Dalitz parameters by fitting the Dalitz plot with the parameterization defined in Eq. (2). All the bins are fitted to the general parameterization with the χ^2 minimization. To find an optimized binning, the Dalitz plot parameters are calculated using the method (a) and (b) for varying number of bins in X and Y axis as shown in Table 1. It is found that the best χ^2/ndf is obtained for 10^5 events with a binning of 18×18 for both methods. It can be seen that a systematics arises due to varying bin size on the parameters a and b in the invariant mass method. In both the methods a binning of 18×18 is thus used to calculate the Dalitz plot parameters, which is higher than the resolution of the Dalitz variables X and Y. A comparison of the parameters obtained using the methods (a), with different photon resolution, and (b), with the generated BESIII parameters, is shown in Fig. 3 along with their errors and χ^2/ndf. It can be seen from the figure that with poorer photon resolution the parameters a and b systematically deviate from the generated value. Though the values of the parameters c and d are consistent as shown in Table 1, it is observed that due to poor photon resolution, deviation of the a and b parameters from the input in the exclusive measurement method is significant compared to the missing 2γ measurement method. Even the deviation of parameter b is higher than that of a, as b is the coefficient of a quadratic Y. The poor photon resolution directly affects the η momentum resolution, which is reflected in the poor Y resolution. The missing measurement method is, therefore, preferable for the Dalitz plot analysis in this case of poor photon resolution.

Since in the missing measurement method the Dalitz variable Y depends on the charge particle resolution, we have carried out another study to examine the variation of the Y resolution with the charged particle resolution. We found that Y varies linearly with the charged particle resolution as shown in Fig. 4. It can be concluded that if the charge particle resolution is equally poor as the photon resolution, the missing analysis method is still a better choice compared to the exclusive measurement method. In the present experimental condition of 1% charged particle resolution and 6% neutral particle resolution, the exclusive measurement method introduces a poorer Y resolution compared to the missing measurement method.

III. Background study

The missing measurement method comes with a cost of additional combinatoric background for the channel $\eta' \to \eta\ \pi^+\ \pi^-$. This combinatoric background arises from π^+ and π^- produced in the decays $\eta' \to \eta\ \pi^+\ \pi^-$ and $\eta \to \pi^+\ \pi^-\ \pi^0$, which leaves the π^+ and π^- in both decays with a similar available energy and hence similar kinematics. We consider a background $\eta' \to \eta\ \pi^0\ \pi^0$ generated with the same Dalitz plot parameters because of isospin symmetry [1] as $\eta' \to \eta\ \pi^+\ \pi^-$ and η is further

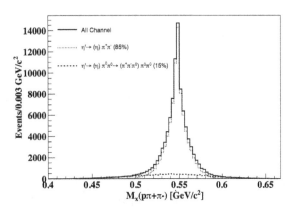

Figure 5: The number of events vs. the missing mass $M_x(\eta',\ \pi+\pi-)$ from the signal and background channel.

decayed into $\eta \to \pi^+\ \pi^-\ \pi^0$. We mixed the signal channel with 5%, 10% and 15% background and then folded it to add resolution. The Dalitz plot parameters from these three set are given in Table 2. The missing mass $M_x(\eta',\ \pi+\pi-)$ in Fig. 5 clearly shows the contribution of the background channel in the η meson mass region due to misidentification. The Dalitz plot parameters a and b related to the Dalitz variable Y are systematically shifted from their generated values due to the presence of this background. Meanwhile, the Dalitz parameters c and d do not change since π^+ and π^- from both channels have similar kinematics, as shown in Fig. 6. However, this combinatoric background can be subtracted by determining the number of background events in each bin of the Dalitz plot. The contribution of background in each bin is calculated from the generated Monte-Carlo samples before implementing the resolution to the particles, which introduces a systematics from the migration of background events to the nearby bins in the Dalitz plot. The general parameterization of Eq. (2) is fitted to the background subtracted Dalitz plot and the parameters are given in Table 3. The systematics from the migration of background events is small in the missing measurement method, and the extracted parameters after background subtraction show that method (b) is a better choice than the exclusive measurement [method (a)] when charge particle resolution is 1% and photon resolution is 6%.

Table 2: Comparison of the Dalitz plot parameters extracted from method (b) with different background contributions.

input pars	method (b) (No bkg)	method (b) (5% bkg)	method (b) (10% bkg)	method (b) (15% bkg)
a (-0.047)	-0.044 ± 0.008	-0.074 ± 0.008	-0.087 ± 0.008	-0.125 ± 0.008
b (-0.069)	-0.063 ± 0.015	-0.078 ± 0.015	-0.130 ± 0.015	-0.119 ± 0.015
c (0.019)	0.014 ± 0.005	0.016 ± 0.005	0.020 ± 0.005	0.015 ± 0.005
d (-0.073)	-0.081 ± 0.009	-0.070 ± 0.009	-0.064 ± 0.009	-0.062 ± 0.009
$\frac{\chi^2}{ndf}$	1.14	0.86	0.94	1.16

Table 3: Comparison of the Dalitz plot parameters extracted from method (a) and the background subtracted Dalitz plot obtained from method (b).

input pars	method (a)	method (b) (5% bkg)	method (b) (10% bkg)	method (b) (15% bkg)
a (-0.047)	-0.062 ± 0.008	-0.045 ± 0.008	-0.045 ± 0.008	-0.042 ± 0.008
b (-0.069)	-0.142 ± 0.015	-0.066 ± 0.016	-0.086 ± 0.016	-0.084 ± 0.017
c $(+0.019)$	0.014 ± 0.005	$+0.018 \pm 0.006$	$+0.022 \pm 0.006$	$+0.019 \pm 0.006$
d (-0.073)	-0.096 ± 0.009	-0.073 ± 0.010	-0.060 ± 0.010	-0.076 ± 0.010
$\frac{\chi^2}{ndf}$	1.04	0.93	1.15	1.02

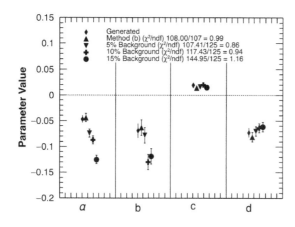

Figure 6: Comparison of the Dalitz plot parameters from generated and missing mass method without any background and with the background added in a proportion of 5%, 10% and 15% of the total number of events.

IV.   Conclusions

In this article, we have described two methods for measuring the Dalitz variables X and Y in the context of the decay $\eta' \rightarrow \eta \, (2\gamma) \, \pi^+ \, \pi^-$. The exclusive measurement method and the invariant mass method, though commonly used for measuring the Dalitz variables, are shown here to be unsuitable in the case of poor photon resolution. In contrast, the missing (2γ) analysis method is found to be more suitable when the photon resolution is poor. The resolution in Y versus charge particle resolution, in the missing analysis method, suggests that, even in the case of equally poor charged particle resolution and photon resolution the missing analysis method is a better choice compared to the exclusive measurement method. The missing analysis method, however, comes with the disadvantage of the combinatoric background and the Dalitz plot parameters deviate systematically from their central values depending on the fraction of background present. Once the background subtraction is implemented, the missing analysis method performs even better and becomes more attractive. In conclusion, this work provides guidance to select a suitable method for the extraction of the Dalitz plot parameters depending on the detector resolution as well as the background present in the signal region and may be extended to the three body decay channels of other mesons when the detector resolution is poor.

[1] B Borasoy, R Nißler, *Hadronic η and η' decays*, Eur. Phys. J. A **26**, 383 (2005).

[2] K Naito, M Oka, M Takizawa, T Umekawa, $U_A(1)$ *breaking effects on the light scalar meson spectrum*, Prog. Theor. Phys. **109**, 969 (2003).

[3] S Weinberg, *The U(1) problem*, Phys. Rev. D **11**, 3583 (1975).

[4] J Bijnens, *η and η′ physics*, Proceedings of the 11th International Conference on Meson-Nucleon Physics and the Structure of the Nucleon, eConf C **070910**, 104 (2007).

[5] N Beisert, B Borasoy, *The η′ → ηππ decay in U(3) chiral perturbation theory*, Nuc. Phys. A **705**, 433 (2002).

[6] S D Bass, *Gluonic effects in η- and η′-nucleon and nucleus interactions*, Acta Phys. Slovaca **56**, 245 (2006).

[7] D Alde *et al.*, *Matrix element of the η′(958) → ηπ⁰π⁰ decay*, Phys. Lett. B **177**, 115 (1986).

[8] A Kupsc, *Decays of η and η′ mesons: An introduction*, Int. J. Mod. Phys. E **18**, 1255 (2009).

[9] A H Fariborz, J Schechter, *η′ → ηππ decay as a probe of a possible lowest lying scalar nonet*, Phys. Rev. D **60**, 034002 (1999).

[10] R Escribano, P Masjuan, J J Sans-Cillero, *Chiral dynamics predictions for η′ → ηππ*, J. High Energy Phys. **2011**, 094 (2011).

[11] V Dorofeev *et al.*, Study of $\eta' \to \eta\pi^+\pi^-$ Dalitz plot, Phys. Lett. B **651**, 22 (2007).

[12] M Ablikim *et al.*, Measurement of the matrix element for the decay $\eta' \to \eta\pi^+\pi^-$, Phys. Rev. D **83**, 012003 (2011).

[13] A M Blik *et al.*, Measurement of the matrix element for the decay $\eta' \to \eta\pi^0\pi^0$ with the GAMS-4π spectrometer, Phys. Atom. Nucl.+ **72**, 231 (2009).

[14] I Fröhlich *et al.*, *Pluto: A Monte Carlo simulation tool for hadronic physics*, PoS(ACAT2007) **076**, (2007).

[15] P Garg, D K Mishra, P K Netrakanti, A K Mohanty, B Mohanty, *Unfolding of event-by-event net-charge distributions in heavy-ion collisions*, J. Phys. G: Nucl. Part. Phys. **40**, 055103 (2013).

[16] M Ullrich, W Khn, Y Liang, B Spruck, M Werner, *Simulation of the BESIII endcap time of flight upgrade*, Nucl. Instrum. Meth. A **769**, 32 (2015).

[17] M Battaglieri, R De Vita, V Kubarovsky, *Pentaquark at JLab: The g11 experiment in CLAS*, AIP Conf. Proc. **792**, 742 (2005).

[18] M Amarian *et al.*, *The CLAS forward electromagnetic calorimeter*, Nucl. Instrum. Meth. A **460**, 239 (2001).

[19] R Brun, F Rademakers, *ROOT: An object oriented data analysis framework*, Nucl. Instrum. Meth. A **389**, 81 (1997).

[20] S Ghosh, *Dalitz plot analysis of* $\eta' \to \eta\pi^+\pi^-$, AIP Conf. Proc. **1735**, 030018 (2016).

17

Pattern formation mechanisms in sphere-forming diblock copolymer thin films

Leopoldo R. Gómez,[1] Nicolás A. García,[1] Richard A. Register,[2] Daniel A. Vega[1]*

The order–disorder transition of a sphere-forming block copolymer thin film was numerically studied through a Cahn–Hilliard model. Simulations show that the fundamental mechanisms of pattern formation are spinodal decomposition and nucleation and growth. The range of validity of each relaxation process is controlled by the spinodal and order–disorder temperatures. The initial stages of spinodal decomposition are well approximated by a linear analysis of the evolution equation of the system. In the metastable region, the critical size for nucleation diverges upon approaching the order–disorder transition, and reduces to the size of a single domain as the spinodal is approached. Grain boundaries and topological defects inhibit the formation of superheated phases above the order–disorder temperature. The numerical results are in good qualitative agreement with experimental data on sphere-forming diblock copolymer thin films.

I. Introduction

Many practical applications of polymers and other soft matter involve the self-assembly of the system into complex multidomain morphologies [1]. The final properties and applications of such materials depend on the ability to control the morphology by adjusting molecular features and macroscopic variables. For example, by appropriate control over their molecular architecture, block copolymers can be designed as rigid and transparent thermoplastics, or as soft and flexible elastomers, depending on the features of their building blocks [1].

The most important features of the block copolymer phase behavior are already captured by the simplest A–B diblock architecture [1, 2]. Here, the unfavorable interactions between blocks, and the constraints imposed

by the connectivity between their constituents, results in a nanophase separation leading to the formation of periodic morphologies. For example, depending on molecular size, temperature, and relative volume fraction of the two blocks, diblock copolymer melts can develop body-centered-cubic (BCC) arrays of spheres, hexagonal patterns of cylinders, gyroids, or lamellar structures. For these systems, the periodicity of the self-assembled pattern is mainly controlled by the average molecular weight, and typically is in the range of 10–100 nm. Also, since the magnitude of the interblock repulsive interaction generally diminishes with temperature, an order–disorder transition can be induced thermally at the order–disorder transition temperature T_{ODT} [1, 2].

During recent decades, the properties of self-assembling copolymers have received great attention because of their potential use in nanotechnology [3–10]. Applications of block copolymer systems include, among others, templates for nanoporous materials, solar cells, and photonic crystals. Perhaps the most pressing application for understanding pattern formation in two-dimensional thin film systems is block copolymer lithography [6–9]. This process uses self-assembled patterns, such as single layers of cylinders or spheres in

[1] Instituto de Física del Sur (IFISUR), Consejo Nacional de Investigaciones Científicas y Técnicas (CONICET), Universidad Nacional del Sur, 8000 Bahía Blanca, Argentina.

[2] Department of Chemical and Biological Engineering, Princeton University, Princeton, New Jersey, 08544, USA.

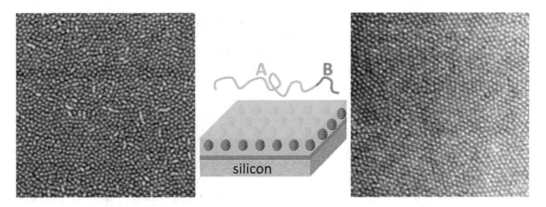

Figure 1: AFM phase images of the PS-PEP block copolymer thin film supported on a silicon substrate after annealing at T = 333 K. Left and right panels show the pattern configuration after 5 minutes and 255 minutes of thermal annealing, respectively. Image size: 1.0 μm x 1.0 μm. The schematic in the central panel shows the block copolymer configuration in the thin film. Here, the minority phase (polystyrene) forms hexagonally-packed arrays of nearly spherical domains (red domains in the scheme). Note the presence of a wetting layer of PS blocks on the silicon substrate.

copolymer thin films, as templates to fabricate devices at the nanometer length scale. However, the use of these materials to obtain lithographic masks requires the production of structures with long-range order. Since the degree of ordering is controlled mainly by the density of topological defects, it is important to determine the physical mechanisms involved in the nanophase separation process and the effects of the thermal history on the pattern morphology.

Experimentally, the time evolution of the density of topological defects and correlation lengths at $T < T_{ODT}$ has been studied through atomic force microscopy (AFM) in block copolymer thin films with different morphologies [11–16]. On the one hand, by comparing the density of disclinations (orientational defects) and the correlation length, it was shown that in cylinders (smectic phase) the dominant mechanism of coarsening involves mostly the annihilation of complex arrays of disclinations [11]. On the other hand, in monolayers of sphere-forming thin films with hexagonal order it was found that the majority of the defects are condensed in grain boundaries [16]. The orientational correlation length was found to grow following a power law, but with a higher exponent than the translational correlation length [16, 17]. However, as the typical exponents observed in the scaling laws of these experimental systems are relatively small ($\sim \frac{1}{5}$ for the translational correlation length and $\frac{1}{4}$ for the orientational correlation length), they are hard to distinguish from logarithmic, glass-like, dynamics [18].

The process of phase separation and the kinetics of ordering in different two-dimensional systems have been studied through different phase field models [17–20]. In particular, simulations of hexagonal systems with a Cahn–Hilliard model were found to be in good agreement with experimental data for block copolymer thin films. Through simulations, it was also shown that the orientational correlation length grows via annihilation of dislocations [17]. In addition, simulations have also shown that triple points, regions where three grains meet, control the dynamics of defect annihilation and can lead to the formation of metastable configurations of domains that slow down the dynamics, with correlation lengths growing logarithmically in time [18].

In block copolymer systems, the degree of order and content of defects depend not only on T_{ODT}, but also on thermal history, depth of quench, and other characteristic temperatures, like the glass transition temperature for each block, the spinodal temperature, and the temperature of crystallization (if any) [1, 2].

In general, in first-order phase transitions, the depth of quench determines whether the system relaxes to equilibrium by means of spinodal decomposition or nucleation and growth [21,22]. Spinodal decomposition is the relaxation process of a system quenched into an unstable state. In this case the phase transition is a spontaneous process, beginning with the amplification of fluctuations which are small in amplitude but large in extent. Nucleation and growth is the physical mechanism of relaxation emerging when the system is quenched into a metastable state. This is an activated process, and a free energy barrier must be overcome in order to relax to the stable phase. In polymeric systems both mechanisms can be inhibited due to crystallization or vitrification of any of the copolymer blocks [1].

In this paper, we study the disorder–order transition and the ordering kinetics in sphere-forming block copolymer thin films as a function of the depth of quench. The dynamics are studied through a Cahn–Hilliard equation and compared with experiments on diblock copolymer thin films. We have organized the paper as follows: Section II presents the experimental system. In section III we present the equations of evolution of the order parameter for diblock copolymer systems and the classical linear instability analysis of the Cahn–Hilliard equation. Section III.i describes the numerical scheme employed to solve the model equations. Results and concluding remarks are presented in sections IV and V, respectively.

II. Experimental System

The sphere-forming diblock copolymer used in this work consists of a chain of polystyrene (PS, 3300 g/mol) covalently bonded to a chain of poly(ethylene-alt-propylene) (PEP, 23100 g/mol). The copolymer was synthesized through sequential living anionic polymerization of styrene and isoprene followed by selective saturation of the isoprene block [23]. Because the two polymer species are immiscible, the minority polystyrene forms spherical microdomains within the majority poly(ethylene-alt-propylene). The bulk morphology of the PS-PEP diblock copolymer consists of arrays of spherical domains of PS packed with a BCC order, with $T_{ODT} = 394 \pm 2$ K, according to small-angle X-ray scattering experiments. The glass transition temperature for the PS block was estimated to be $T_g^{PS} \sim 320$ K while the glass transition temperature of the PEP block is well below room temperature [24].

The block copolymer was deposited (film thickness ca. 30 nm) on silicon substrates via spin coating from a disordered state in toluene, a good solvent for both blocks, to produce a quasi-two-dimensional periodic hexagonal lattice of polystyrene spheres within a matrix of poly(ethylene-alt-propylene). The film thickness was measured using ellipsometry (Gaertner Scientific LS116S300, $\lambda = 632.8$ nm). Order was induced through vacuum annealing above the glass transition temperature of both blocks and below the T_{ODT} of the

block copolymer. In this system, the PEP block wets the air interface, while PS wets the silicon oxide substrate [25, 26], i.e., there is an asymmetric wetting condition (see schematic in Fig. 1).

Samples were imaged with a Veeco Dimension 3000 AFM in tapping mode, using phase contrast imaging. The contrast is provided by the difference in elastic modulus between the hard polystyrene spheres and the softer poly(ethylene-alt-propylene) blocks. The repeat spacing for the block copolymer is 25 nm (as measured on thin films by AFM). The spring constant of the tip (uncoated Si) was ~ 40 N/m and its resonant frequency 300 kHz.

During spin coating, most of the toluene evaporates rapidly (within a few seconds), so the block copolymer thin film suffers a relatively quick quench well below the glass transition temperature of the PS domains, inhibiting the relaxation of the early nanophase separated structure towards equilibrium. The deep and quick quench below the spinodal and order–disorder temperatures forms a nanodomain structure that contains a high density of defects. Figure 1 shows tapping-mode AFM phase images of the PS-PEP system at a very early stage of annealing. Note in this figure the large content of defects and that the spherical domains are not well defined. On the other hand, after ~ 4 hours of annealing above the glass transition temperature of the PS block, the pattern shows a higher degree of order and a well-defined structure of spherical domains. In order to better quantify the degree of order in this system, here we calculate the orientational order parameter ξ_6. Using standard image tools to identify the position of the individual spheres and applying a Delaunay triangulation [13, 16] it is possible to determine the inter-sphere bond orientation $\theta(r)$, with regard to a reference axis. Then, we can evaluate the local orientational order parameter at a position $\mathbf{r}$: $\Phi(\mathbf{r}) = \exp[6i(\theta(\mathbf{r}_1) - \theta(\mathbf{r}_2))]$, where $\mathbf{r} = \mathbf{r}_1 - \mathbf{r}_2$ [17, 18]. The azimuthally-averaged correlation function $g_6(r) = \langle \Phi(\mathbf{r}) \rangle$ was then calculated and the correlation length ξ_6 was measured by fitting $g_6(r)$ with an exponential $\exp(-r/\xi_6)$. Figure 2 shows the azimuthally-averaged correlation function $g_6(r)$ for the two pattern configurations shown in Fig. 1. During annealing, the correlation length of the film increases from $\lesssim 10$ nm to 80 nm upon increasing the annealing time from 5 to 255 minutes.

III. Model

The phase transition and the dynamics of nanophase separation for a diblock copolymer can be described through the Cahn–Hillard model [1]. A convenient order parameter for the diblock copolymer can be defined in terms of the local volume fractions for each block as $\psi = \phi_A - \phi_B - (1 - 2f)$. Here ϕ_A and ϕ_B are the local densities for the A and B blocks, respectively, and f is the volume fraction of one block in the copolymer. The dynamics can be described by the following time-dependent equation for a conserved order parameter [27]

$$\frac{\partial \psi}{\partial t} = M \, \nabla^2 \frac{\delta F\{\psi\}}{\delta \psi}. \tag{1}$$

In this equation, M is a phenomenological mobility coefficient, and $F\{\psi\}$ is the free energy functional. In this model, all the dynamics are rescaled by the mobility coefficient M, which fixes the characteristic timescale of the diffusive phenomena that drive the dynamics towards equilibrium. In block copolymer systems, the diffusion process depends on several parameters, including the monomeric friction coefficients of the individual blocks, molecular weight distribution, composition, symmetry of the mesophase structure, and degree of segregation between blocks [28]. In addition, the diffusion also depends on the molecular architecture

and degree of entanglement, that dictate whether the relaxation occurs by a Rouse, reptation, or arm retraction mechanism [28–30]. Thus, a detailed description of the dynamics involves an enormous complexity and consequently, at present, there is no phase field model equation like Eq. (1) able to capture all of the controlling parameters. However, as the experiments here are conducted at temperatures well above the glass transition temperatures of both blocks, we found that a constant mobility provides a good approximation to the dynamic response.

In the mean field approximation, the free energy functional for a diblock copolymer can be decomposed into a sum of short-range and long-range terms [31, 32]

$$F\{\psi\} = F_S\{\psi\} + F_L\{\psi\}. \tag{2}$$

The short-range contribution F_S has the form of a Landau free energy and can be expressed as

$$F_S\{\psi\} = \int d\mathbf{r}\{W(\psi) + \frac{1}{2} D \, (\nabla \psi)^2\}, \tag{3}$$

where $W(\psi)$ represents the mixing free energy of the homogeneous blend of disconnected A and B homopolymers, the term containing the gradient represents the free energy penalty generated by the spatial variations of ψ (interfacial energy), and D is a parameter related to the Kuhn segment length a_0 and the number fraction of A monomers per chain f as: $D = \frac{b^2}{48f(1-f)}$ [33].

The free energy $W(\psi)$ has the form of a non-symmetrical double well

$$W(\psi) = -\frac{1}{2}[\tau - a(1 - 2f)^2] \, \psi^2 + \frac{v}{3} \, \psi^3 + \frac{u}{4} \, \psi^4. \tag{4}$$

Here, the parameter τ is related to temperature by means of the Flory-Huggins parameter χ through [34, 35]

$$\tau = 8f(1-f)\rho_0 \chi - \frac{2 \, s(f)}{f(1-f) \, N}, \tag{5}$$

where ρ_0 is the monomer density, $N = N_A + N_B$ is the total number of monomers in the diblock copolymer chain, $s(f)$ is a constant of order one [31], $v = \nu(1 - 2f)$ and $b = \frac{9}{[2Na_0(1-f)]^2}$. Here, a, ν, and u are phenomenological constants derived by Leibler using the random phase approximation [31, 33].

The Flory-Huggins interaction parameter χ measures the incompatibility between the two monomer units and depends on the absolute temperature T as $\chi =$

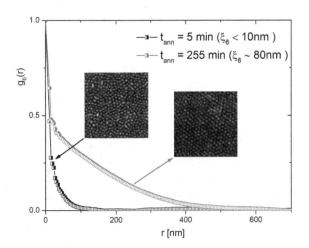

Figure 2: Orientational correlation function $g_6(r)$ for the two patterns shown in Fig. 1. As annealing time increases, there is an increasing order in the system, as shown by the slower decay in $g_6(r)$ with r and the increase in the correlation length ξ_6.

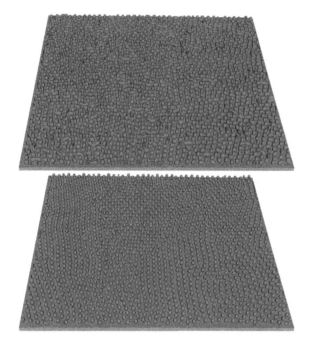

Figure 3: Phase separation process after 4.5×10^4 time steps for a sphere-forming diblock copolymer monolayer quenched within the spinodal region. Top and bottom panels correspond to $\tau_r = \tau/\tau_s - 1 = 0.34$ and $\tau_r = 0.025$, respectively. In order to describe the structure of domains developed by the minority phase, here we employ a contour plot at a fixed value of the order parameter ($\psi = -0.22$).

$A + B/T$, where A and B are phenomenological constants [1]. Thus, note that here τ increases as temperature decreases.

The term $F_L\{\psi\}$, representing the long-range contribution that accounts for the connectivity between blocks, can be expressed as [32]

$$F_L\{\psi\} = \frac{b}{2} \int d\mathbf{r}_1 \int d\mathbf{r}_2 \, G(\mathbf{r}_1 - \mathbf{r}_2) \, \psi(\mathbf{r}_1) \, \psi(\mathbf{r}_2),$$

(6)

where G is the solution of $\nabla^2 G(\mathbf{r}) = -\delta(\mathbf{r})$.

Then, using Eqs. (2)–(6), Eq. (1) takes the form

$$\frac{1}{M} \frac{\partial \psi}{\partial t} = \nabla^2 (\tilde{f}(\psi) - D\nabla^2 \psi) - b\psi,$$

(7)

where $\tilde{f}(\psi)$ is given by

$$\tilde{f}(\psi) = -[\tau - a (1 - 2f)^2] \psi + v\psi^2 + u\psi^3.$$

(8)

Equation (7) is actually a coarse-grained phase field model. In recent years, such models have been used to study a variety of systems under different conditions, including shear and other external fields, curvature and patterned substrates, and pattern formation in 3D [20, 33, 36–40]. In this work, the free energy parameters were selected to capture the desired symmetry and segregation strength of the block copolymer hexagonal phase [20, 34, 35]. More details about the equilibrium phase diagram and the different spinodals can be found elsewhere [34, 35, 41–44].

i. Numerical methods

In this work, Eq. (7) was numerically solved under confinement between impenetrable walls that represent the interfaces with the air and the silicon wafer. To describe the confinement interactions we consider an external field that couples with the two components of the block copolymer system (see details in Ref. [45]). The film thickness was fixed at the monolayer value. We employed periodic boundary conditions along the (x,y)-axis. The evolution equation, Eq. (7), was solved through the cell dynamics method, which has been extensively used in non-equilibrium studies of this kind of system [35]. One of the main advantages of this model is that it is efficient over the time scales involved in the dynamics of defect annihilation, and is thus appropriate to describe the dynamics of coarsening (see Ref. [46] for an extensive and detailed analysis of the cell dynamics method and a comparison with the Cahn–Hilliard–Cook model). Another remarkable advantage of this model is its matricial nature (continuum discretization), which allows the implementation of a transparent parallelization into a GPU code. Here, we solved this model using a dual buffering scheme (Ping-Pong technique) and an optimized use of the different GPU memories [47,48].

We study how the process of pattern formation varies with the parameter related to the temperature τ, while keeping the rest of the parameters fixed at the following values: $M = 1.0$, $D = 0.1$, $a = 1.5$, $v = 2.3$, $u = 0.38$ and $b = 0.01$. The initial homogeneous (disordered) state is simulated by a random noise distribution of the order parameter.

Figure 3 shows two snapshots of typical pattern configurations obtained through simulations of monolayers, at different conditions of annealing when an initially disordered system is quenched into the spinodal region ($\tau_s < \tau < \tau_{\text{ODT}}$).

To obtain a better comparison with the experimental AFM phase images shown in Fig. 1, rather than considering a contour plot to describe the individual PS spheres, in Fig. 4 we integrate the order parameter describing the density fluctuations along the direction perpendicular to the substrate. In this way, the interfaces are smoothed out, facilitating the comparison with AFM. Note the qualitative agreement between experimental data (Fig. 1) and the numerical results shown in Fig. 4 for systems at different annealing conditions.

ii. Linear Instability Analysis

Due to the strong degree of confinement in the monolayer, in general the formation of ordered structures and patterns can be well approximated as a 2D process. In this section we consider temperatures in the vicinity of the order–disorder transition ($T \lesssim T_{\text{ODT}}$) and study the pattern evolution of a modulated structure in the plane $z = 0$, whose order parameter profile can be described by the following sum of Fourier modes

$$\psi(\mathbf{r}, t) = \sum_{\mathbf{k}} \psi_{\mathbf{k}} \, exp \, (i \, \mathbf{k} \cdot \mathbf{r} + \lambda t), \qquad (9)$$

where $\psi_{\mathbf{k}}$ is the Fourier coefficient at $t = 0$ (we consider that $\mathbf{k}$ is restricted to the thin film's middle plane and neglect the spatial variation of ψ along the direction perpendicular to the thin film surfaces).

The stability of the solution $\psi = 0$ (high temperature phase) for a system quenched into an unstable state can be studied by linearizing Eq. (7) around $\psi = 0$ [49,50]. Substituting Eq. (9) into the linearized Eq. (7), it can be easily shown that the amplification factor λ satisfies

$$\lambda(k) = -D \, k^4 + [\tau - a \, (1 - 2 \, f)^2] \, k^2 - b, \qquad (10)$$

where $k = \|\mathbf{k}\|$. The spinodal/nucleation and growth transition line can be calculated as the lowest value of τ for which an extended mode can grow, i.e., $\lambda > 0$ for some k

$$\tau_s = 2 \sqrt{D \, b} + a \, (1 - 2 \, f)^2. \qquad (11)$$

Thus, for $\tau < \tau_s$, no extended mode can be amplified and the state $\psi = 0$ is a stable or metastable configuration. If $\tau > \tau_s$, some modes grow exponentially with time. These growing modes are constrained according to

$$k_1^2 \leq k^2 \leq k_2^2, \qquad (12)$$

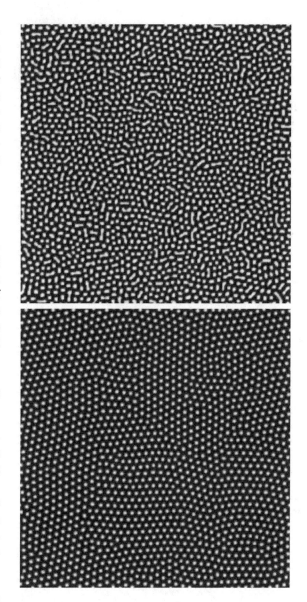

Figure 4: Two-dimensional pattern configurations for the 3D simulation data shown in Fig. 3. Observe the similarities between these patterns and those in the experimental system shown in Fig. 1.

where

$$k_1^2 = \frac{\Gamma - \sqrt{\Gamma^2 - 4 \, D \, b}}{2 \, D},$$
$$\qquad (13)$$
$$k_2^2 = \frac{\Gamma + \sqrt{\Gamma^2 - 4 \, D \, b}}{2 \, D},$$

with $\Gamma = \tau - a \, (1 - 2 \, f)^2$. Moreover, the most unstable mode (where λ is maximal) is

$$k_s = \sqrt{\frac{\Gamma}{2\,D}}. \qquad (14)$$

If $\tau \to \tau_s$, this has the form $k_s \to (b/D)^{1/4}$, and is the only mode which is unstable.

Note that, in general, the spinodal temperature does not coincide with the order–disorder transition temperature, T_{ODT}. If a homogeneous system is quenched between these temperatures, it can remain in a metastable state, which can relax only by nucleation and growth. For block copolymers the two temperatures coincide only for symmetric lamellar morphologies, $f = \frac{1}{2}$ [1].

Figure 5 shows the range of unstable modes as given by Eq. (12). Note that the distribution of unstable modes is not symmetrically distributed around k_s, the wave vector at the spinodal. This skewness becomes progressively larger as the depth of quench increases. Consequently, for systems quenched in the neighborhood of the spinodal, one can expect a very strong length scale selectivity and a strong free energy penalty for elastic distortions which shift the system from the optimum k_s. On the other hand, deep quenches stabilize the presence of different elastic distortions and thus, more disordered patterns can be expected. In addition, due to the skewness of the distribution, phase-separated systems generated at deep quenches are characterized by an average wave vector $k > k_s$.

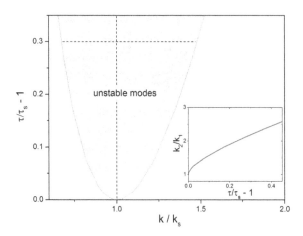

Figure 5: Range of unstable modes as a function of the reduced temperature. Inset: k_2/k_1 as a function of the reduced temperature. Note that the unstable modes are not evenly distributed around k_s.

It is known that the linear instability analysis describes only the initial stage (short times) of the spinodal decomposition mechanism. As time proceeds, the dynamics become highly nonlinear and higher-order wave numbers emerge [51–53]. We will discuss these other stages of spinodal decomposition in the next section.

IV. Results

i. Spinodal Decomposition

Spinodal decomposition is the process of relaxation of thermodynamically unstable states. Early studies on spinodal decomposition date back to the 1960s, with the pioneering works of Cahn and Hilliard [53], Hillert [54], and Lifshiftz and Slyozov [55]. Such studies were mainly focused on the mechanism of phase formation and macroscopic phase separation in solid binary alloys and fluid binary mixtures.

At present, it is well known that spinodal decomposition in binary mixtures has three different regimes [56]. Initially, some modulations present in the homogeneous phase grow exponentially with time, following the early linear evolution dynamics. With time, the nonlinear coupling between these growing modes slows their growth. In this second stage, the pattern has well-defined interfaces delimiting domains of the different stable phases. At longer times, the average domain size increases in order to reduce the total interfacial area. The asymptotic long-time state consists of two macroscopic domains, one for each phase.

In block copolymers, most of the studies of pattern formation and ordering investigate the later stage of spinodal decomposition (the coarsening stage), focusing on the kinetic exponents of evolution, and comparison with other related systems, like the self-assembled structures observed in Rayleigh-Benard convection [17, 18, 46, 56–60]. Here, we consider the early stages of spinodal decomposition in sphere-forming block copolymer systems under confinement in monolayers. We have observed two leading factors that control the degree of order during annealing below the spinodal. For $\tau > \tau_s$ the system is nanophase-separated but disordered and the kinetics of ordering are completely inhibited, while for $\tau \gtrsim \tau_s$ the strong length scale selectivity observed in Fig. 5 leads to well-defined hexagonal patterns with a lower density of defects and faster ordering kinetics. The differences in mode selectivity as a function of τ can be visualized in Fig. 3, where it can be appreciated that deep quenches lead to patterns with poorly defined symmetry.

In order to explore the dynamics and to facilitate comparison with the experimental results, here we have computed the circularly averaged structure factor for both experiments and simulations. The circularly averaged structure factor is defined as

$$S_k = \langle \tilde{\psi}(\mathbf{k})\, \tilde{\psi}^*(\mathbf{k}) \rangle, \tag{15}$$

where $\tilde{\psi}(\mathbf{k})$ is the Fourier transform of the order parameter.

For the experimental data, we have calculated S_k through the phase fields obtained by imaging the block copolymer with AFM. Although the density maps obtained by AFM are correlated with the local elasticity of the system, rather than with density fluctuations, they adequately describe the main pattern features: symmetry, defects, and degrees of local and long-range order.

Figure 6 shows S_k for systems quenched and annealed at different conditions, as obtained from the simulations. Here, if the high-temperature phase is deeply quenched into the spinodal region ($\tau_r = \tau/\tau_s - 1 = 0.2$), the length-scale selectivity imposed by the radius of gyration of the molecule leads to a nanophase-separated system with poorly defined symmetry. The insets in Fig. 6 show the 2D scattering function for this system under two different conditions. The broad halo

of intensity is consistent with a poorly defined symmetry and liquid-like order. In this case, there is not only one mode which can grow, but rather a continuous range of modes delimited by an annulus determined by $k_1 \leq k \leq k_2$ (see Fig. 5). As the temperature is low and a wide spectrum of modes is stable, the kinetics of coarsening are frozen. Observe in Fig. 6 that even after 3×10^5 time steps, S_k remains unaffected when $\tau_r = 0.2$. Similar results were found by Yokojima and Shiwa, and by Sagui and Desai, in a related system [61–63]. By including thermal fluctuations it has been shown that the system can move towards equilibrium via defect annihilation and grain growth [17,64].

When the same system is subjected to a shallow quench, $\tau_r = 3 \times 10^{-4}$, S_k shows the typical features of hexagonal patterns, with a sharper main peak at $k_{\max}$ and well-defined higher-order peaks at $\sqrt{3}k_{\max}$, $\sqrt{4}k_{\max}$ and $\sqrt{7}k_{\max}$. The rings of nearly uniform intensity are also a signature of isotropy. However, in this case the isotropy emerges as a consequence of the polycrystalline structure and not a liquid-like order, as in the case of systems deeply quenched below the spinodal temperature T_S.

The dominant features of the scattering function during spinodal decomposition are in good qualitative agreement with the experimental data shown in Fig. 7. In the experimental system, the early patterns are char-

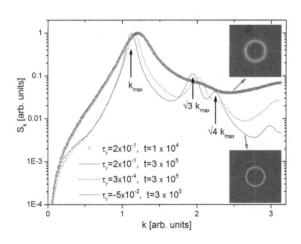

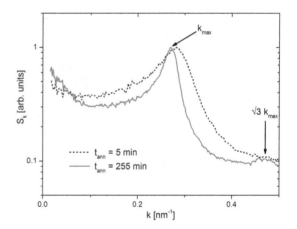

Figure 6: Circularly averaged scattering function S_k from simulations, at different depths of quench and annealing times. The insets show the 2D patterns of S_k. The higher-order peaks, located at $\sqrt{3}k_{\max}$, $\sqrt{4}k_{\max}$ and $\sqrt{7}k_{\max}$, are the signature of hexagonal order.

Figure 7: Circularly averaged scattering function S_k for the experimental data at two different annealing times. Annealing temperature: T=333 K. Note that as annealing time increases, the main peak sharpens, shifts towards lower values of k, and develops a second-order peak at $\sqrt{3}k_{\max}$, characteristic of a hexagonal pattern.

acterized by a strong length-scale selectivity, but short-range order (see also Fig. 1). As the annealing time increases, there is a shift of the main peak in S_k towards lower values of k. In addition, the main peak sharpens, and a weak higher-order peak located at $\sqrt{3}k_{\max}$ can be clearly identified. These features are in good agreement with the results of Figs. 1 and 2, which show increasing order with annealing time.

Although there is qualitative agreement between experiments and simulations, it must be emphasized that the model employed here cannot capture the detailed dynamics of the system as a function of temperature. In block copolymers, the chain mobility depends on an effective monomeric friction coefficient that is strongly dependent on temperature, the glass transition and/or crystallization temperatures of the individual blocks, the degree of segregation between blocks, the symmetry of the self-assembled structure, and the degree of entanglement. Although some of these properties can be qualitatively captured through an effective mobility coefficient [M in Eq. (1)], at present it is not quantitatively clear how these factors affect the dynamics.

ii. Nucleation and growth

When the high-temperature phase is quenched to temperatures slightly above the spinodal (quenches with $\tau \lesssim \tau_s$), all the modes decrease exponentially in time and the order parameter goes to zero across the whole system. That is, the initially randomly distributed order parameter $\psi(\mathbf{r})$, characterized by $\langle\psi(\mathbf{r})\rangle = 0$ and $\langle\psi(\mathbf{r})\psi(\mathbf{r'})\rangle \neq 0$, dies off. However, in the same range of temperatures, an initial crystalline hexagonal pattern is completely stable, indicating a region of metastability $\tau_s > \tau > \tau_{\mathrm{ODT}}$, where τ_{ODT} denotes the limit of metastability.

In addition, within this metastable region, a polycrystalline structure can improve its degree of order via the annihilation of defects, with a similar mechanism to the one observed for quenches only slightly below the spinodal. Figure 6 also shows S_k for a system quenched into this metastable region ($\tau_s > \tau > \tau_{\mathrm{ODT}}$, $\tau_r = -5 \times 10^{-2} < 0$). Note that at the same time of annealing (3×10^5 timesteps), as compared with the system quenched below the spinodal ($\tau_r = 3 \times 10^{-4} > 0$), this system shows an improved order, as the main peak sharpens and higher-order peaks become better defined.

The most distinctive feature of the metastable region is that only the nuclei or grains above a critical size can grow to develop the equilibrium phase, while smaller

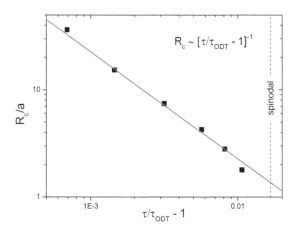

Figure 8: Critical nucleus size R_c vs. the reduced temperature $\tau/\tau_{\mathrm{ODT}} - 1$. Here a is the lattice constant. Note that $R_c \sim a$ for $\tau \sim \tau_s$.

nuclei collapse due to the surface free energy. According to the classical picture of nucleation and growth in 2D, the variation in the free energy ΔF due to the formation of a nucleus of radius R can be expressed as [21, 22]

$$\Delta F = 2\sigma\pi R - \pi|\Delta F_0|R^2, \qquad (16)$$

where σ represents the line tension and $\Delta F_0 \propto (\tau_{\mathrm{ODT}} - \tau)$ is the difference in the local free energies of the high-temperature (disordered) and the low-temperature (ordered) phases, which drives the transition. Due to the competition between these surface and volume contributions, only those nuclei whose size exceeds a critical value R_c can propagate to form the equilibrium phase.

In order to obtain the dependence of the critical size for nucleation R_c on the depth of quench, we explore the stability of crystalline nuclei of different sizes.

To study the stability of a crystal seed, we change τ back and forth around the critical value to obtain a defect-free crystal patch. The partial melting and recrystallization involved in this process removes defects and long-wavelength elastic distortions and yields a crystal seed that is then used as an initial condition in the numerical solution of Eq. (1). Figure 8 shows the critical size for nucleation R_c as a function of τ. Around the order–disorder temperature τ_{ODT} we found that the critical size for grain growth R_c diverges as $R_c \sim 1/(\tau/\tau_{\mathrm{ODT}}-1)$, in agreement with classical the-

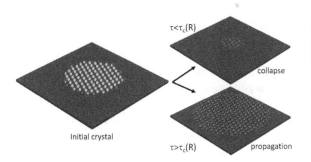

Figure 9: Time evolution of a crystalline domain as a function of temperature for a system quenched to $\tau_{\text{ODT}} < \tau < \tau_s$. Here the left panel corresponds to a hexagonal grain of initial size $R_0/a = 4.6$, where a is the lattice constant. The right panels correspond to a grain that collapses [top, $\tau < \tau_c(R)$] or propagates [bottom, $\tau > \tau_c(R)$].

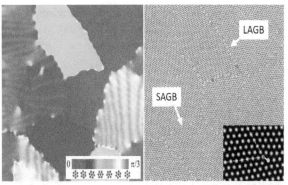

Figure 10: Orientational map $\theta(\mathbf{r})$ (left panel) and defect structure (right panel) of the hexagonal lattice determined by a Delaunay triangulation. Spheres with seven neighbors are indicated with a green dot, those with five neighbors in red. Dislocations are formed by a pair of 5-7 disclinations separated by one lattice constant, and are indicated by a connecting yellow line segment (see also the inset with a dislocation). A comparison between the defect structure and local orientation indicates that grain boundaries are decorated with linear arrays of dislocations. The bottom of the figure for $\theta(\mathbf{r})$ shows the color scale used to indicate the local orientation. Here the patterns correspond to $\tau_r = 1.64 \times 10^{-2}$ and 9×10^4 time steps.

ories of nucleation and growth [22]. Note also that as the spinodal temperature is approached, the critical size for nucleation becomes on the order of the lattice constant a. This result indicates that the process of spinodal decomposition could be inhibited, as the nucleation and growth process precedes spinodal decomposition [22].

Figure 9 shows typical snapshots of the evolution of nuclei as a function of temperature for quenches within the metastable region ($\tau_{\text{ODT}} < \tau < \tau_s$). For an initial crystalline nucleus of a given size R, and with $R_c \propto (\tau_{\text{ODT}} - \tau)^{-1}$, if the temperature is lowered to a value where R is larger than R_c, grain growth occurs. Otherwise, the crystal collapses since line tension overcomes the bulk contribution.

iii. Melting

A perfect crystalline structure may get trapped in a superheated state, such that the system may remain ordered when annealing above τ_{ODT}. However, we observed that for polycrystalline structures, topological defects trigger the phase transition, inhibiting superheating and disordering the system when $\tau \gtrsim \tau_{\text{ODT}}$. Figure 10 shows a polycrystalline structure obtained during annealing at $\tau_r = 1.64 \times 10^{-2}$. This figure shows the local orientation of the different hexagonal grains and also the defect structure. In 2D crystals with hexagonal symmetry, the elementary defects, named disclinations, are the domains which have a number of neighbors different from six (a five-coordinated domain is a positive disclination and a seven-coordinated domain is a negative disclination). In general, these de-

fects are too energetic to be found in isolation: they couple in pairs to form dipoles, also known as dislocations (see Fig. 10). Note in Fig. 10 that dislocations are piled up in linear arrangements, decorating the grain boundaries. Figure 11 shows two snapshots of the system during annealing at temperatures above T_{ODT}. Note in this figure the strong correlation between the liquid-like phase ($\psi = 0$) and the position of the initial grain boundaries. In addition, it can also be observed that melting does not start uniformly at all grain boundaries, but depends on the orientational mismatch between neighboring crystals.

The average distance between the dislocations located along a grain boundary depends on the orientation mismatch between neighboring grains, $\Delta\theta$, as $d_{ds} \propto a/\Delta\theta^{-1}$, where a is the lattice constant. Note that the largest possible mismatch in orientation between crystals in a hexagonal pattern is 30 Degrees. Compared with small-angle grain boundaries (SAGB), large-angle grain boundaries (LAGB, those for which the orientational mismatch is in the range 10–30 Degrees), are more energetic and contain a larger number of dislocations per unit length. Consequently, the phase transition is initially triggered at LAGB, where the crystal is more disordered. We also observe that the kinetics of disordering are affected by the initial defect structure. Note in Fig. 12 that a polycrystal with small crystal size, and thus a higher content of dislocations, leads to a larger fraction of the liquid phase at the same annealing time.

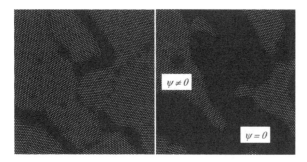

Figure 11: Upon increasing the temperature above T_{ODT}, the crystals of Fig. 10 melt first at LAGB (left panel). As time proceeds, the liquid phase ($\psi = 0$) propagates through the system. $\tau = 0.938\tau_s$. Left panel: $t = 1.5 \times 10^3$ time steps. Right panel: $t = 5 \times 10^4$ time steps.

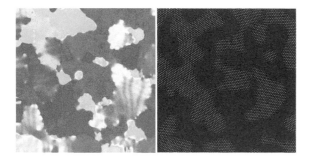

Figure 12: Melting of the crystal phase for a system containing a relatively large initial content of topological defects(left panel). Although the temperature and annealing time are the same as in the left panel of Fig. 11, due to the larger initial content of defects, there is a larger fraction of liquid phase here. $\tau = 0.938\tau_s$, $t = 1.5 \times 10^3$ time steps.

V. Conclusion

In this work, we studied the different mechanisms leading to the self-assembly of sphere-forming block copolymer thin films. Using a Cahn–Hilliard free energy model valid for diblock copolymers, we followed the evolution of the system with numerical simulations for different temperatures, and presented a unified view of the phase transition process. The results are consistent with a first-order transition. Below a spinodal temperature T_{S}, the initially homogeneous disordered state relaxes towards equilibrium by spinodal decomposition, by the spontaneous growth of characteristic modes. Above this temperature, the system can only self-assemble into an ordered phase by nucleation and growth ($T_{\mathrm{S}} < T < T_{\mathrm{ODT}}$). Above the order–disorder temperature T_{ODT} a well-ordered system can in principle remain superheated without disordering. However, dislocations and grain boundaries trigger the melting of the structure, such that defective states cannot be superheated above T_{ODT}.

For systems quenched into the metastable region, we found that the critical size for nucleation and growth diverges as the disorder–order temperature is approached, following the law: $R_c \sim 1/(\tau/\tau_{\mathrm{ODT}} - 1)$, in agreement with classical theories of nucleation and growth.

Acknowledgments

We gratefully acknowledge financial support from the National Science Foundation MRSEC Program through the Princeton Center for Complex Materials (DMR-1420541), Universidad Nacional del Sur and the National Research Council of Argentina (CONICET). The PS-PEP block copolymer was hydrogenated by John Sebastian.

[1] I W Hamley, *The physics of block copolymers*, Oxford University Press (1998).

[2] F S Bates, G H Fredrickson, *Block copolymers-designer soft materials*, Phys. Today **52**, 32 (1999).

[3] I W Hamley, *Nanostructure fabrication using block copolymers*, Nanotechnology **14**, R39 (2003).

[4] T P Lodge, *Block copolymers: past successes and future challenges*, Macromol. Chem. Phys. **204**, 265 (2003).

[5] A V Ruzette, L Leibler, *Block copolymers in tomorrow's plastics*, Nat. Mater. **4**, 19 (2005).

[6] R A Segalman, *Patterning with block copolymer thin films*, Mat. Sci. Eng. R, **48**, 191 (2005).

[7] S B Darling, *Directing the self-assembly of block copolymers*, Prog. Polym. Sci. **32**, 1152 (2007).

[8] H C Kim, S M Park, W D Hinsberg, *Block copolymer based nanostructures: Materials, processes, and applications to electronics*, Chem. Rev. **110**, 146 (2010).

[9] C M Bates, M J Maher, D W Janes, C J Ellison, C G Willson, *Block copolymer lithography*, Macromolecules **47**, 2 (2014).

[10] M J Fasolka, A M Mayes, *Block copolymer thin films: Physics and applications,* Annu. Rev. Mater. Res. **31**, 323 (2001).

[11] C Harrison, D H Adamson, Z D Cheng, J M Sebastian, S Sethuraman, D A Huse, R A Register, P M Chaikin, *Mechanisms of ordering in striped patterns*, Science **290**, 1558 (2000).

[12] C Harrison, Z D Cheng, S Sethuraman, D A Huse, P M Chaikin, D A Vega, J M Sebastian, R A Register, D H Adamson, *Dynamics of pattern coarsening in a two-dimensional smectic system*, Phys. Rev. E **66**, 011706 (2002).

[13] M L Trawick, M Megens, C Harrison, D E Angelescu, D A Vega, P M Chaikin, R A Register, D H Adamson, *Correction for piezoelectric creep in scanning probe microscopy images using polynomial mapping*, Scanning **25**, 1 (2003).

[14] A P Marencic, M W Wu, R A Register, P M Chaikin, *Orientational order in sphere-forming block copolymer thin films aligned under shear*, Macromolecules **40**, 7299 (2007).

[15] A P Marencic, D H Adamson, P M Chaikin, R A Register, *Shear alignment and realignment of sphere-forming and cylinder-forming block-copolymer thin films*, Phys. Rev. E **81**, 011503 (2010).

[16] C Harrison, D E Angelescu, M Trawick, Z D Cheng, D A Huse, P M Chaikin, D A Vega, J M Sebastian, R A Register, D H Adamson, *Pattern coarsening in a 2D hexagonal system*, Europhys. Lett. **67**, 800 (2004).

[17] D A Vega, C K Harrison, D E Angelescu, M L Trawick, D A Huse, P M Chaikin, R A Register, *Ordering mechanisms in two-dimensional sphere-forming block copolymers*, Phys. Rev. E **71**, 061803 (2005).

[18] L R Gómez, E M Vallés, D A Vega, *Lifshitz-Safran coarsening dynamics in a 2D hexagonal system*, Phys. Rev. Lett. **97**, 188302 (2006).

[19] N A García, R A Register, D A Vega, L R Gómez, *Crystallization dynamics on curved surfaces*, Phys. Rev. E **88**, 012306 (2013).

[20] W Li, F Qiu, Y Yang, A C Shi, *Ordering dynamics of directed self-assembly of block copolymers in periodic two-dimensional fields*, Macromolecules **43**, 1644 (2010).

[21] P M Chaikin, T C Lubensky, *Principles of condensed matter physics*, Cambridge University Press (1995).

[22] P G Debenedetti, *Metastable liquids*, Princeton University Press (1996).

[23] J M Sebastian, C Lai, W W Graessley, R A Register, *Steady-shear rheology of block copolymer melts and concentrated solutions: Disordering stress in body-centered-cubic systems*, Macromolecules **35**, 2707 (2002).

[24] D E Angelescu, C K Harrison, M L Trawick, R A Register, P M Chaikin, *Two-dimensional melting transition observed in a block copolymer*, Phys. Rev. Lett. **95**, 025702 (2005).

[25] D E Angelescu, J H Waller, D H Adamson, P Deshpande, S Y Chou, R A Register, P M Chaikin, *Macroscopic orientation of block copolymer cylinders in single-layer films by shearing*, Adv. Mater. **16**, 1736 (2004).

[26] D E Angelescu, J H Waller, R A Register, P M Chaikin, *Shear-induced alignment in thin films of spherical nanodomains*, Adv. Mater. **17**, 1878 (2005).

[27] A Adland, Y Xu, A Karma, *Unified theoretical framework for polycrystalline pattern evolution*, Phys. Rev. Lett. **110**, 265504 (2013).

[28] G H Fredrickson, F S Bates, *Dynamics of block copolymers: Theory and experiment* Annu. Rev. Mater. Sci. **26**, 501 (1996).

[29] M Doi, S F Edwards, *The theory of polymer dynamics*, Clarendon Press (1986).

[30] D A Vega, J M Sebastian, Y L Loo, R A Register, *Phase behavior and viscoelastic properties of entangled block copolymer gels*, J. Polym. Sci. Part B: Polym. Phys. **39**, 2183 (2001).

[31] L Leibler, *Theory of microphase separation in block copolymers*, Macromolecules **13**, 1602 (1980).

[32] T Ohta, K Kawasaki, *Equilibrium morphology of block copolymer melts*, Macromolecules **19**, 2621 (1986).

[33] M Serral, M Pinna, A V Zvelindovsky, J B Avalos, *Cell dynamics simulations of sphere-forming diblock copolymers in thin films on chemically patterned substrates*, Macromolecules **49**, 1079 (2016).

[34] I W Hamley, *Cell dynamics simulations of block copolymers*, Macromol. Theory Simul. **9**, 363 (2000).

[35] S R Ren, I W Hamley, *Cell dynamics simulations of microphase separation in block copolymers*, Macromolecules **34**, 116 (2001).

[36] N Provatas, K Elder, *Phase-field methods in material science and engineering*, Wiley (2010).

[37] L R Gómez, D A Vega, *Amorphous precursors of crystallization during spinodal decomposition*, Phys. Rev. E **83**, 021501 (2011).

[38] Y Guo, J Zhang, B Wang, H Wu, M Sun, J Pan, *Microphase transitions of block copolymer/homopolymer under shear flow*, Condens. Matter Phys. **18**, 23801 (2015).

[39] A D Pezzutti, L R Gómez, D A Vega, *Smectic block copolymer thin films on corrugated substrates*, Soft Matter **11**, 2866 (2015).

[40] S K Mkhonta, K R Elder, Z F Huang, *Emergence of chirality from isotropic interactions of three length scales*, Phys. Rev. Lett. **116**, 205502 (2016).

[41] T Ohta, Y Enomoto, J L Harden, M Doi, *Anomalous rheological behavior of ordered phases of block copolymers. 1*, Macromolecules **26**, 4928 (1993).

[42] M Nonomura, *Stability of the fcc structure in block copolymer systems*, J. Phys.: Condens. Matter **20**, 465104 (2008).

[43] K Yamada, S Komura, *The dynamics of order-order phase separation*, J. Phys.: Condens. Matter **20**, 155107 (2008).

[44] R Choksi, X Ren, *On the derivation of a density functional theory for microphase separation of diblock copolymers*, J. Stat. Phys. **113**, 151 (2003).

[45] A A Abate, G T Vu, A D Pezzutti, N A García, R L Davis, F Schmid, R A Register, D A Vega, *Shear-aligned block copolymer monolayers as seeds to control the orientational order in cylinder-forming block copolymer thin films*, Macromolecules **49**, 7588 (2016).

[46] G J A Sevink, *Rigorous embedding of cell dynamics simulations in the Cahn–Hilliard–Cook framework: Imposing stability and isotropy*, Phys. Rev. E **91**, 053309 (2015).

[47] J Nickolls, I Buck, M Garland, K Skadron, *Scalable parallel programming with CUDA*, ACM Queue **6**, 40 (2008).

[48] NVIDIA, *CUDA Programming Guide 8.0*, http://docs.nvidia.com/cuda/ (2017).

[49] D A Vega, L R Gómez, *Spinodal-assisted nucleation during symmetry-breaking phase transitions*, Phys. Rev. E **79**, 051607 (2009).

[50] A D Pezzutti, L R Gómez, M A Villar, D A Vega, *Defect formation during a continuous phase transition*, Europhys. Lett. **87**, 66003 (2009).

[51] J W Cahn, *The later stages of spinodal decomposition and the beginnings of particle coarsening*, Acta Met. **14**, 1685 (1966).

[52] J S Langer, M Bar-on, H D Miller, *New computational method in the theory of spinodal decomposition*, Phys. Rev. A **11**, 1417 (1975).

[53] J W Cahn, J E Hilliard, *Free energy of a nonuniform system. I Interfacial free energy*, J. Chem. Phys. **28**, 258 (1958).

[54] M Hillert, *A solid-solution model for inhomogeneous systems*, Acta Met. **9**, 525 (1961).

[55] I M Lifshitz, V V Slyozov, *The kinetics of precipitation from supersaturated solid solutions*, J. Phys. Chem. Solids **19**, 35 (1961).

[56] A J Bray, *Theory of phase-ordering kinetics*, Adv. Phys. **43**, 357 (1994).

[57] Y Oono, S Puri, *Computationally efficient modeling of ordering of quenched phases*, Phys. Rev. Lett. **58**, 836 (1987).

[58] Y Oono, M Bahiana, *2/3-Power law for copolymer lamellar thickness implies a 1/3-power law for spinodal decomposition*, Phys. Rev. Lett. **61**, 1109 (1988).

[59] F Liu, N Goldenfeld, *Dynamics of phase separation in block copolymer melts*, Phys. Rev. A **39**, 4805 (1989).

[60] M Bahiana, Y Oono, *Cell dynamical system approach to block copolymers*, Phys. Rev. A **41**, 6763 (1990).

[61] Y Yokojima, Y Shiwa, *Hydrodynamic interactions in ordering process of two-dimensional quenched block copolymers*, Phys. Rev. E **65**, 056308 (2002).

[62] C Sagui, R C Desai, *Kinetics of topological defects in systems with competing interactions*, Phys. Rev. Lett. **71**, 3995 (1993).

[63] C Sagui, R C Desai, *Late-stage kinetics of systems with competing interactions quenched into the hexagonal phase*, Phys. Rev. E **52**, 2807 (1995).

[64] L R Gómez, E M Vallés, D A Vega, *Effect of thermal fluctuations on the coarsening dynamics of 2D hexagonal system*, Physica A **386**, 648 (2007).

18

Dilute antiferromagnetism in magnetically doped phosphorene

A. Allerdt,[1] A. E. Feiguin[1]*

We study the competition between Kondo physics and indirect exchange on monolayer black phosphorous using a realistic description of the band structure in combination with the density matrix renormalization group (DMRG) method. The Hamiltonian is reduced to a one-dimensional problem via an exact canonical transformation that makes it amenable to DMRG calculations, yielding exact results that fully incorporate the many-body physics. We find that a perturbative description of the problem is not appropriate and cannot account for the slow decay of the correlations and the complete lack of ferromagnetism. In addition, at some particular distances, the impurities decouple forming their own independent Kondo states. This can be predicted from the nodes of the Lindhard function. Our results indicate a possible route toward realizing dilute anti-ferromagnetism in phosphorene.

I. Introduction

Phosphorene, a single layer of black phosphorus, is one of the many allotropes of the element. Others include red, white, and velvet phosphorus. Of these, black is the most thermodynamically stable and least reactive [1]. Its crystalline structure resembles that of graphite, whose 2D form is famously known as graphene, and can similarly be fabricated into 2D layers. The main qualitative distinction lies in the fact that phosphorene has a "puckered" hexagonal structure which is responsible for opening a gap in the band structure. In addition, the bonding structure is vastly different, since phosphorene does not have sp^2 bonds, but is composed of $3p$ orbitals [2, 3].

Successful production of monolayer black phosphorus has been achieved only in the last few years, with the first publications concerning its single layer properties appearing in 2014 [4–7]. While being a semi-conductor with a small direct band gap in bulk form (~ 0.3 eV) [8], the gap increases as the number of layers is de-creased.

Since its appearance, there have been many proposed promising applications and exotic properties. Besides being a semi-conductor with a band gap in the optical range, it has ample flexibility and high carrier mobility [4, 9]. Further, it also possesses interesting optical properties such as absorbing light only polarized in the armchair direction, indicating a possible future as a linear polarizer [10]. Its stable excitons present possible applications in optically driven quantum computing [11]. Interest on its applications continues to grow. For a comprehensive review, we refer to Ref. [11].

Our understanding of magnetic doping in phosphorene is still in its infancy. A thorough study of metal adatoms adsorbed on phosphorene was performed in Ref. [12] where a variety of structural, electronic and magnetic properties emerge. The authors show the binding energies are twice the amount in graphene. A DFT study has proposed possible chemical doping by means of adsorption of different atoms, ranging from n-type to p-type, as well as transition metals with finite magnetic moment [13]. Another conceivable method of moving the Fermi level is to induce strain on the lattice, causing the level to move into the conduction band [14]. The case of magnetic impurities is considerably

*a.feiguin@northeastern.edu

[1] Department of Physics, Northeastern University, Boston, Massachusetts MA 02115, USA.

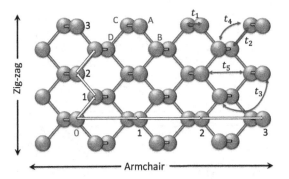

Figure 1: Phosphorene lattice showing the directions chosen to place the impurities for calculations shown in Figs. 3, 4, and 5. The numbering of lattice sites refers to the impurity separations in the same figures. Also shown are the five hopping parameters borrowed from Ref. [5]. The labels A, B, C and D refer to the four sublattices.

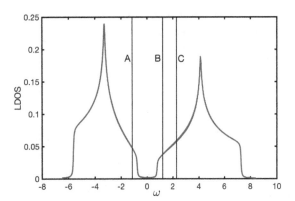

Figure 2: Local density of states of phosphorene. Vertical lines show positions of the different Fermi energies used in the calculations.

non-trivial, since one has to account for the Kondo effect [15] with the impurity being screened by the conduction spins in the metallic substrate, and the RKKY interaction [16–18], an effective indirect exchange between impurities mediated by the conduction electrons. These two phenomena are expected to be present and compete in phosphorene when the Fermi level is not sitting in the gap. The RKKY interaction in phosphorene with the inclusion of mechanical strain is investigated in Ref. [14]. However, all their calculations are based on second order perturbation theory and ignore the effects that Kondo physics can induce. In this work, we present numerical results for two Kondo impurities in phosphorene that capture the full many-body physics and we discuss the competition between Kondo and indirect exchange.

II. Model and numerical method

The Hamiltonian studied in this work is the two-impurity Kondo model, generically written as:

$$H = H_{band} + J_K \left(\vec{S}_1 \cdot \vec{s}_{\mathbf{r}_1} + \vec{S}_2 \cdot \vec{s}_{\mathbf{r}_2} \right), \quad (1)$$

where H_{band} is the non-interacting part describing the band structure of the material while the impurities are locally coupled to the substrate at positions $\mathbf{r}_1$ and $\mathbf{r}_2$ via an interaction J_K. Here, $\vec{S}_i$ represent the impurities, and $\vec{s}_{\mathbf{r}_i}$ is the spin of the conduction electrons at site $\mathbf{r}_i$. This problem has been theoretically studied in other

materials using a variety of approaches. This work, however, will take a non-perturbative approach revealing important subtle points. Traditionally, the RKKY interaction is characterized by the Lindhard function in second order perturbation theory, which takes the form:

$$\chi_{ij} = 2\Re \left[\sum_{E_\alpha > E_f > E_\beta} \frac{\langle i|\alpha\rangle \langle \alpha|j\rangle \langle j|\beta\rangle \langle \beta|i\rangle}{E_\alpha - E_\beta} \right], \quad (2)$$

where $|i, j\rangle$ represent lattice sites and $|\alpha, \beta\rangle$ are eigenstates of H_{band}. This function oscillates and changes sign with a period that depends on the density of the electrons. When the full geometry of the problem is taken into account, the picture becomes more complex, since a system could have a complex Fermi surface even with more than one band crossing the Fermi level in some cases [19]. However, it has been shown by the authors in previous studies [19–21] that this does not uncover the full picture.

We adapt the four-band model introduced in Ref. [5] to obtain a tight-binding description of the phosphorene band structure. Five hopping parameters are employed to closely reproduce the bands near the Fermi level with a gap of approximately 1.6 eV. These are shown schematically in Fig. 1, and their values are restated in Table 1. A plot of the density of states is shown in Fig. 2. All energies in the calculations are in units of eV. To see the band dispersions, we refer to Ref. [5].

In order to solve the interacting problem we first perform a unitary transformation to map the quadratic part of the Hamiltonian H_{band} onto an equivalent one-dimensional one, as described in detail in Refs. [20,22].

t_1	-1.220
t_2	3.665
t_3	-0.205
t_4	-0.105
t_5	-0.055

Table 1: Tight binding hopping parameters in eV used in the calculations throughout this work.

The full many-body calculation can in turn be carried out using the density matrix renormalization group (DMRG) algorithm [23–25] with high accuracy and without approximations. For all DMRG calculations, the total system size is $L = 124$ (including impurities), and we fix the truncation error to be smaller than 10^{-7}. It is known that phosphorene nano-ribbons can host quasi-flat edge states whose emergence is topologically similar to graphene [26]. A study of their even/odd properties under application of an electric field has recently been performed [27], revealing a gap opening in these edge states. Edge states can introduce notable finite-size effects [21]. However, due to the geometry of the lattices produced by the Lanczos transformation employed in this work, these edge states are not present. In addition, we shall focus our study in a regime away from half-filling.

III. Results

We start by calculating the Lindhard function as a reference using Eq. (2). Figure 3 shows results for two different Fermi energies, one in the conduction band, and one in the valence band. Notice the different period of oscillation, which is to be expected from conventional RKKY theory [16–18]. We also show, in different panels, the DMRG result for the impurity spin-spin correlations $\langle S_1^z S_2^z \rangle$. Since the system is $SU(2)$ symmetric, only the z-component is displayed. For small values of J_K we find that he RKKY correlations saturate at -1/4. This is an indication that the two impurities are decoupled forming "free moments". When this happens, the spins can be pointing in either direction yielding a four-fold degenerate ground state. Since we force $S^z = 0$ in our calculations, the impurities have no choice but to align anti-ferromagnetically. This is a finite size effect that is expected when the interaction J_K is of the order of the level spacing in the non-interacting bands. Apart from short distances, the spin correlations roughly follow the oscillation patterns predicted

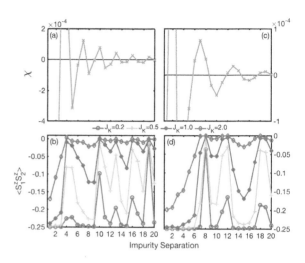

Figure 3: (a) Lindhard function and (b) spin-spin correlations along the zig-zag direction with the Fermi level at position A in Fig. 2. (c) and (d) are the same quantities with the Fermi level at position C.

by the Lindhard function with one striking difference: ferromagnetism is non-existent. This lack of ferromagnetism is also seen in previous works [19–21]. However, in doped graphene, for example [21], the correlations depart drastically from the Lindhard function, and Kondo physics dominates after just a few lattice spacings. Here, we find that RKKY physics plays the dominant role, with correlations persisting even for quite large coupling (J_K=2.0 eV), which is greater than the band gap.

Figure 4 again the Lindhard function and spin correlations, but for a different Fermi energy and in an extended range. This is to highlight the fact that the RKKY oscillations survive at large distances with little signs of decay. Ferromagnetism is again absent in contrast with the predictions of perturbation theory, exemplifying the need for exact numerical techniques that can capture the many-body physics. The impurities, however, do appear to form their independent Kondo singlets at some particular distances. This results in the impurities being completely uncorrelated $\langle S_1^z S_2^z \rangle = 0$ and typically occurs at positions where the Lindhard function has a node, as observed in Ref. [20].

To show that these effects are not due to the particular directions chosen, calculations were done along other paths with qualitatively similar results. As an example, we show in Fig. 5 the correlations along the perpendicular direction with the Fermi level at posi-

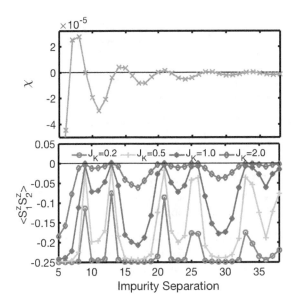

Figure 4: Lindhard function (top) and spin-spin correlations (bottom) along the zig-zag direction with the Fermi level at position B in Fig. 2. Results are shown with an extended range, highlighting the persistent oscillations even for large J_K.

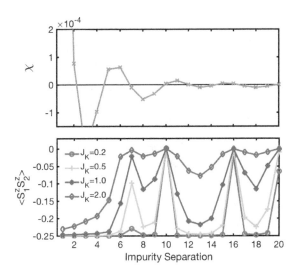

Figure 5: Lindhard function (top) and spin-spin correlations (bottom) along the armchair direction with the Fermi level at position B in Fig. 2.

tion B. As can be seen from the lattice structure, single layer phosphorene contains four sublattices, as labeled in Fig. 1. Assuming the first impurity is placed on an 'A' site, along the zig-zag direction the impurities follow an "$A - A, A - B, A - A, A - B...$" pattern. Alternatively, the second impurity could be placed on the other "layer" along the zig-zag direction resulting in a "$A - C, A - D, A - C, A - D...$" pattern. It is clear that $A - D$ is equivalent to $B - C$. The armchair direction, as it has been defined, consists of impurities always on the same sublattice. In all cases (not shown), ferromagnetism is absent and we find dominant antiferromagnetic RKKY interactions.

IV. Conclusions

By means of a canonical transformation and exact numerical calculations using the DMRG method, we have studied the competition between Kondo and RKKY physics on phosphorene. The method is numerically exact, and even though it is limited to finite systems, these can be very large, of the order of a hundred lattice sites and more. Our results highlight the non-perturbative nature of the RKKY interaction and the non-trivial ab-

sence of ferromagnetism. This remains an outstanding question and should stimulate more research in this direction. It is possible that by adding a repulsive interaction between conduction electrons, the system may acquire a net magnetic moment, a behavior of this type has been observed in graphene doped with hydrogen defects [28–30]. According to Lieb's theorem [31], this is expected to occur for bi-partite lattices when two impurities are on the same sublattice. Even though phosphorene has four sublattices (and hence, the theorem does not rigorously apply), the system may still realize similar physics. Unfortunately, our approach can only describe non-interacting/quadratic Hamiltonians and we cannot prove this conjecture. On the other hand, the dominant anti-ferromagnetism at all dopings and distances (more robust than in graphene [21]) indicates a route toward realizing dilute 2D anti-ferromagnetism with phosphorene.

Acknowledgements - The authors are grateful to the U.S. Department of Energy, Office of Basic Energy Sciences, for support under grant DE-SC0014407.

[1] R B Jacobs, *Phosphorus at high temperatures and pressures*, J. Chem. Phys. **59**, 945 (1937).

[2] L Pauling, M Simonetta, *Bond orbitals and bond energy in elementary phosphorus*, J. Chem. Phys. **20**, 29 (1952).

[3] R R Hart, M B Robin, N A Kuebler, *3p orbitals, bent bonds, and the electronic spectrum of the p4 molecule*, J. Chem. Phys. **42**, 3631 (1965).

[4] H Liu *et al.*, *Phosphorene: An unexplored 2d semiconductor with a high hole*, ACS Nano **8**, 4033 (2014).

[5] A N Rudenko, M I Katsnelson, *Quasiparticle band structure and tight-binding model for single- and bilayer black phosphorus*, Phys. Rev. B **89**, 201408 (2014).

[6] A S Rodin, A Carvalho, A H Castro Neto, *Strain-induced gap modification in black phosphorus*, Phys. Rev. Lett. **112**, 176801 (2014).

[7] L Li *et al.*, *Black phosphorus field-effect transistors*, Nat. Nanotechnol. **9**, 372 (2014).

[8] R W Keyes, *The electrical properties of black phosphorus*, Phys. Rev. **92**, 580 (1953).

[9] X Ling, H Wang, S Huang, F Xia, M S Dresselhaus, *The renaissance of black phosphorus*, P. Natl. Acad. Sci. USA **112**, 4523 (2015).

[10] V Tran, R Soklaski, Y Liang, L Yang, *Layer-controlled band gap and anisotropic excitons in few-layer black phosphorus*, Phys. Rev. B **89**, 235319 (2014).

[11] A Carvalho *et al.*, *Phosphorene: From theory to applications*, Nat. Rev. Mater. **1**, 16061 (2016).

[12] V V Kulish, O I Malyi, C Persson, P Wu, *Adsorption of metal adatoms on single-layer phosphorene*, Phys. Chem. Chem. Phys. **17**, 992 (2015).

[13] P Rastogi, S Kumar, S Bhowmick, A Agarwal, Y S Chauhan, *Effective doping of monolayer phosphorene by surface adsorption of atoms for electronic and spintronic applications*, IETE J. Res. **63**, 205 (2017).

[14] H-J Duan *et al.*, *Anisotropic RKKY interaction and modulation with mechanical strain in phosphorene*, New J. Phys. **19**, 103010 (2017).

[15] A C Hewson, *The Kondo Problem to Heavy Fermions*, Cambridge University Press, New York (1983).

[16] K Yosida, *Magnetic properties of Cu-Mn alloys*, Phys. Rev. **106**, 893 (1957).

[17] M A Ruderman, C Kittel, *Indirect exchange coupling of nuclear magnetic moments by conduction electrons*, Phys. Rev. **96**, 99 (1954).

[18] T Kasuya, *A theory of metallic ferro- and antiferromagnetism on Zener's model*, Prog. Theor. Phys. **16**, 45 (1956).

[19] A Allerdt, R Žitko, A E Feiguin, *Nonperturbative effects and indirect exchange interaction between quantum impurities on metallic (111) surfaces*, Phys. Rev. B **95**, 235416 (2017).

[20] A Allerdt, C A Büsser, G B Martins, A E Feiguin, *Kondo versus indirect exchange: Role of lattice and actual range of RKKY interactions in real materials*, Phys. Rev. B **91**, 085101 (2015).

[21] A Allerdt, A E Feiguin, S D Sarma, *Competition between Kondo effect and RKKY physics in graphene magnetism*, Phys. Rev. B **95**, 104402 (2017).

[22] C A Büsser, G B Martins, A E Feiguin, *Lanczos transformation for quantum impurity problems in d-dimensional lattices: Application to graphene nanoribbons*, Phys. Rev. B **88**, 245113 (2013).

[23] S R White, *Density matrix formulation for quantum renormalization groups*, Phys. Rev. Lett. **69**, 2863 (1992).

[24] S R White, *Density-matrix algorithms for quantum renormalization groups*, Phys. Rev. B **48**, 10345 (1993).

[25] A E Feiguin, *The density matrix renormalization group method and its time-dependent variants*, AIP Conf. Proc. **1419**, 5 (2011).

[26] M Ezawa, *Topological origin of quasi-flat edge band in phosphorene*, New J. Phys. **16**, 115004 (2014).

[27] B Zhou, B Zhou, X Zhou, G Zhou, *Even–odd effect on the edge states for zigzag phosphorene nanoribbons under a perpendicular electric field*, J. Phys. D: Appl. Phys. **50**, 045106 (2017).

[28] O V Zazyev, L Helm, *Defect-induced magnetism in graphene,* Phys. Rev. B **75**, 125408 (2007).

[29] S Casolo, O M Løvvik, R Martinazzo, G F Tantardini, *Understanding adsorption of hydrogen atoms on graphene,* J. Chem. Phys. **130**, 054704 (2009).

[30] H González-Herrero *et al., Atomic-scale control of graphene magnetism by using hydrogen atoms,* Science **352**, 437 (2016).

[31] E H Lieb, *Two theorems on the Hubbard model,* Phys. Rev. Lett. **62**, 1201 (1989).

19

Smartphones on the air track: Examples and difficulties

Manuel Á. González,[1*] Alfonso Gómez,[1] Miguel Á. González[1]

In this paper we describe a classical experiment with an air track in which smartphones are used as experimental devices to obtain physical data. The proposed experiment allows users to easily observe and measure relationships between physical magnitudes, conservation of momentum in collisions and friction effects on movement by utilizing the users' own mobile devices.

I. Introduction

Smartphones and tablets have sensors that can be used to redesign physics experiments by substituting laboratory equipment with those devices [1]. This technique can enrich data availability in an experiment and also reduces its costs. The work in the laboratory under controlled conditions can also help experimenters to learn the use and limitations of smartphones and to apply them in experiments outside the laboratory. In this paper, we show some results obtained in a classical experiment, pointing out basic concepts that can be explored and analyzed by using smartphones, as well as some difficulties of the work.

II. Using the smartphone in an air-track experiment

A simple and usual mechanics experiment consists in studying linear movement on an air track without friction. On it, positions, speeds and accelerations can be measured by using measuring tapes, chronometers or more sophisticated photoelectric cells. Here, we discuss how smartphones can be used to analyze the movement of a body in three different configurations of the experiment: elastic collisions between two carts, accelerated movement of a cart pulled by a falling body via pulley, and movement of a cart when the air pump is switched off and it is stopped by friction.

As experimental devices in our work we used two smartphones, a Samsung Galaxy S3 mini and a Samsung Galaxy S4. These smartphones were placed horizontally on the moving carts with their Y-axis pointed along the direction of the movement and their Z-axis vertically. The carts can hold additional weights so that we could study interactions between bodies with same or different masses. To access the data recorded using the sensors of the smartphones, we used two Android apps: Sensor-Mobile [2] and Physics Toolbox [3], so that the advantages or disadvantages, accuracy and numerical noise, of different apps can also be analyzed.

i. Elastic collisions between carts in a horizontal air track

This is a simple experiment that is done by using two carts, each with a smartphone measuring its acceleration, allowing us to check the momentum conservation in a collision between two bodies. In our experiments, we used different configurations

*E-mail: manuelgd@termo.uva.es

[1] Universidad de Valladolid, 47011 Valladolid, Spain.

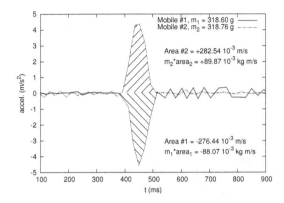

Figure 1: Accelerations of two bodies of equal mass during a collision. A spreadsheet can be useful to calculate the area under the acceleration curves and check momentum conservation readily.

that can be explored: both carts moving on the air track in the same or in opposite directions, or one cart at rest while the other moves towards it. Moreover, by adding different masses to the carts we have analyzed collisions between bodies with the same or with different masses.

Figure 1 shows the results of one of such experiments corresponding to an elastic collision between two bodies (cart plus smartphone) of similar masses. In this case only one of the bodies was moving before the impact. Once the acceleration of each body is measured by the smartphones, the CSV files generated by the apps can be transferred to a computer and analyzed easily using a spreadsheet program. Here the spreadsheet was used to obtain the area under the curve of each acceleration using a simple numerical method like the trapezoidal rule. As can be seen in Fig. 1, when the masses are nearly equal the areas calculated from the smartphone data are also similar (the difference is a little larger than the 2%) considering the experimental noise. And multiplying those areas by the masses of the bodies the conservation of linear momentum is checked, as shown in Fig. 1.

ii. Frictionless movement of a cart pulled by a falling body through pulley

Another dynamics problem studied in the first courses of physics is the movement of a body, on a frictionless horizontal or inclined plane, connected to a cable that passes over a pulley and then is fastened to a falling object. Here, we study such movement by placing a smartphone on the moving cart and measuring its acceleration.

Figure 2 shows some results of two measurements of this type. They correspond to the acceleration suffered by a body (cart plus smartphone) of mass $m_c = 318.76$ g on a horizontal plane when the falling body has different masses in two different experiments, $m_b = 20.02$ g and 29.99 g. According to the theory studied in the classroom, assuming that there is no friction, the acceleration of the cart is $a = \frac{m_b}{m_c + m_b} g$ that for our conditions gives accelerations $a_{(m_b=20.02)} = 0.579$ m/s^2 and $a_{(m_b=29.99)} = 0.843$ m/s^2. As can be seen these results agree well with the average acceleration of the cart as measured by the smartphone on it. The users could also add a second smartphone on the falling body and check that the acceleration of the cart and that of the falling body are the same within the experimental accuracy, according to this simple model. From the accelerometer measurements, the dependence of the travelled space with acceleration and time in the uniformly accelerated movement can be checked. If the length of the cable is the same in two measurements with different falling masses, then the ratio of travelling times $\Delta t_2 / \Delta t_1$ must be equal to the square root of the ratio of the average accelerations measured by the smartphones $\sqrt{a_2 / a_1}$. As can be seen from the measurements in the figure, $\Delta t_2 / \Delta t_1 \approx 1.199$, while for the accelerations $\sqrt{a_2 / a_1} \approx 1.207$, so that the the theoretical dependence $s = \frac{1}{2} a t^2$ is easily proved.

iii. Movement with friction of a cart pulled by a falling body through pulley

In this section, we show some results that can be obtained if the air pump of the track is set off and the cart moves with friction. For this experiment the lower part of the cart should be protected, for example with duct tape, to avoid damaging the air track with scratches, which will also vary the friction coefficient as the experiment is performed. The experiment is the same as in the section ii., except for the existence of friction. The acceleration of the cart is lower than the one without friction, the cart decelerates, and finally stops if the cable is long enough for the falling object to reach the floor

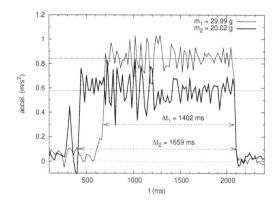

Figure 2: Measurements of the acceleration of a cart on an air track when pulled by different masses in two experiments. The average values of acceleration, in the two experiments, marked with lines in the figure, were $a_1 \approx 0.84$ m/s^2 and $a_2 \approx 0.58$ m/s^2

before the cart gets to the track end.

Figure 3 shows an example of the results with the smartphone accelerometer in such experiment. That figure shows acceleration data of three independent measurements under the same conditions to illustrate the repeatability of the results. Some comments are necessary about the results shown in the figure. As can be seen there, the acceleration is positive when the movement is accelerated, and changes its sign when the falling body reaches the floor and the friction decelerates the movement of the cart. A very remarkable result that immediately calls our attention is the oscillations in the accelerated part of the movement. As can be seen in Fig. 3 (and in Fig. 4), these oscillations appear consistently in all the experiments and then, cannot be considered an experimental artifact. In our opinion, these oscillations appear due to the combined effect of the push on the string of the falling body and the pull due to the friction on the cart attached to the end of the string. These oscillations do not appear when the air pump is working and there is no friction. From our point of view, these oscillations reflect variations in the acceleration of the cart due to changes in the string tension. This is a problem that we consider very interesting to see. When this experiment is done using classical measurement devices the usual approximation that considers inextensible strings and constant tension seems to hold true, but smartphone sensors are sen-

sible to small variations in them, as can be seen here. This observation can help the experimenters to consider the limitations of the physical model and the influence of the string characteristics. Figure 3 shows in an inset the experimental points of the three measurements together with a damped oscillation with period $T \approx 0.105$ s. For the theoretical behaviour we have tested accelerations for pure Coulomb and pure viscous damping but the experimental results seems to be a mixing of both behaviours, what makes more difficult its analysis. The inset in Fig. 3 shows a comparison of the experimental points with a theoretical curve for the acceleration. From this comparison, we have an estimation of the initial elongation of the string to be around $1.2 \cdot 10^{-3}$ m (the length of the string used in this experiment was approximately 1.75 m). In addition, a qualitative analysis of the oscillations was acomplished by using strings with different elasticity. We observed that these oscillations were less noticeable for strings with lower elasticities, and that even for low elastic strings these oscillations were masked by the experimental noise. On the other hand, for more rigid strings, such as metallic strings, the amplitude of the oscillations was smaller due to their lower deformation. Anyway, both the mixing of effects in this result and its experimental difficulties could make an exact analysis harder than in the classical version of the experiment.

Other results can be obtained easily by the users. From the masses of the cart ($m_c = 318.8$ g) and of the falling body ($m_b = 358.05$ g) they can calculate the theoretical acceleration without friction, resulting $a_{teo} = 5.18$ m/s^2, that is clearly higher than the experimental value in the accelerated part of the movement, whose average value is $a_{exp} = 2.386$ m/s^2 (see Fig. 3). Using this experimental value, the students can obtain the friction coefficient between the duct tape protecting the cart and the track with $\mu = (m_b g - (m_c + m_b)a_{exp})/(m_c g)$, resulting $\mu = 0.61$. Once μ is known the friction deceleration $a_{friction} = \mu g = 5.94$ m/s^2 can be calculated, that, as shown in Fig. 3, agrees well with the experimental results within the experimental noise. From the experimental data the impulse-momentum relationship can also be observed. As the cart started and ended its movement with null speed, the areas under the acceleration and deceleration curves in Fig. 4 must have the same value

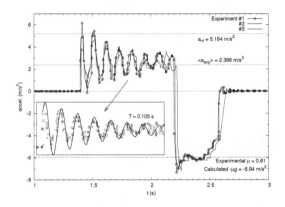
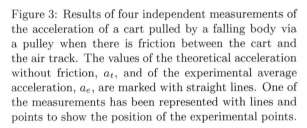
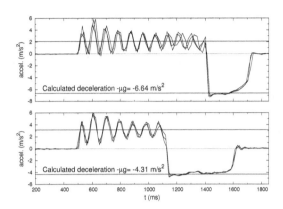

Figure 3: Results of four independent measurements of the acceleration of a cart pulled by a falling body via a pulley when there is friction between the cart and the air track. The values of the theoretical acceleration without friction, a_t, and of the experimental average acceleration, a_e, are marked with straight lines. One of the measurements has been represented with lines and points to show the position of the experimental points.

Figure 4: Accelerations recorded by the smartphone in an experiment like that in Fig. 3 but two different friction coefficients: painted Al cart on Al track (above) and duct tape covered cart on Al track (below). The cart and the falling body masses were $m_c = 317.3$ g and $m_b = 359.1$ g, respectively. The experimental average accelerations of each case have been marked with horizontal lines, as well as the decelerations once the falling body reaches the floor.

but opposite signs. By using a spreadsheet program and the numerical trapezoidal rule, those areas can be easily obtained.

Finally, Fig. 4 shows another result that can be discussed. Results for two different conditions of friction are shown in it: an aluminium cart on an aluminium track and a duct tape covered cart on an aluminium track. Due to different friction coefficients, the cart accelerations are different and, consequently, their travelled times. As for Fig. 3, the users can calculate the dynamic friction coefficients and the corresponding decelerations, and compare them with the experimental results. A final result that can be discussed is the influence of different friction coefficients on the stopping times.

III. Conclusions

Smartphone sensors are very useful tools that permit to do measurements in the laboratory and outside it. The use of these devices can also increase the interest in physics and facilitate the understanding of the physical phenomena. We have shown some results of an air track experiment using a smartphone. Some of the concepts reinforced with this experiment are acceleration, collisions, momentum, friction, friction coefficient, impulse-

momentum theorem and even elasticity. These experiments can also be useful to learn the importance of sensors accuracy and measuring frequency, as well as the reliability of the used applications.

Acknowledgements - This work has been supported by the University of Valladolid within its Teaching Innovation Program under grants PID2016_64 and PID2016_67.

[1] R Vieyra, C Vieyra, P Jeanjacquot, A Martí, M Monteiro, *Turn your smartphone into a science laboratory*, The Science Teacher **82**, 32 (2015).

[2] SensorMobile `https://play.google.com/store/apps/details?id=com.sensor.mobile`, last accessed September 13, 2017.

[3] PhysicsToolbox `https://play.google.com/store/apps/details?id=com.chrystianvieyra.physicstoolboxsuite`, last accessed September 13, 2017.

20

Smartphone audio port data collection cookbook

Kyle Forinash,[1*]   Raymond Wisman[1†]

The audio port of a smartphone is designed to send and receive audio but can be harnessed for portable, economical, and accurate data collection from a variety of sources. While smartphones have internal sensors to measure a number of physical phenomena such as acceleration, magnetism and illumination levels, measurement of other phenomena such as voltage, external temperature, or accurate timing of moving objects are excluded. The audio port cannot be only employed to sense external phenomena. It has the additional advantage of timing precision; because audio is recorded or played at a controlled rate separated from other smartphone activities, timings based on audio can be highly accurate. The following outlines unpublished details of the audio port technical elements for data collection, a general data collection recipe and an example timing application for Android devices.

I. Audio port technical elements for data collection

The audio port physical interface to the smartphone is a 4-pole jack connecting to some external device. Table 1 presents the American Headset Jack (AHJ) standard used by many smartphones for connections to the 4-pole jack. Figures 1 and 2 present two similar measurement circuits based upon the same resistor-capacitor network, Fig. 1 circuit for temperature measurements [1] and Fig. 2 circuit with a pair of photoresistors acting as gates for timing measurements of a moving object [2].

Although timing and temperature measures are obviously different, the circuits and data collection methods are quite similar. In each circuit, a variable resistor connects the speaker audio out-

*E-mail: kforinas@ius.edu
†E-mail: rwisman@ius.edu

[1] Indiana University Southeast, 4201 Grantline Road, New Albany, Indiana 47150, USA.

Table 1: Audio connections to 4-pole jack.

Tip	Left speaker
1	Right speaker
2	Ground
3	Microphone

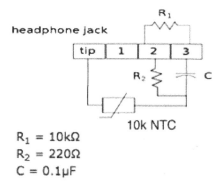

$R_1 = 10k\Omega$
$R_2 = 220\Omega$
$C = 0.1\mu F$

Figure 1: Temperature measurement circuit.

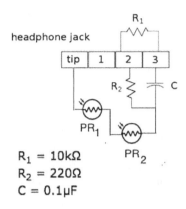

$R_1 = 10k\Omega$
$R_2 = 220\Omega$
$C = 0.1\mu F$

Figure 2: Time measurement circuit.

put to the microphone input; the speaker output is a wave which peak amplitude at the microphone input varies with the variable resistor, greater resistance resulting in lower peak amplitude. Figure 3 illustrates the effect of a photoresistor on the microphone signal amplitude while a moving object momentarily blocks and reduces the illumination reaching the photoresistor - as the illumination decreases, resistance increases and peak amplitude decreases.

The microphone input amplitude is digitally sampled (44100 samples per second is common) over time as displayed in Fig. 4. Each digital value represents the speaker output amplitude as changed by the circuit, then received and digitized as the microphone input.

A recipe for data collection with Fig. 1 and 2 type circuits generally is:

1. Choose a variable resistor to measure some phenomena such as force, humidity, etc.

2. Generate a wave of fixed frequency and peak amplitude through the speaker output and sample it as the microphone input as illustrated in Fig. 4.

3. Detect the significant elements of the microphone input. Significant for timing applications, wave peaks can be detected by the simple method described in Fig. 4.

4. Convert the significant elements to corresponding data. For example, by recording a

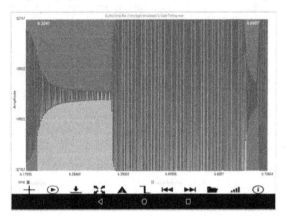

Figure 3: Microphone input signal amplitude versus time graph from circuit in Fig. 2 when measuring an object falling from 1 meter. The amplitude of the wave input at the microphone drops as the object entering the first photoresistor (PR1) at time 6.2041 s and again at time 6.6607 s when entering the second photoresistor (PR2). The measured time of 0.456 s compares favorably with the predicted time of 0.451 s.

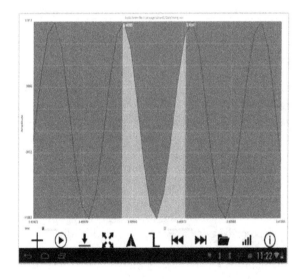

Figure 4: Speaker output of a 4410 Hz sine wave sampled at 44100 samples per second at the microphone, producing the highlighted 10 samples per wave. Simple peak detection is possible by comparing three consecutive samples (x0, x1, x2) where: $0 < x0 < x1 > x2 > 0$ the wave peak occurs at sample x1.

range of peak amplitudes produced by the circuit of Fig. 1 and the corresponding temperatures, an equation fitted to that amplitude vs. temperature data can convert the circuit amplitude output to temperature.

II.   Timing data collection recipe

For timing of a moving object, some form of gate is needed to detect entry and exit. Ideally, gates are binary, they only open or close with no intermediate states, unfortunately reality does not represent the ideal case. For timing applications, the circuit of Fig. 2 includes one or more photoresistors (gates) connected serially so that when all are illuminated (closed) resistance is at its minimum and the peak signal amplitude input to the microphone is at its maximum; hence, when illumination of any one of the photoresistors is reduced, circuit resistance increases and the signal amplitude reduces.

Because the photoresistor resistance does not change instantaneously when illumination changes, as illustrated in Fig. 3 by the gradual peak amplitude drop, and in the case where one photoresistor is illuminated differently than another producing a different amplitude result for each, a peak amplitude threshold must be determined. Above the threshold a gate is considered closed (exited) and below the threshold a gate is considered open (entered). To determine the threshold, the peak amplitude is initially calibrated by illuminating (closing) all gates and incrementally increasing the speaker volume until the microphone peak amplitude approaches its maximum, M. After fixing the volume for M, the microphone peak amplitude for the circuit when any single gate is open (entered), L, is determined by blocking (entering) each gate in turn and recording its lowest peak amplitude. L is set to the greatest peak amplitude of any blocked gate to ensure the threshold is crossed when any gate is blocked. The maximum M and minimum L then set the upper and lower output limits of the circuit peak amplitude. The threshold amplitude, TA, the peak amplitude level that determines when an object enters or exits a gate is: $TA=(M+L)/2$, the amplitude midpoint between M and L. An amplitude crossing the threshold from above is defined as gate entered, while crossing it from below is defined as gate exited.

With the threshold defined, deriving timing data from the microphone audio is both straightforward and accurate. From the general recipe, the practicalities of timing a moving object reduces to:

1. Choose a photoresistor to measure illumination.

2. Generate a sine wave of 4410 Hz and volume setting at M through the speaker and sample at the microphone at 44100 samples per second which corresponds to 10 samples per wave.

3. Detect the wave peaks by the method described in Fig. 4.

4. Convert wave peaks to corresponding timing data by noting the sample number at threshold crossings.

Microphone input sampling occurs at 44100 samples per second or one sample every 0.000023 s (equals 1 sample/44100 samples/second). Accurate timing between events is simply a matter of counting samples between threshold crossings. Gates are entered when a wave peak crosses the threshold from above and the sample number of the peak is recorded at this time. Similarly, gates are exited when a wave peak crosses the threshold from below. The time between entering two gates is then:

$$t = \frac{g_2 - g_1}{\frac{44100 \ samples}{second}}, \qquad (1)$$

where t=time between Gate 1 and 2, g_1 = Gate 1 entry sample number, and g_2 = Gate 2 entry sample number.

Figure 3 illustrates the microphone input while timing the one meter free fall of an object when entering two photoresistor gates. The amplitude of the wave input at the microphone drops as the object entering the first photoresistor (PR1) at time 6.2041 s and again at time 6.6607 s when entering the second photoresistor (PR2). The measured time of 0.456 s = 20110 samples/44100 samples/second compares favorably with the predicted time of 0.451 s.

The prescribed recipe above was followed in the design of the *GateTiming* [3] timing application that measures times of entry and exit at multiple gates. Figure 5 presents the times recorded

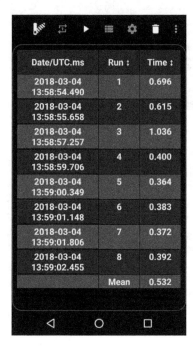

Figure 5: Sample timing results from *GateTiming* application.

Table 2: Times of a block falling one meter.

	Time [s]
	0.453
	0.457
	0.471
	0.451
	0.448
Mean	0.456

during multiple runs through two gates. The *Date/UTC.ms* column is the clock time on entering the first gate, the *Time* column is the time between entering the first and second gate; a more detailed display of each gate entry and exit times is a viewing option. Table 2 lists the times as recorded by the application of a block falling one meter. The variation of measurements is primarily due to slight differences in the height when dropped above the first gate. For example, dropping from 1/2 inch (1.27 cm) above the first gate produced a timing in the range of 0.42 s, the time is lower than the one predicted because the block is traveling a little faster through the gates (about 0.5 mph or 0.80 km/h at the first gate). Other timing variations are likely due to slight changes to the block profile relative to the gates, not always maintaining a stable orientation.

III. Conclusions

Although only timing was closely examined in this paper, a range of audio port data collection applications are possible, including our own Android applications: *AudioTime+* [4], a general purpose tool for signal display and analysis that can examine the behavior of a variable resistor in the circuit of Fig. 1 and 2, *GateTiming* [3], to measure the time an object passes between or through a series of gates, *DCVoltmeter* [5], measurement of 0-10VDC via voltage-to-frequency conversion of microphone input, then converting the frequency to a corresponding voltage value; a simple, active voltage controlled oscillator circuit is required for the conversion. Other applications for data collection from a variety of sources using the basic recipe are also possible. By using this method, data can be collected from almost any sensor based on variable resistance.

[1] K Forinash, R Wisman, *Smartphones - Experiments with an External Thermistor Circuit*, The Physics Teacher **50**, 566 (2012).

[2] K Forinash, R Wisman, *Photogate Timing with a Smartphone*, The Physics Teacher **53**, 234 (2015).

[3] R Wisman, K Forinash, *Mobile Science - GateTiming*, December 2017. Google Play: https://play.google.com/store/apps/details?id=edu.ius.rwisman.gatetiming

[4] R Wisman, K Forinash, *Mobile Science - AudioTime+*, November 2013. Google Play: https://play.google.com/store/apps/details?id=edu.ius.audiotimeplus

[5] R Wisman, K Forinash, *Mobile Science - DCVoltmeter*, January 2016. Google Play: https://play.google.com/store/apps/details?id=edu.ius.dcvoltmeter

21

Temperature-dependent transport measurements with Arduino

A. Hilberer,[1] G. Laurent,[1] A. Lorin,[1] A. Partier,[1] J. Bobroff,[2] F. Bouquet,[2]* C. Even,[2]
J. M. Fischbach,[1] C. A. Marrache-Kikuchi,[3]† M. Monteverde,[2] B. Pilette,[1] Q. Quay[2]

The current performances of single-board microcontrollers render them attractive, not only
for basic applications, but also for more elaborate projects, amongst which are physics
teaching or research. In this article, we show how temperature-dependent transport measurements can be performed by using an Arduino board, from cryogenic temperatures
up to room temperature or above. We focus on two of the main issues for this type of
experiments: the determination of the sample temperature and the measurement of its
resistance. We also detail two student-led experiments: evidencing the magnetocaloric
effect in Gadolinium and measuring the resistive transition of a high critical temperature
superconductor.

I.   Introduction

The development of single-board microcontrollers
and single-board computers has given physicists
access to a large variety of inexpensive experi-
mentation that can be used either to design simple test benches, to put together set-ups for class
demonstration or to devise student practical work.
Moreover, specifications of single-board components are now such that, although they cannot rival
with state-of-the-art scientific equipment, one can
nonetheless derive valuable physical results from
them.

*E-mail: frederic.bouquet@u-psud.fr
†E-mail: claire.marrache@u-psud.fr

[1]   Magistère de Physique Fondamentale, Département
de Physique, Univ. Paris-Sud, Université Paris-Saclay,
91405 Orsay Campus, France.

[2]   Laboratoire de Physique des Solides, CNRS, Univ. Paris-
Sud, Université Paris-Saclay, 91405 Orsay Campus,
France.

[3]   CSNSM, Univ. Paris-Sud, CNRS/IN2P3, Université
Paris-Saclay, 91405 Orsay, France.

We will here focus on the Arduino microcontroller board [1]. Let us note that other boards,
such as MBED, Hawkboard, Rasberry Pi, or
Odroid to cite but a few, exist which may be
cheaper and/or have better characteristics than Arduino. In our case, we have employed Arduino
boards to take advantage of the important user's
community. This has been an important selling
point for the students with whom we are working.

Indeed, our experience with Arduino is primarily based on undergraduate project-based physics
labs [2] we have initiated within the Fundamental Physics Department of Université Paris Sud for
students to gain a first hands-on practice of experimental physics. In these practicals, students are
asked to choose a subject they want to study during
a week-long project. They then have to design and
build the experiment with the equipment available
in the lab. The aim is not only to study a physical phenomenon, but to do so by using inexpensive
materials and low-cost boards.

In this article, we will describe two projects that
have been developed by third year students. The
first one aimed at quantifying the magnetocaloric

effect in Gadolinium and the second one, which has been popular amongst students, consisted in measuring the resistive transition of a high critical temperature superconductor (HTCS). However, the techniques to do so can more generally be used for any experiment involving the measurement of a low voltage while varying the set-up temperature. In particular, they could be applied to simple transport characterization of samples in research laboratories.

In the following, we will focus on two important issues for this kind of measurements: thermometry and thermal anchoring on the one hand, and measuring resistances on the other.

II. Determining the temperature of an object using microcontrollers

One of the experimental control parameters that is most commonly used to make a physical system properties vary is temperature. There is a wide variety of temperature sensors, depending on the temperature range of interest. The aim of this paper is not to list those, but rather, to focus on the most ordinarily found sensors compatible with an Arduino read out. We will also review some basic techniques to ensure a proper thermal contact between the sample and the thermometer.

i. Sensors types

a. Built-in Arduino sensors

There are a number of temperature sensors that are generally sold with standard Arduino kits. The Arduino Starter Kit, for instance, comes with a TMP36 low voltage temperature sensor [3] (available for about $1 if purchased separately) whose operating principle is based on the temperature-dependence of the voltage drop across a diode.

The advantage of this type of thermometer is that it can be directly plugged into Arduino without any additional electrical circuit. Furthermore, provided that the corresponding library is downloaded, the temperature is straightforwardly read via the computer interface in °C, so that no calibration is needed.

However, these sensors are limited in accuracy and operation: the TMP36 sensor for example has a ±2°C precision over the −40°C to +125°C range

where it can operate. If they are extremely convenient for non-demanding temperature read-outs, such as students atmospheric probes for example [4], they are not adapted to the precision needed for most research lab experiments.

b. Thermocouple

Thermocouples are cheap and robust thermal sensors that are industrially available for about $15, and which cover a wide range of temperatures (for example from −200°C to +1250°C for a type K thermocouple [5]). They are one of the few thermometers that are reliable at temperatures much higher than room temperature.

Thermocouples are also extremely convenient to measure the temperature of small-sized samples. Indeed, only the hot junction between the two metals needs to be in contact with the region where the temperature is to be monitored. On the other hand, the quality of the readings will strongly depend on how thermally stable the cold junction is, and the temperature measurement is less precise than using a thermistor. Indeed, the voltage to be measured is small: the sensitivity of a thermocouple is of the order of tens of microvolts per Kelvin, and it decreases when the temperature decreases (a type-K thermocouple has a sensitivity of 40 μV.K^{-1} at room temperature but a sensitivity of 10 μV.K^{-1} at liquid nitrogen temperature).

Let us note that an amplification of the voltage signal is then needed to read the temperature with Arduino. Some chips provide a ready-to-use thermocouple amplifier for microcontrollers (such as the MAX31856 breakout with a resolution of a quarter of a kelvin when using the Adafruit library [6] and an accuracy of a few kelvins). Better sensitivity could be achieved with a home-made amplifier (see below) and some care.

c. Platinum thin resistive films

Platinum thin films are practical and very reliable resistive thermometers typically working from 20 K to 700 K [7]. They are therefore suited for cryogenic applications – at least down to liquid nitrogen temperatures – as well as for moderate heating. The advantage of this sensor is that its response is entirely determined by the value of its resistance at 0°C [8]. The most commonly used platinum resistance is the so-called Pt100 which has a resistance of 100 Ω at 0°C and costs approximately \$3 to \$5. These thermal sensors have a typical precision of about 20 mK up to 300 K and about 200 mK above room temperature. Moreover, their magnetic field-dependent temperature errors are well-known [7].

It is possible to mount those resistances on a dedicated Arduino resistance-to-temperature converter such as MAX31865 [9], but it is often simpler to plainly measure the resistance with a dedicated electrical circuit as will be explained in section III. This is particularly convenient for low or high temperature measurements for which the Arduino board cannot be at the same temperature as the thermometer and the sample are.

ii. Thermal anchoring

For the temperature measurement to be relevant, the thermometer must be in good thermal contact with the sample. How to achieve a good thermal anchoring is a subject of investigation in itself, but in this section we will outline a few standard techniques, focusing on the low temperature case.

To cool down a sample at low temperatures, one could use a Peltier module, but the simplest – and not so expensive – way is to use liquid nitrogen. Some basic safety measures have to be taken to manipulate this cryogenic fluid: use protection glasses, gloves, work in a well-ventilated room and, above all, ensure that it is poured into a vessel that is not leak-tight to allow natural evaporation of liquid nitrogen and avoid pressure build-up in the vessel. Once these precautions are observed, the manipulation is relatively safe.

To ensure that the thermometer indeed probes the sample temperature, the most obvious technique is to solidly attach the thermometer to the sample using good thermal conductors. The thermal sensor can, for instance, be glued onto a copper sample holder. The glue then has to retain its properties at the probed temperature range. In the low temperature case, one frequently uses GE 7031 varnish which sustains very low temperatures and can easily be removed with a solvent. Alternatively, the thermometer could be mechanically fixed with a spring-shaped material whose elasticity is maintained at a low temperature, such as CuNi sheets. Upon cooling, the spring-shaped material will continue to apply pressure onto the thermometer, thus

ensuring a good mechanical and thermal contact with the sample holder.

Another method is to thermally insulate the thermometer and the sample from the outside world, while putting them in contact with a common thermal bath. This can be done by inserting them into a container filled with glass beads of a few millimeters in diameter [10] (inset of Fig. 7), or alternatively, sand. These materials provide a good thermal insulation of the {sample+thermometer} system from the outside world while allowing for an important thermal inertia. Moreover, when working at temperatures close to 77 K, they limit the liquid nitrogen evaporation so that the temperature increases back to room temperature only very slowly: typically for a volume 1 L of beads that is initially immersed in liquid nitrogen, the temperature reaches back 300 K in 3 to 4 hours. The heat exchange between the sensor and the sample is then guaranteed through the evaporated N$_2$ gas, thus ensuring the temperature is homogeneous within the entire volume. An alternative method for achieving good thermal contact between the sensor and the sample through gas exchange is explained in Ref. [11].

III. Measuring resistances with microcontrollers

Microcontroller inputs give a reading of electric potentials. Measuring resistances is then slightly more complicated than plugging a resistance into an ohmmeter. For educational purposes, this is actually rather valuable since it gives students the opportunity to experiment with the notion of resistance and to realize that even the simplest measurement may present some challenge. In the following, we will present standard methods to measure resistances and we will particularly focus on the low-resistance case.

i. Current-Voltage measurement

The simplest set-up for measuring a standard resistance is the voltage divider set-up represented in Fig. 1: the resistance of interest R_0 is put in series with a reference resistance R_{ref}. The voltage drop across both resistances is controlled by the board 5 V output. The potential V_1 can be read by one

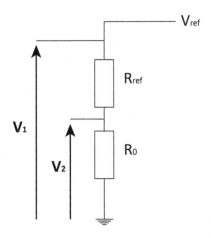

Figure 1: Schematic representation of the current-voltage set-up.

of the microcontroller's inputs and should be close to 5 V. The potential V_2 – read by a second input – corresponds to the voltage drop across the unknown resistance. R_0 can then be determined through the simple relation:

$$R_0 = \frac{V_2}{V_1 - V_2} R_{ref} \qquad (1)$$

The monitoring of V_1 allows for a better precision through a direct monitoring of the current. Arduino's 5 V output sometimes varies in time. To have a better stabilization of the voltage, it may be useful to use an external power source for the microcontroller and not use the computer's USB output. Let us note that, if $R_{ref} \gg R_0$, the current through the circuit can be considered to be constant, which is often very convenient when the resistance measurement does not require a large precision.

This method presents several drawbacks when dealing with small values of R_0: since the ultimate resolution of an Arduino UNO board is of about 1 mV with $V_{ref} = 1.1$ V, one cannot measure R_0 smaller than about $2 \times 10^{-4} R_{ref}$. In the case of a standard commercial HTCS sample for instance, the normal state resistance is often of the order of a few tens of mΩ. To observe the resistance drop across the critical temperature T_c of a superconductor, R_{ref} should then be of the order of a few Ω. Such resistances are commercially available or, alternatively, can be custom-made with a relatively

good precision (of the order of a few mΩ) by using a long string of copper wire (commercially available Cu wires of 0.2 mm in diameter have a resistance of about 0.5 Ω/m for example). However, unless V_2 is amplified, the precision of the measurement is not optimal. Moreover, using this method to measure small resistances leads the circuit current to exceed the maximum current allowed at the microcontroller's output. In the following, we will see another method to measure small resistances.

ii. Wheatstone bridge

Another resistance determination method, which can achieve a good precision, is the Wheatstone bridge. The principle of the measurement is illustrated in Fig. 2. R_1 and R_3 are fixed value resistances, while R_2 is a tunable resistance and R_0 the resistance of interest. The potentials V_1 and V_2 are then related by:

$$V_2 - V_1 = \left(\frac{R_2}{R_1 + R_2} - \frac{R_0}{R_0 + R_3} \right) V_e \qquad (2)$$

The bridge is co-called "balanced" when R_2 is tuned such that V_1 and V_2 are equal. The resistances are then related through:

$$R_0 = \frac{R_2 R_3}{R_1} \qquad (3)$$

The precision that can be achieved through this method and when using a microcontroller is about the same as for the current-voltage measurement method. However, this method is not very practical when dealing with resistances R_0 that vary, since the bridge has to be maintained close to balance at each measurement point. In particular, it is not well suited for the measurement of a superconductor's resistive transition.

iii. Voltage amplifier

The most practical solution for the measurement of small voltages – and hence small resistances – is the amplification of the potential difference across the resistance. This can be done via standard voltage amplification set-ups using operational amplifiers, either in single-ended or differential input configurations. In the single-ended case, illustrated in Fig. 3, the output potential is given by:

$$V_{out} = 1 + \frac{R_2}{R_1} V_{in} \qquad (4)$$

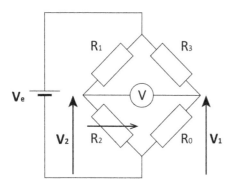

Figure 2: Schematic representation of the Wheatstone bridge: R_1 and R_3 are fixed resistances, R_2 is a tunable resistance, and R_0 is the resistance of interest.

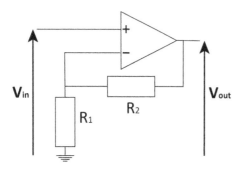

Figure 3: Voltage amplification.

The input voltage V_{in} can then be amplified at will, depending on the ratio $\frac{R_2}{R_1}$. The output voltage V_{out} can then be read by the microcontroller.

This amplification method has a much larger precision than the previously mentioned methods, it does not require tuning at each data point and can be used to measure any small voltage: the voltage drop across a superconductor, but also the difference of potential across a thermocouple, or to derive thermoelectric coefficients (Seebeck or thermopower).

Going beyond this simple amplification method requires substantially more work. One possible method is to fabricate a microcontroller-based lock-in amplifier, as demonstrated in Ref. [12].

iv. Using another ADC than Arduino's

The specifications of Arduino Digital-Analog Converter are often the main limitation in the above measurements. As already mentioned, the Arduino ADC provides – at best – 10 bits on the 1.1 V internal reference voltage, and can only measure a voltage in single-ended configurations.

One alternative would be to use another microcontroller, with a better ADC. For example, the low-cost FRDM-KL25Z from NXP [14] provides a ADC that can measure a voltage either in single-ended or in differential input configurations with 16 bits on 3.3 V.

The ease-of-use and the large users community can be a strong motivation to keep Arduino as your board of choice. In which case, a second solution would be to use an external ADC when better resolution or a differential mode configuration is needed. For example, we have tested the ADS1115 chip [15]: this external ADC can measure 4 single channels or 2 differential channels with 16 bits on 4.1 V. The possibility of a preamplification up to 16 times brings the resolution down to 8 μV per bit instead of the standard 5 mV (or 1 mV with the 1.1 V internal reference).

The possibility of measuring a voltage in a differential mode configuration with a resolution better than 10 μV are two important advantages that open many interesting possibilities for physics measurements: for instance, measuring a strain gauge or a resistance in a four-wire configuration, or measuring directly a thermocouple or the resistance of a superconductor across the transition.

The main drawback to this method is that it is not as easy as using the Arduino ADC: a library should be installed first (but good tutorials can be found online, see for example Ref. [15]). Also, an external ADC is generally not as robust as the Arduino's ADC, and the user should carefully monitor the voltage input so as not to damage the ADC.

IV. Evidencing a magnetocaloric effect with microcontrollers

To illustrate these methods, let us detail the magnetocaloric effect that we have measured. This effect consists in the temperature change occurring when a magnetic material is placed in a varying

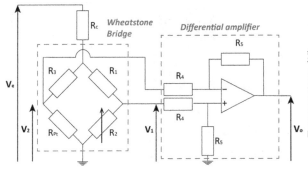

Figure 4: Measurement of the resistance of a Pt resistive thermometer with a Wheatstone bridge and amplified by an opamp-based circuit.

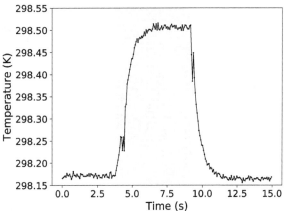

Figure 5: Magnetocaloric effect in a Gd sample submitted to a 0.51 T magnetic field (blue background) before going back to the zero-field situation (white background). Each data point corresponds to the average of 50 measurements. The noise level is of the order of 10 mK.

magnetic field. A more detailed explanation of this the phenomenon can be found in Ref. [16].

In our case, Gadolinium (Gd) was chosen for its paramagnetic properties and its Curie temperature close to room temperature ($T_{Curie} = 292$ K). At a temperature of about 298 K, a 2.242 g Gd sample was submitted to the magnetic field created by a neodymium magnet of maximum value 0.51 T.

In this experiment, the challenge was to measure the small temperature difference induced by the application of a magnetic field. To this effect, a Pt100 thermistor was put in good thermal contact with the Gd sample via thermal paste. The resistance change was measured by a Wheatstone bridge with the following characteristics: $R_1 = R_3 = 100$ Ω, R_2 has been set at 108 Ω to be close to balance at the considered temperature and $V_e = 5$ V via Arduino's internal source. An additional resistance $R_c = 800$ Ω was placed in series to limit the current going through the Pt100, thus avoiding heating the thermometer. V_e is then replaced by $\frac{R_1+R_2}{R_1+R_2+2R_c}V_e$ in Eq. (2). The off-balance difference of potential $V_2 - V_1$ was differentially amplified with a gain of 100 ($R_4 = 1.5$ kΩ and $R_5 = 150$ kΩ). The voltage V_{out} was then read by the board (Arduino Mega in this case) using 2.56 V as Arduino's ADC reference voltage [17]. The overall read-out circuit is schematically shown in Fig. 4. The temperature is then inferred knowing that, in the [273 K - 323 K] range, the Pt100 response can be linearly fitted by:

$$T[\text{K}] = 2.578R_{Pt}[\Omega] + 15.35 \qquad (5)$$

As illustrated in Fig. 5, the magnetocaloric ef-

fect is clearly visible with an amplitude of about $\Delta T \simeq 0.33 \pm 0.01$ K and a time scale of a few seconds. The resolution of the setup corresponds to 50 mK (18 mΩ). Each data point in Fig. 5 corresponds to an average of 50 measurements so that the effective noise that can be observed is of about 10 mK, or about 5 mΩ in resistance. This yields a relative precision for the measurement of a few 10^{-5}, which is remarkable given the simplicity of the apparatus . When the magnet is taken away from the Gd sample, the temperature decreases back to its initial value, as predicted by the isentropic character of the magnetocaloric effect.

V. Measuring a superconducting resistive transition with microcontrollers

For the second experiment, we would like to detail is the measurement of the superconducting resistive transition of a HCTS. Indeed, in such compounds, the critical temperature T_c below which the sample is superconducting and exhibits zero resistance is larger than 77 K. The transition can therefore easily be observed by cooling the sample down to liquid nitrogen temperature and warming it back up to

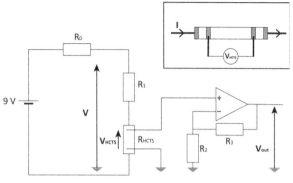

Figure 6: Amplification of the voltage drop across a superconducting sample. Inset: geometry of the HCTS sample.

room temperature.

In the present case, the HCTS is a commercial $Bi_2Sr_2Ca_2Cu_3O_{10}$ sample which specifications indicate a critical temperature $T_c = 110$ K at midtransition point and a room temperature resistivity of 1 mΩ.cm [18]. In this case, the experimental challenge is therefore to measure very small resistances with good precision. To achieve this, it is essential to adopt a four-wire measurement configuration for the superconductor, as schematized in the inset of Fig. 6. Indeed, in this way, no contact resistances or connection wires contribute to the measured resistance. Moreover, the voltage drop across the superconductor has been amplified by a factor of 480 by a single-ended operational amplifier set-up as shown in Fig. 6. The current going through the superconductor is fixed by $R_0 = 110.0$ $\Omega \gg R_1, R_{HCTS}$ and is experimentally measured via the potential V read at the extremity of a home-made resistance $R_1 = 1.18$ Ω, made out of copper wire.

The temperature has been measured with a Pt100 resistive thermometer using the set-up shown in Fig. 1 with $R_{ref} = 217.3$ Ω and a reference voltage of $V_{ref} = 3.3$ V provided by one of Arduino UNO's internal sources.

Both the sample and the thermometer have been attached to a printed circuit board and have been wrapped in cotton to ensure temperature homogeneity. The ensemble was placed in a polystyrene container filled with glass beads (inset of Fig. 7). Liquid nitrogen was then poured into the container

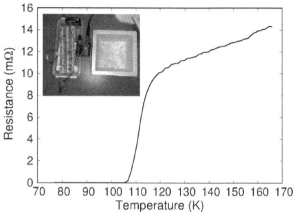

Figure 7: Resistive transition of a superconductor measured with a voltage amplification. Inset: experimental setup. The blue polystyrene container is filled with glass beads. Both the superconductor and the Pt100 thermometer are immersed inside with liquid nitrogen.

and the temperature of the ensemble was let to increase back to room temperature while recording the data.

In this manner, we have measured the resistive transition given in Fig. 7. The experimental data have been averaged by a convolution with a Gaussian of half width 0.4 K to take into account the error in the temperature measurement. As it can be seen, the resolution of the measurement is of the order of 0.1 mΩ for the superconductor's resistance. The latter is actually dominated by the thermal gradient that may exist between the sample and the thermometer if the operator does not carefully check that both are close to one another or if the container is forcefully warmed-up (with a hair dryer for instance). Nonetheless, the precision of the measurement is good (< 1% relative uncertainty) and the measured mid-point T_c is of 112 K ± 2K, very close to the value given by the specifications.

VI.   Conclusion

In conclusion, we have shown that standard temperature and resistance measurement methods could be adapted to microcontrollers. The performances that are then attainable are sufficient to probe with reasonable sensitivity a large range of physics phenomena such as thermoelectric effects, temperature-dependence of the resistivity, Hall effect, magnetocaloric effect, etc. We have illustrated this with the measurements of magnetocaloric effect in Gadolinium and of the resistive transition of a high critical temperature superconductor. We believe that the scope of inexpensive, transportable and easy-to-build experiments that are accessible through the use of single-board microcontrollers is continuously expanding and, in some cases, can now even replace standard characterization methods in research laboratories. Furthermore, they provide a large range of opportunities to devise innovative teaching activities that enhances students involvement.

Acknowledgements - We thank all the students who have participated in the Arduino-based labworks. We thankfully acknowledge Patrick Puzo for welcoming this project-based teaching within the Magistère de Physique d'Orsay curriculum. This work has been supported by a "Pédagogie Innovante" grant from IDEX Paris-Saclay.

Author contributions - A. H. and G. L. designed and conducted the measurements of the magnetocaloric effect. A. L. and A. P. designed and conducted the measurements of the superconducting resistive transition. J. B., F. B., C. E., J. M. F., C. A. M-K., M. M., B. P. and Q. Q. have supervised and coordinated the studies. All authors have contributed to the redaction of the paper.

[1] *Arduino Project Website,*
https://www.arduino.cc/

[2] F Bouquet, J Bobroff, M Fuchs-Gallezot, L Maurines, *Project-based physics labs using low-cost open-source hardware,* Am. J. Phys. **85**, 216 (2017).

[3] *TMP36 specifications,*
https://www.arduino.cc/en/uploads/Main/TemperatureSensor.pdf

[4] V K Merhar, R Capuder, T Marošević, S Artač, A Mozer, M Štekovič, *Vič goes to near space,* The Physics Teacher **54**, 482 (2016).

[5] *Thermocouple operation temperature range,*
https://www.omega.co.uk/techref/colorcodes.html

[6] *Thermocouple amplifier,*
https://www.adafruit.com/product/3263

[7] *Characteristics of Pt thermometers are available at the following website,*
https://www.lakeshore.com/Documents/LSTC_Platinum_l.pdf

[8] *Platinum RTD sensor resistance to temperature conversion tables are available at the following websites,*
https://www.omega.com/techref/pdf/z252-254.pdf,
www.intech.co.nz/products/temperature/typert/RTD-Pt100-Conversion.pdf

[9] *RTD-to-Digital converter,*
https://www.maximintegrated.com/en/products/sensors/MAX31865.html

[10] G Ireson, *Measuring the transition temperature of a superconductor in a pre-university laboratory,* Physics Education **41**, 556 (2006).

[11] L M León-Rossano, *An inexpensive and easy experiment to measure the electrical resistance of high-Tc superconductors as a function of temperature,* Am. J. Phys. **65**, 1024 (1997).

[12] K D Schultz, *Phase-sensitive detection in the undergraduate lab using a low-cost microcontroller,* Am. J. Phys. **84**, 557 (2016).

[13] R Henaff, G Le Doudic, B Pilette, C Even, J M Fischbach, F Bouquet, J Bobroff, M Monteverde, C A Marrache-Kikuchi, *A study of kinetic friction: The Timoshenko oscillator,* Am. J. Phys. **86**, 174 (2018).

[14] https://os.mbed.com/platforms/KL25Z/

[15] https://www.adafruit.com/product/1085

[16] V Percharsky, K A Gscheider Jr, *Magnetocaloric effect and magnetic refrigeration,* J. Magn. Magn. Mater. **200**, 44 (1999).

[17] https://www.arduino.cc/reference/en/language/functions/analog-io/analogreference/

[18] https://shop.can-superconductors.com/index.php?id_product=11&controller=product

22

Calculation of almost all energy levels of baryons

Mario Everaldo de Souza [1*]

It is considered that the effective interaction between any two quarks of a baryon can be approximately described by a simple harmonic potential. The problem is firstly solved in Cartesian coordinates in order to find the energy levels irrespective of their angular momenta. Then, the problem is also solved in polar cylindrical coordinates in order to take into account the angular momenta of the levels. Comparing the two solutions, a correspondence is made between the angular momenta and parities for almost all experimentally determined levels. The agreement with the experimental data is quite impressive and, in general, the discrepancy between calculated and experimental values is below 5%. A couple of levels of Δ, N, $\Sigma^{\pm}$, and Ω present discrepacies between 6.7% and 12.5% [$N(1655)$, $N(1440)$, $N(1675)$, $N(1685)$, $N(1700)$, $N(1710)$, $N(1720)$, $N(1990)$, $N(2600)$, $\Delta(1700)$, $\Delta(2000)$, $\Delta(2300)$, $\Sigma^{\pm}(1189)$, $\Lambda(1520)$, $\Omega(1672)$ and $\Omega(2250)$].

I. Introduction

There are several important works that deal with the calculation of the energy levels of baryons. One of the most important ones is the pioneering work of Gasiorowicz and Rosner [1] which has calculation of baryon energy levels and magnetic moments of baryons using approximate wavefunctions. Another important work is that of Isgur and Karl [2] which strongly suggests that non-relativistic quantum mechanics can be used in the calculation of baryon spectra. Other very important attempts towards the understanding of baryon spectra are the works of Capstick and Isgur [3], Bhaduri et al. [4] Murthy et al. [5], Murthy et al. [6] and Stassat et al. [7]. Still another important work that attempts to describe baryon spectra is the recent work of Hosaka, Toki and Takayama [8] that makes use of a non-central harmonic potential (called by the authors *the deformed oscillator*) and is able to describe many levels. This present work describes many more levels and is more consistent in the characterization of angular momenta and parities of levels. It is an updated version of the pre-print of Ref. [9].

II. The approximation for the effective potential

The effective potential between any two quarks of a baryon is not known and thus a couple of different potentials can be found in the literature. Of course, the effective potential is the result of the attractive and repulsive forces of QCD and is completely justified because, as it is well known that the strong

*E-mail: mariodesouza.ufs@gmail.com

[1] Universidade Federal de Sergipe, Departamento de Física, Av. Marechal Rondon, s/n, Campus Universitário, Jardim Rosa Elze 49100-000, São Cristovão, Brazil.

force becomes repulsive for very short distances, and thus repulsion and attraction can form a potential well that can be approximated with a harmonic potential about the equilibrium point. Taking into consideration the work of Isgur and Karl [2] about the use of non-relativistic quantum mechanics, and considering that the three quarks of a baryon are always on a plane, we consider that the system can be approximately described by three non-central and non-relativistic linear harmonic potentials. This is a calculation quite different from those found in the literature and explains almost all energy levels of baryons.

III. Calculation in Cartesian coordinates and comparison with experimental data

The initial calculation, in which we have used Cartesian coordinates, does not, of course, consider the angular momentum of the system, that is, it does not take into account the symmetries of the system. This calculation is important for the iden-

tification of the energy levels given by the experimental data, and for the assignment of the angular momenta later on. Also, it allows the prediction of many energy levels. Since each oscillator has two degrees of freedom, the energy of the system of 3 quarks is given by [10]

$$
\begin{aligned}
E_{n,m,k} =& h\nu_1(n+1) + h\nu_2(m+1) \\
& + h\nu_3(k+1)
\end{aligned}
\tag{1}
$$

where $n, m, k = 0, 1, 2, 3, 4, \ldots$ Of course, we identify $h\nu_1$, $h\nu_2$, $h\nu_3$ with the ground states of the corresponding energy levels of baryons, and thus $h\nu_1$, $h\nu_2$, $h\nu_3$ are equal to the masses of constituent quarks. Since we do not take isospin into account, we cannot distinguish between N and Δ states, or between Σ and Λ states. The experimental values for the baryon levels were taken from Particle Data Group (Nakamura et al. [11]). The masses of constituent quarks are taken as $m_u = m_d = 0.31$ GeV, $m_s = 0.5$ GeV, $m_c = 1.7$ GeV, $m_b = 5$ GeV, and $m_t = 174$ GeV. We have, thus, the following formulas (see Table 1) for the energy levels of all known baryons up to now:

Baryons	Formulas for the energy levels (in GeV)
N, Δ^-, Δ^{++}	$E_{n,m,k} = 0.31(n+m+k+3)$
Λ^0, Σ^+, Σ^0, Σ^-	$E_{n,m,k} = 0.31(n+m+2) + 0.5(k+1)$
Ξ^0, Ξ^-	$E_{n,m,k} = 0.31(n+1) + 0.5(m+k+2)$
Ω^-	$E_{n,m,k} = 0.5(n+m+k+3)$
Λ_c^+, Σ_c^+, Σ_c^{++}, Σ_c^0	$E_{n,m,k} = 0.31(n+m+2) + 1.7(k+1)$
Ξ_c^0, Ξ_c^+	$E_{n,m,k} = 0.31(n+1) + 0.5(m+1) + 1.7(k+1)$
Ω_c^0	$E_{n,m,k} = 0.5(n+m+2) + 1.7(k+1)$
X_{cc}	$E_{n,m,k} = 0.31(n+1) + 1.7(m+k+2)$
Λ_b^0	$E_{n,m,k} = 0.31(n+m+2) + 5(k+1)$
Ξ_b^0, Ξ_b^-	$E_{n,m,k} = 0.31(n+1) + 0.5(m+1) + 5(k+1)$
Ω_b^-	$E_{n,m,k} = 0.5(n+m+2) + 5(k+1)$

Table 1: Formulas for most energy levels of all baryons.

In Tables 1 to 11, E_C is the calculated value by the above formulas, E_M is the measured value and the error is given by $Error = 100\% \times |E_M - E_C|/E_C$.

Within the scope of our simple calculation, many levels are degenerate, of course. Further calculations, taking into account spin-orbit and spin-spin

effects, should lift part of the degeneracy. We notice that these effects are quite complex. States such as $1.70(N)D_{13}$ and $1.70(\Delta)D_{33}$ clearly show that isospin does not play an important role in the splitting of the levels. In general, the $Error$ is below 5%.

State(n,m,k)	E_C (GeV)	E_M (GeV)	$Error(\%)$	$L_{2I,2J}$	Parity
$0,0,0$	0.93	0.938(N)	0.9	P_{11}	$+$
$n+m+k=1$	1.24	1.232(Δ)	0.6	P_{33}	$+$
$n+m+k=2$	1.55	1.44(N)	7.1	P_{11}	$+$
$n+m+k=2$	1.55	1.52(N)	1.9	D_{13}	$-$
$n+m+k=2$	1.55	1.535(N)	1.0	S_{11}	$-$
$n+m+k=2$	1.55	1.6(Δ)	3.1	P_{33}	$+$
$n+m+k=2$	1.55	1.62(Δ)	4.5	S_{31}	$-$
$n+m+k=2$	1.55	1.655(N)	6.7	S_{11}	$-$
$n+m+k=2$	1.55	1.675(N)	8.1	D_{15}	$-$
$n+m+k=2$	1.55	1.685(N)	8.7	F_{15}	$+$
$n+m+k=2$	1.55	1.70(N)	9.7	D_{13}	$-$
$n+m+k=2$	1.55	1.70(Δ)	9.7	D_{33}	$-$
$n+m+k=2$	1.55	1.72(N)	11.0	P_{13}	$+$
$n+m+k=3$	1.86	1.71(N)	8.1	P_{11}	$+$
$n+m+k=3$	1.86	1.90(N)	2.2	P_{13}	$+$
$n+m+k=3$	1.86	1.90(Δ)	2.2	S_{31}	$-$
$n+m+k=3$	1.86	1.905(Δ)	2.4	F_{35}	$+$
$n+m+k=3$	1.86	1.91(Δ)	2.7	P_{31}	$+$
$n+m+k=3$	1.86	1.92(Δ)	3.2	P_{33}	$+$
$n+m+k=3$	1.86	1.93(Δ)	3.8	D_{35}	$-$
$n+m+k=3$	1.86	1.94(Δ)	4.3	D_{33}	$-$
$n+m+k=3$	1.86	2.0(N)	7.5	F_{15}	$+$
$n+m+k=4$	2.17	1.95(Δ)	10.1	F_{37}	$+$
$n+m+k=4$	2.17	1.99(N)	8.3	F_{17}	$+$
$n+m+k=4$	2.17	2.00(Δ)	7.8	F_{35}	$+$
$n+m+k=4$	2.17	2.08(N)	4.1	D_{13}	$-$
$n+m+k=4$	2.17	2.09(N)	3.7	S_{11}	$-$
$n+m+k=4$	2.17	2.10(N)	3.2	P_{11}	$+$
$n+m+k=4$	2.17	2.15(Δ)	0.9	S_{31}	$-$
$n+m+k=4$	2.17	2.19(N)	0.9	G_{17}	$-$
$n+m+k=4$	2.17	2.20(N)	1.4	D_{15}	$-$
$n+m+k=4$	2.17	2.20(Δ)	1.4	G_{37}	$-$
$n+m+k=4$	2.17	2.22(N)	2.3	H_{19}	$+$
$n+m+k=4$	2.17	2.225(N)	2.5	G_{19}	$-$
$n+m+k=4$	2.17	2.3(Δ)	6.0	H_{39}	$+$
$n+m+k=5$	2.48	2.35(Δ)	5.2	D_{35}	$-$
$n+m+k=5$	2.48	2.39(Δ)	3.6	F_{37}	$+$
$n+m+k=5$	2.48	2.40(Δ)	3.2	G_{39}	$-$
$n+m+k=5$	2.48	2.42(Δ)	2.4	$H_{3,11}$	$+$
$n+m+k=6$	2.79	2.60(N)	6.8	$I_{1,11}$	$-$
$n+m+k=6$	2.79	2.70(N)	3.2	$K_{1,13}$	$+$
$n+m+k=6$	2.79	2.75(Δ)	1.4	$I_{3,13}$	$-$
$n+m+k=7$	3.10	2.95(Δ)	4.8	$K_{3,15}$	$+$
$n+m+k=7$	3.10	3.10(N)	0	$L_{1,15}$	$-$
$n+m+k=8$	3.21	?	?	?	?

Table 2: Energy levels of baryons N and Δ.

State (n,m,k)	E_C (GeV)	E_M (GeV)	$Error$ (%)	$L_{2I,2J}$	Parity
$0,0,0$	1.12	$1.116(\Lambda)$	0.4	P_{01}	$+$
$0,0,0$	1.12	$1.189(\Sigma^\pm)$	6.2	P_{11}	$+$
$0,0,0$	1.12	$1.193(\Sigma^0)$	6.5	P_{11}	$+$
$n+m=1, k=0$	1.43	$1.385(\Sigma)$	3.2	P_{13}	$+$
$n+m=1, k=0$	1.43	$1.405(\Lambda)$	1.7	S_{01}	$-$
$n+m=1, k=0$	1.43	$1.48(\Sigma)$	3.5	?	?
$0,0,1$	1.62	$1.52(\Lambda)$	6.2	D_{03}	$-$
$0,0,1$	1.62	$1.56(\Sigma)$	3.7	?	$+$
$0,0,1$	1.62	$1.58(\Sigma)$	2.5	D_{13}	$-$
$0,0,1$	1.62	$1.60(\Lambda)$	1.2	P_{01}	$+$
$0,0,1$	1.62	$1.62(\Sigma)$	0	S_{11}	$-$
$0,0,1$	1.62	$1.66(\Sigma)$	2.5	P_{11}	$+$
$0,0,1$	1.62	$1.67(\Lambda)$	3.1	S_{01}	$-$
$n+m=2, k=0$	1.74	$1.67(\Sigma)$	4.0	D_{13}	$-$
$n+m=2, k=0$	1.74	$1.69(\Lambda)$	2.9	D_{03}	$-$
$n+m=2, k=0$	1.74	$1.69(\Sigma)$	2.9	?	?
$n+m=2, k=0$	1.74	$1.75(\Sigma)$	0.6	S_{11}	$-$
$n+m=2, k=0$	1.74	$1.77(\Sigma)$	1.7	P_{11}	$+$
$n+m=2, k=0$	1.74	$1.775(\Sigma)$	2.0	D_{15}	$-$
$n+m=2, k=0$	1.74	$1.80(\Lambda)$	3.4	S_{01}	$-$
$n+m=2, k=0$	1.74	$1.81(\Lambda)$	4.0	P_{01}	$+$
$n+m=2, k=0$	1.74	$1.82(\Lambda)$	4.6	F_{05}	$+$
$n+m=2, k=0$	1.74	$1.83(\Lambda)$	5.2	D_{05}	$-$
$n+m=1, k=1$	1.93	$1.84(\Sigma)$	4.7	P_{13}	$+$
$n+m=1, k=1$	1.93	$1.88(\Sigma)$	2.6	P_{11}	$+$
$n+m=1, k=1$	1.93	$1.89(\Lambda)$	2.1	P_{03}	$+$
$n+m=1, k=1$	1.93	$1.915(\Sigma)$	0.8	F_{15}	$+$
$n+m=1, k=1$	1.93	$1.94(\Sigma)$	0.5	D_{13}	$-$
$n+m=3, k=0$	2.05	$2.00(\Lambda)$	2.5	?	?
$n+m=3, k=0$	2.05	$2.00(\Sigma)$	2.5	S_{11}	$-$
$n+m=3, k=0$	2.05	$2.02(\Lambda)$	1.5	F_{07}	$+$
$n+m=3, k=0$	2.05	$2.03(\Sigma)$	1.0	F_{17}	$+$
$n+m=3, k=0$	2.05	$2.07(\Sigma)$	1.0	F_{15}	$+$
$n+m=3, k=0$	2.05	$2.08(\Sigma)$	1.5	P_{13}	$+$
$0,0,2$	2.12	$2.10(\Sigma)$	0.9	G_{17}	$-$
$0,0,2$	2.12	$2.10(\Lambda)$	0.9	G_{07}	$-$
$0,0,2$	2.12	$2.11(\Lambda)$	0.5	F_{05}	$+$
$n+m=2, k=1$	2.24	$2.25(\Sigma)$	0.5	?	?
$n+m=4, k=0$	2.36	$2.325(\Lambda)$	1.5	D_{03}	$-$
$n+m=4, k=0$	2.36	$2.35(\Lambda)$	0.4	H_{09}	$+$
$n+m=1, k=2$	2.43	$2.455(\Sigma)$	1.0	?	?
$n+m=3, k=1$	2.55	$2.585(\Lambda)$	1.4	?	?

Table 3: Energy levels of Σ and Λ.

State (n,m,k)	E_C (GeV)	E_M (GeV)	$Error$ (%)	$L_{2I,2J}$	Parity
$0,0,3$	2.62	$2.62(\Sigma)$	0	?	?
$n+m=5, k=0$	2.67	?	?	?	?
$n+m=2, k=2$	2.74	?	?	?	?
$n+m=4, k=1$	2.86	?	?	?	?
$n+m=1, k=3$	2.93	?	?	?	?
$n+m=6, k=0$	2.98	$3.00(\Sigma)$	0.7	?	?
$n+m=3, k=2$	3.05	?	?	?	?
$n+m=0, k=4$	3.12	?	?	?	?
$n+m=5, k=1$	3.17	$3.17(\Sigma)$	0	?	?
$n+m=2, k=3$	3.24	?	?	?	?
$n+m=2, k=3$	3.29	?	?	?	?

Table 3 (Cont.): Energy levels of Σ and Λ.

State (n,m,k)	E_C (GeV)	E_M (GeV)	$Error$ (%)	$L_{2I,2J}$	Parity
$0,0,0$	1.31	$1.315(\Xi^0)$	0.5	P_{11}	+
$0,0,0$	1.31	$1.321(\Xi^-)$	0.8	P_{11}	+
$1,0,0$	1.62	1.53	5.6	P_{13}	+
$1,0,0$	1.62	1.62	0	?	?
$1,0,0$	1.62	1.69	4.3	?	?
$n=0, m+k=1$	1.81	1.82	0.6	D_{13}	−
$2,0,0$	1.93	1.95	1.0	?	?
$n=1, m+k=1$	2.12	2.03	4.2	?	?
$n=1, m+k=1$	2.12	2.12	0	?	?
$n=3, m=k=0$	2.24	2.25	0.5	?	?
$n=0, m+k=2$	2.31	2.37	2.6	?	?
$n=2, m+k=1$	2.43	?	?	?	?
$n=4, m=k=0$	2.55	2.5	2.0	?	?
$n=1, m+k=2$	2.62	?	?	?	?

Table 4: Energy levels of Ξ.

State (n,m,k)	E_C (GeV)	E_M (GeV)	$Error$ (%)
$0,0,0$	1.5	1.672	11.17
$n+m+k=1$	2.0	2.25	12.5
$n+m+k=2$	2.5	2.38	4.8
$n+m+k=2$	2.5	2.47	1.2
$n+m+k=3$	3.0	?	?

Table 5: Energy levels of Ω.

State (n,m,k)	E_C (GeV)	E_M (GeV)	$Error$ (%)
$0,0,0$	2.32	2.285	1.5
$n+m=1, k=0$	2.63	2.594	0.1
$n+m=1, k=0$	2.63	2.625	0.2
$n+m=2, k=0$	2.94	?	?

Table 6: Energy levels of Λ_c.

State (n, m, k)	E_C (GeV)	E_M (GeV)	$Error$ (%)
$0, 0, 0$	2.51	$2.46(\Xi_c^+)$	2.0
$0, 0, 0$	2.51	$2.47(\Xi_c^0)$	1.6
$1, 0, 0$	2.82	2.79	1.1
$1, 0, 0$	2.82	2.815	0.2
$0, 1, 0$	3.01	2.93	2.7
$0, 1, 0$	3.01	2.98	1.0
$0, 1, 0$	3.01	3.055	1.5
$2, 0, 0$	3.13	3.08	1.6
$2, 0, 0$	3.13	3.123	0.2
$1, 1, 0$	3.32	?	?
$3, 0, 0$	3.44	?	?

Table 7: Energy levels of Ξ_c.

State (n, m, k)	E_C (GeV)	E_M (GeV)	$Error$ (%)
$0, 0, 0$	2.7	2.704	0.2
$n + m = 1, k = 0$	3.2	?	?
$n + m = 2, k = 0$	3.7	?	?

Table 8: Energy levels of Ω_c.

State (n, m, k)	E_C (GeV)	E_M (GeV)	$Error$ (%)
$0, 0, 0$	5.62	5.6202	0.004
$n + m = 1, k = 0$	5.93	?	?
$n + m = 2, k = 0$	6.24	?	?

Table 9: Energy levels of Λ_b^0.

State (n, m, k)	E_C (GeV)	E_M (GeV)	$Error$ (%)
$0, 0, 0$	5.81	$5.79(\Xi_b^0)$	0.2
$0, 0, 0$	5.81	$5.79(\Xi_b^-)$	0.2
$1, 0, 0$	6.12	?	?
$0, 1, 0$	6.31	?	?

Table 10: Energy levels of Ξ_b.

State (n, m, k)	E_C (GeV)	E_M (GeV)	$Error$ (%)
$0, 0, 0$	6.0	6.071	1.2
$n + m = 1, k = 0$	6.5	?	?
$n + m = 2, k = 0$	7.0	?	?

Table 11: Energy levels of Ω_b.

We can predict the energy levels of many heavy baryons, probably already found by the LHC or to be found in the near future. There are, for example, the baryon levels (in GeV):

- scc, $E_{n,m,k} = 0.5(n+1) + 1.7(m+k+2)$;

- ccc, $E_{n,m,k} = 1.7(n+m+k+3)$;

- ccb, $E_{n,m,k} = 1.7(n+m+2) + 5(k+1)$;

- cbb, $E_{n,m,k} = 1.7(n+1) + 5(m+k+2)$;

- etc.

IV.   Calculation in polar cylindrical coordinates and comparison with experimental data

In order to take into account angular momentum and parity, we have to use spherical or polar coordinates. Since the 3 quarks of a baryon are always in a plane, we can use polar coordinates and choose the Z axis perpendicular to this plane. Now the eigenfunctions are eigenfunctions of the orbital angular momentum. Thus, we have three oscillators in a plane and we consider them to be independent. Using again the non-relativistic approximation, the radial Schrödinger equation for the stationary states of each oscillator is given by [12,13]

$$\left[-\frac{\hbar^2}{2\mu}\left(\frac{\partial^2}{\partial\rho^2} + \frac{1}{\rho}\frac{\partial}{\partial\rho} - \frac{m_z}{\rho^2}\right) + \frac{1}{2}\mu\omega^2\rho^2\right] R_{Em}(\rho) = E R_{Em}(\rho) \quad (2)$$

where m_z is the quantum number associated with L_z, μ is the reduced mass of the oscillator, and ω is the oscillator frequency. Therefore, we have three independent oscillators with orbital angular momenta $\vec{L}_1$, $\vec{L}_2$ and $\vec{L}_3$ whose Z components are L_{z1}, L_{z2} and L_{z3}. Of course, the system has total orbital angular momentum $\vec{L} = \vec{L}_1 + \vec{L}_2 + \vec{L}_3$ and each $\vec{L}_i$ has a quantum number l_i associated with it. The eigenvalues of the energy levels are given by [12,13]

$$E = (2r_1 + |m_{z1}| + 1)h\nu_1 + (2r_2 + |m_{z2}| + 1)h\nu_2 + (2r_3 + |m_{z3}| + 1)h\nu_3 \quad (3)$$

in which $r_i = 0,1,2,\ldots$ and it is a radial quantum number, and $|m_{zi}| = 0,1,\ldots,l_i$. Comparing equation (3) with equation (1), we have $n = 2r_1 + |m_{z1}|$; $m = 2r_2 + |m_{z2}|$; $k = 2r_3 + |m_{z3}|$. Let us recall that if we have three angular momenta $\vec{L}_1$, $\vec{L}_2$ and $\vec{L}_3$ associated to the quantum numbers l_1, l_2 and l_3, the total orbital angular momentum $\vec{L}$ is described by the quantum number L given by

$$l_1 + l_2 + l_3 \geq L \geq ||l_1 - l_2| - l_3| \quad (4)$$

where $l_1 \geq |m_{z1}|$; $l_2 \geq |m_{z2}|$; $l_3 \geq |m_{z3}|$.

Because the three quarks are on a plane, only r_i and m_{zi} are good quantum numbers, that is, l_i are not good quantum numbers and their possible values are found indirectly by means of m_{zi} due to the condition $l_i \geq m_{zi}$. This means that the upper values of l_i cannot be found from the model, and as a consequence, the upper value of L cannot be found either. We only determine the values of L comparing the experimental results of the energies of the baryon states with the energy values calculated by E_{nmk}. This is a limitation of the model. The other models have many limitations too. For example, in the Deformed Oscillator Model some quantum numbers are not good either and are only approximate and there is not a direct relation between N and L where N is the total quantum number. In a certain way, a baryon is a tri-atomic molecule of three quarks and thus some features of molecules may show up and that is indeed the case.

Taking into account spin, we form the total angular momentum $\vec{J} = \vec{L} + \vec{S}$ whose quantum numbers are $J = L \pm s$ where $s = 1/2, 3/2$. As we will see, we will be able to describe almost all baryon levels.

As in the case of the rotational spectra of tri-atomic molecules [14], due to the couplings of the different angular momenta, it is expected that there should exist a minimum value of $J = K$ for the total angular momentum and, thus, J should have the possible values $J = K, K+1, K+2, K+3, \ldots$. But in the case of baryons, this feature does not always appear to happen.

i.   Baryons N and Λ

We will classify the levels according to Table 1 and take $J = L \pm 1/2$ or $J = L \pm 3/2(\Delta)$.

a. Level $(n = m = k = 0; 0.93\ GeV)$

The first state of N is the state $(n = m = k = 0)$ with energy 0.93 GeV. Therefore, in this case $l_1 = l_2 = l_3 = 0$ and thus $L = 0$. This is the positive parity state P_{11}

L	N	Δ	Parity
0	$0.938P_{11}$		+

b. Level $(n = m = k = 1; 1.24\ GeV)$

This is the first state of Λ. As $n + m + k = 1$, we have $2r_1 + |m_{z1}| + 2r_2 + |m_{z2}| + 2r_3 + |m_{z3}| = 1$, and thus $|m_{z1}| + |m_{z2}| + |m_{z3}| = 1$, and $l_1 + l_2 + l_3 \geq 1$, and we can choose the sets $|m_{z1}| = 1$, $|m_{z2}| = |m_{z3}| = 0$; $|m_{z2}| = 1$, $|m_{z1}| = |m_{z3}| = 0$; $|m_{z3}| = 1$, $|m_{z1}| = |m_{z2}| = 0$, and $l_1 = 1, l_2 = l_3 = 0$ or $l_2 = 1, l_1 = l_3 = 0$ or still $l_3 = 1, l_1 = l_2 = 0$ which produce $L \geq 0$ (ground state) and thus the level

L	N	Δ	Parity
0		$1.232P_{33}$	+

c. Level $(n = m = k = 2; 1.55\ GeV)$

In this case $n = m = k = 2 = 2r_1 + |m_{z1}| + 2r_2 + |m_{z2}| + 2r_3 + |m_{z3}|$. This means that $|m_{z1}| + |m_{z2}| + |m_{z3}| = 2, 0$ and we have the sets of possible values of l_1, l_2, l_3

l_1, l_2, l_3	2,0,0	0,2,0	0,0,2	1,1,0
L	2	2	2	0,1,2

l_1, l_2, l_3	1,0,1	0,1,1	0,0,0
L	0,1,2	0,1,2	0

in which the second row presents the values of L that satisfy the condition $l_1 + l_2 + l_3 \geq 2, 0$. As 2 is a lower bound, we can also have $L = 3$. There are, therefore, the following possible states

L	N	Δ	Parity
0	$1.44P_{11}$	$1.6P_{33}$	+
1	$1.535S_{11}; 1.655S_{11}$	$1.62S_{31}$	−
	$1.52D_{13}; 1.7D_{13}$		−
2	$1.72P_{13}$	?	+
	$1.685F_{15}$		
3	$1.675D_{15}$	$1.70D_{33}$	−

d. Level $(n = m = k = 3; 1.86\ GeV)$

Since $n = m = k = 3 = 2r_1 + |m_{z1}| + 2r_2 + |m_{z2}| + 2r_3 + |m_{z3}|$, $|m_{z1}| + |m_{z2}| + |m_{z3}| = 3, 1$, and thus $l_1 + l_2 + l_3 \geq 3, 1$. We have, therefore, the possibilities $L = 4, 3, 2, 1, 0$ because of the condition $L \geq ||l_1 - l_2| - l_3|$ and we can arrange the levels in the form

L	N	Δ	Parity
0	$1.71P_{11}$	$1.91P_{31}$	+
1		$1.90S_{31}$	−
		$1.93D_{35}$	−
2	$1.90P_{13}$	?	+
	$2.0F_{15}$		+
3		$1.92P_{33}, 1.94D_{33}$	−
4		$1.905F_{35}$	+

e. Level $(n = m = k = 4; 2.17\ GeV)$

This energy level is split in many close levels. Following what we have done above $n = m = k = 4 = 2r_1 + |m_{z1}| + 2r_2 + |m_{z2}| + 2r_3 + |m_{z3}|$, which yields $|m_{z1}| + |m_{z2}| + |m_{z3}| = 4, 2, 0$, and thus $l_1 + l_2 + l_3 \geq 4, 2, 0$. We have therefore for L the possible values $L = 6, 5, 4, 3, 2, 1, 0$ and the following assignments

L	N	Δ	Parity
0	$2.10P_{11}$	?	+
1	$2.09S_{11}$	$2.15S_{31}$	−
	$2.08D_{13}$		
2		$1.95F_{37}$	+
3	$2.20D_{15}$	?	−
	$2.19G_{17}$		−
4	$2.22H_{19}$	$2.00F_{35}$	+
5	$2.225G_{19}$	$2.20G_{37}$	−
6		$2.30H_{39}$	+

f. Level $(n = m = k = 5; 2.48\ GeV)$

Doing as above $n = m = k = 5 = 2r_1 + |m_{z1}| + 2r_2 + |m_{z2}| + 2r_3 + |m_{z3}|$, and thus $|m_{z1}| + |m_{z2}| + |m_{z3}| = 5, 3, 1$. That is, $l_1 + l_2 + l_3 \geq 5, 3, 1$, and so we may have $L = 5, 4, 3, 2, 1, 0$ because of the conditions $L \geq ||l_1 - l_2| - l_3|$ and $l_i \geq |m_{zi}|$. Experimentally, though, we note that $K = 1$, and hence we have the possible arrangement of levels

L	N	Δ	Parity
1		$2.35D_{35}$	$-$
2		$2.39F_{37}$	$+$
3		$2.40G_{39}$	$-$
4		$2.42H_{3,11}$	$+$
5		?	$-$

g. Level ($n = m = k = 6$; 2.79 GeV)

We have $n = m = k = 6 = 2r_1 + |m_{z1}| + 2r_2 + |m_{z2}| + 2r_3 + |m_{z3}|$ and so $|m_{z1}| + |m_{z2}| + |m_{z3}| = 6, 4, 2, 0$ and thus $l_1 + l_2 + l_3 \geq 6, 4, 2, 0$, and L can be $L = 6, 5, 4, 3, 2, 1, 0$, but, from the experimental values, we note that $K = 5$ and so there are the possible states

L	N	Δ	Parity
5	$2.60I_{1,11}$	$2.75I_{3,13}$	$-$
6	$2.70K_{1,13}$	?	$+$

h. Level ($n = m = k = 7$; 3.10 GeV)

From $n = m = k = 7 = 2r_1 + |m_{z1}| + 2r_2 + |m_{z2}| + 2r_3 + |m_{z3}|$ we obtain $|m_{z1}| + |m_{z2}| + |m_{z3}| = 7, 5, 3, 1$ and hence $l_1 + l_2 + l_3 \geq 7, 5, 3, 1$, and thus the possible values for L are $L = 7, 6, 5, 4, 3, 2, 1, 0$, but we note that $K = 7$. Therefore, we have the list of states

L	N	Δ	Parity
6		$2.95K_{3,15}$	$+$
7	$3.10L_{1,15}$	?	$-$

ii. Baryons Σ and Λ

We will classify the levels according to Table 2. Again $J = L \pm 1/2$.

a. Level ($n = m = k = 0$; 1.12 GeV)

In this state $l_1 = l_2 = l_3 = 0$ and thus $L = 0$ and we have the state

L	Σ	Λ	Parity
0	$1.189(\Sigma^{\pm})P_{11}$; $1.193(\Sigma^0)P_{11}$	$1.116P_{01}$	$+$

b. Level ($n = m = k = 1$; 1.43 GeV)

From $n = m = 1, k = 0$ we obtain $2r_1 + |m_{z1}| + 2r_2 + |m_{z2}| = 1$ and $2r_3 + |m_{z3}| = 0$ which make $|m_{z1}| + |m_{z2}| = 1$ and $|m_{z3}| = 0$. That is, we have the condition $l_1 + l_2 \geq 1, l_3 \geq 0$ which allows us to have the possibilities $L = 0, 1, 2$ and the states

L	Σ	Λ	Parity
0			$+$
1	?	$1.405S_{01}$	$-$
2	$1.358P_{13}$	?	$+$

Thus, the most probable values of L for the state $1.48(\Sigma)$ are $L = 0, 1$. Maybe $K = 1$ in this case and thus $L = 0$ may be suppressed.

c. Level ($0, 0, 1$; 1.62 GeV)

For $n = m = 0$ and $k = 1$ we have $|m_{z1}| = |m_{z2}| = 0$ and $|m_{z3}| = 1$. That is, we have the condition $l_1 \geq 0, l_2 \geq 0, l_3 \geq 1$ which allows us to choose $l_1 = l_2 = 0, l_3 = 1$; $l_1 = l_3 = 1, l_2 = 0$; $l_1 = 0, l_2 = l_3 = 1$, and thus $L \geq 0, 1, 2$, and the states

L	Σ	Λ	Parity
0	$1.66P_{11}$	$1.60P_{01}$	$+$
1	$1.62S_{11}$	$1.67S_{01}$	$-$
	$1.58D_{13}$	$1.52D_{03}$	$-$
2	?	?	$+$

d. Level ($n + m = 2, k = 0$; 1.74 GeV)

In this case $n + m = 2 = 2r_1 + |m_{z1}| + 2r_2 + |m_{z2}|$ and $k = 2r_3 + |m_{z3}| = 0$, and thus we obtain $|m_{z1}| + |m_{z2}| = 2, 0$ and $|m_{z3}| = 0$. Thus, we have the conditions $l_1 + l_2 \geq 2, 0$ and $l_3 \geq 0$. We can then choose $l_1 = l_2 = l_3 = 0$; $l_1 = l_2 = 1, l_3 = 0$; $l_1 = 2, l_2 = l_3 = 0$; $l_2 = 2, l_1 = l_3 = 0$; $l_1 = 3, l_2 = l_3 = 0$, and we may have thus $L = 0, 1, 2, 3$ and the assignments

L	Σ	Λ	Parity
0	$1.77P_{11}$	$1.81P_{01}$	$+$
1	$1.75S_{11}$	$1.80S_{01}$	$-$
	$1.67D_{13}$	$1.69D_{03}$	$-$
2	?	$1.82F_{05}$	$+$
3	$1.775D_{15}$	$1.83D_{05}$	$-$

We can then say that the state $1.69(\Sigma)$ is probably a F_{15} state.

e.   Level $(n + m = 1, k = 1; 1.93~GeV)$

We have $n + m = 1 = 2r_1 + |m_{z1}| + 2r_2 + |m_{z2}|$ and $k = 2r_3 + |m_{z3}| = 1$, from which we obtain $|m_{z1}| + |m_{z2}| = 1$ and $|m_{z3}| = 1$. Hence, we have the condition $l_1 + l_2 \geq 1$ and $l_3 \geq 1$. We can then have the sets $l_1 = 1, l_2 = 0, l_3 = 1$; $l_1 = 0, l_2 = 1, l_3 = 1$. Both yield $L \geq 2, 1, 0$ and thus we identify the states

L	Σ	Λ	Parity
0	$1.88P_{11}$	?	+
1	$1.94D_{13}$	?	−
2	$1.84P_{13}$	$1.89P_{03}$	+
	$1.915F_{15}$		+

f.   Level $(n + m = 3, k = 0; 2.05~GeV)$

With $n + m = 3 = 2r_1 + |m_{z1}| + 2r_2 + |m_{z2}|$ and $k = 2r_3 + |m_{z3}| = 0$ we obtain $|m_{z1}| + |m_{z2}| = 3, 1$ and $|m_{z3}| = 0$, and thus the conditions $l_1 + l_2 \geq 3, 1$ and $l_3 \geq 0$ which yield $L \geq 4, 3, 2, 1, 0$, and the possible identification taking into account that maybe $K = 1$

L	Σ	Λ	Parity
1	$2.00S_{11}$	?	−
2	$2.08P_{13}$	?	+
	$2.07F_{15}$		
3	?	?	−
4	$2.03F_{17}$	$2.02F_{07}$	+

g.   Level $(0, 0, 2; 2.12~GeV)$

In this case $n = 0 = 2r_1 + |m_{z1}|$, $m = 0 = 2r_2 + |m_{z2}|$ and $k = 2 = 2r_3 + |m_{z3}|$ and thus $|m_{z1}| = |m_{z2}| = 0$ and $|m_{z3}| = 2, 0$. Hence, we have the condition $l_1 \geq 0, l_2 \geq 0$ and $l_3 \geq 2, 0$. We can then choose the sets $l_1 = l_2 = l_3 = 0$; $l_1 = 0, l_2 = 0, l_3 = 2$; $l_1 = 0, l_2 = 0, l_3 = 3$; $l_1 = l_2 = l_3 = 1$; $l_1 = l_2 = 1, l_3 = 0$ which make $L \geq 3, 2, 1, 0$, and probably $K = 2$. Hence, we have the possible states

L	Σ	Λ	Parity
2	?	$2.11F_{05}$	+
3	$2.10G_{17}$	$2.10G_{07}$	−

h.   Level $(n + m = 4, k = 0; 2.36~GeV)$

From $n + m = 4 = 2r_1 + |m_{z1}| + 2r_2 + |m_{z2}|$ and $k = 2r_3 + |m_{z3}| = 0$ we obtain $|m_{z1}| + |m_{z2}| = 4, 2, 0$ and

$|m_{z3}| = 0$, and thus the conditions $l_1 + l_2 \geq 4, 2, 0$ and $l_3 \geq 0$ which produce $L \geq 4, 3, 2, 1, 0$, and the possible identification

L	Σ	Λ	Parity
0			
1		$2.325D_{03}$	−
2		?	
3		?	
4		$2.35H_{09}$	+

Probably in this case $K = 1$ and the levels with $L = 2, 3$ are missing just because of a lack of experimental data. It appears that there is no state of Σ.

V.   Discussion and conclusion

One can immediately ask about the spin degrees of freedom of the three quarks since the spin-spin interaction makes a contribution to the mass. We can say that we took care of part of it because the formulas of the energy levels depend on the three parameters $h\nu_1$, $h\nu_2$ and $h\nu_3$ which are assigned according to the masses of the constituent quarks which have already taken into account the spin-spin interaction because the masses of constituent quarks are in perfect agreement with the ground state levels of baryons. Of course, the spin-spin interaction contribution depends on the energy level as is well known from the bottomonium spectrum, for example. But, as it is seen in the spectrum of bottomonium, the spin-spin contribution diminishes as the energy of the level increases. In bottomonium, the difference between the energies of $\eta_b(1S)$ and $\Upsilon(1S)$ is about 69.4 MeV, while between $\eta_b(2S)$ and $\Upsilon(2S)$ it is about 36.3 MeV, and between $\eta_b(3S)$ and $\Upsilon(3S)$ it is about 25.2 MeV, where we have used, for the energies of $\eta_b(2S)$ and $\eta_b(3S)$, the predicted values from reference [15], 9987.0 MeV and 10330 MeV, respectively. In the case of baryons, the spin-spin interaction varies from 15 MeV to 30 MeV for levels of N, Σ, Ξ and Λ [16]. Therefore, we observe that the spin-spin interaction is of the order of magnitude of the splitting beween neighboring levels. For example, the measured energy of the D_{13} level of N is 1.52 MeV, while our calculated value is 1.55 MeV, and thus the difference is 0.03 GeV= 30 MeV which is of the order of the spin-spin interaction. And that is why there are large discrepancies in the calculation of the lowest levels of Ω because in this case all quark spins are parallel and thus, the total spin-spin contribution is larger than in other baryons in which two spins are up and the other spin is down. For the lowest state of Ω, the discrepancy is about 1.672 GeV $-$ 1.5 GeV $=$ 0.172 GeV $=$ 172 MeV. This is actually the worse calculation. But we either consider the mass of constituent quarks or we try to find tentative values for the masses of quarks like is done in QCD models which use a quite different range of arbitrary quark masses. The use of the constituent quark mass is completely justifiable in our case because we do not attempt to calculate at all the splitting between neighboring baryon levels. Such calculation can be made in the future upon improving the present model.

We only addressed the angular momenta of N, Δ, Σ and Λ due to a lack of experimental data for the other baryons. Of course, the state $n = m = l = 0$ is missing for the Δ particle because this corresponds to the ground state of the nucleon.

We notice that the simple model above describes almost all energy levels of baryons. The splitting for a certain L is quite complex. Sometimes, there is almost no dependence on spin, such as, for example, the states of Σ with $L = 2$, $2.08P_{13}$ and $2.07F_{15}$. On the other hand, the states of Σ with the same $L = 2$, $1.84P_{13}$ and $1.915F_{15}$, present a strong spin-orbit dependence. It can just be a matter of obtaining more accurate experimental results.

It is important to observe that part of the splitting is primarily caused by the spin-orbit interaction and is very complex because, in some cases, it appears to be the normal spin-orbit and, in other cases, it appears to be the inverted (negative) spin-orbit. In the simple model above, the oscillators were considered approximately independent but there may exist some coupling among them and this can contribute to the splitting of levels. As we have discussed in the first paragraph, part of the splitting should be attributed to the spin-spin interaction which was not taken into account in a detailed way. Of course, part of it was considered inside the values of the three parameters $h\nu_1$, $h\nu_2$ and $h\nu_3$ which are taken as the masses of the three constituent quarks of a given baryon. It is important to observe that the discrepancy between calculated and measured values diminishes as the energy increases. This fact shows that the splitting

is mainly caused by the spin-spin interaction.

Another important conclusion is that with the simple model above we cannot calculate the values of K, and from the above results we note that it is a quite difficult task because there appears to exist no pattern with respect to this. As in the case of triatomic molecules, the values of K are found from the experimental data.

As we notice, in the above tables the increase in the energy of levels allows the existence of higher values of L (and J). This is an old fact and is so because equation (3) has a linear dependence on $|l_{zi}|$.

For experimentalists, the classifications above are very important and can help them in the prediction of energies and angular momenta of levels. An old version of this work that appeared in Ref. [9] predicted the energies of *all levels* which have lately been reported, and this is a very important fact. For example, for Ξ_c it predicted the levels (on page 8 of [9]) with energies 2.82, 3.01 and 3.13, and since 2002 the following corresponding levels of Ξ_c have been found: 2.815, (2.93; 2.98; 3.055); (3.08; 3.123).

As it is well known, the first order correction term of anharmonicity in an oscillator for each degree of freedom is of the form

$$\Delta E = A \left(p + \frac{1}{2} \right)^2 \qquad (5)$$

where A is a constant and p is a non-negative integer ($p = 0, 1, 2, 3, \ldots$). Therefore, the calculated energies of levels with high quantum numbers would be away from the experimental values. This is not observed above and, thus, the anharmonicity should be quite low. For example, for $n + m + k = 7$ of N we obtain that the experimental and calculated values are the same (3.10 GeV). In the case of Σ, we have the same kind of behavior because for $(n + m = 5, k = 1)$ we also have the same calculated and experimental value for Σ (3.17 GeV).

The assignments of the angular momenta for some few levels are only reasonable attempts. It is the case, for example, of the level $2.0F_{15}$ of N which can belong to either the $(n + m + k = 3)$ or to the $(n + m + k = 4)$ levels. We chose the former because 2.0 is closer to 1.86 than to 2.17. For the level $1.99F_{17}$ we chose the $(n + m + k = 4)$ level because it appears that the highest value of J

for the level $(n + m + k = 3)$ is 5. It is a strange feature that the level $(n + m + k = 5)$ only contains $\Delta's$. Having in mind what has been justified above, we chose the $2.35 D_{35}$ level of Δ belonging to $(n + m + k = 5)$ as 2.35 is closer to 2.48 than to 2.17. The level $2.60 I_{1,11}$ of N was assigned as belonging to $(n + m + k = 6)$ because its energy is between 2.55 GeV and 2.75 GeV. The level $1.74 D_{13}$ of Σ was chosen as belonging to $(n + m = 2, k = 0)$ because $(0, 0, 1)$ already has a D_{13} level for $\Sigma(1.58)$. We made similar considerations in the choice of the levels $1.83(\Lambda) D_{05}$, $1.84(\Sigma) P_{13}$ and $1.94(\Sigma) D_{13}$. These ambiguities will be settled either with data with smaller widths or with a more improved model.

Some levels are not described by the simple approximation above. It is the case, for instance, of $\Xi(1530) P_{13}$ which is probably a composite of $\Xi(0, 0, 0) \equiv \Xi(1.31)$ with a pion excitation (that is, it is a hadronic molecule). Its decay is actually $\Xi(1.31)\pi$. In the same way, the state $\Sigma_c(2455)$ appears to be a composite state of $\Lambda_c^+(2285)$ and a pion excitation. The same appears to hold for the other known states of Λ_c.

As a whole, the model describes quite well the baryonic spectra but it is far from describing the detailed splitting which appears to be quite complex and may depend on the spin-spin interaction. It does not provide a way of calculating the values of K. With the acquisition of more data from other baryons, we may be able to find more patterns and to improve the model. Due to the complexity of the problem, we will probably have to go back and forth several times in the improvements of the model as it has been done in the description of the molecular spectrum of molecules. But it is still the only model that describes almost all levels of baryons in a consistent way and is able to predict the energies of levels yet to be found experimentally.

Acknowledgements - I thank the comments of the referee Prof. José Muñoz.

---

[1] S Gasiorowicz, J L Rosner, *Hadron spectra and quarks*, Am. J. Phys. **49**, 954 (1981).

[2] N Isgur, G Karl, *P-wave baryons in the quark model*, Phys. Rev. D **18**, 4187 (1978).

[3] S Capstick, N Isgur, *Baryons in a relativized quark model with chromodynamic*, Phys. Rev. D **34**, 2809 (1986).

[4] R K Bhaduri, B K Jennings, J C Waddington, *Rotational bands in the baryon spectrum*, Phys. Rev. D **29**, 2051 (1984).

[5] M V N Murthy, M Dey, R K Bhaduri, *Rotational bands in the baryon spectrum. II*, Phys. Rev. D **30**, 152 (1984).

[6] M V N Murthy, M Brack, R K Bhaduri, B K Jennings, *The spin-orbit puzzle in the spectra and deformed baryon model*, Z. Phys. C **29**, 385 (1985).

[7] P Stassart, F Stancu, J-M Richard, L Theußl, *On the scalar meson exchange in the baryon spectra*, J. Phys. G: Nucl. Partic. **26**, 397 (2000).

[8] A Hosaka, H Toki, M Takayama, *Baryon spectra in deformed oscillator quark model*, Mod. Phys. Lett. **13**, 1699 (1998).

[9] M E de Souza, *Calculation of the energy levels and sizes of baryons with a noncentral harmonic potential*, arXiv:hep-ph/0209064v1 (2002).

[10] M E de Souza, *The energies of baryons*, In: Proceedings of the XIV Brazilian National Meeting of the Physics of Particles and Fields, Eds. A J da Silva, A Suzuki, C Dobrigkeit, C Z de Vasconcelos, C Wotzacek, R F Ribeiro, S R de Oliveira, S A Dias, Pag. 331, Sociedade Brasileira de Física, São Paulo (1993).

[11] K Nakamura et al.(Particle Data Group), *Review of particle physics*, J. Phys. G: Nucl. Partic. **37**, 075021 (2010).

[12] R Shankar, *Principles of quantum mechanics*, Plenum Press, New York (1994).

[13] W Greiner, J A Maruhn, *Nuclear models*, Springer, Berlin (1996).

[14] G Herzberg, *Infrared and Raman spectra*, Van Nostrand Heinhold Company, New York (1945).

[15] L. Bai-Qing and C. Kuang-Ta, *Bottomonium spectrum with screeened potential*, Commun. Theor. Phys. **52**, 653 (2009).

[16] H Hassanabadi, A A Rajabi, *Determination of the potential coefficients of the baryons and the effect of spin and isospin potential on their energy*, In: 11th Internation Conference on meson-Nucleon Physics and Structure of the Nucleon, Eds. H Machner, S Krewald, Pag. 128, IKP, Forschungzentrum Jülich (2007).

23

Beltrami flow structure in a diffuser: Quasi-cylindrical approximation

Rafael González,[1,2*] Ricardo Page,[3] Andrés S. Sartarelli[1]

We determine the flow structure in an axisymmetric diffuser or expansion region connecting two cylindrical pipes when the inlet flow is a solid body rotation with a uniform axial flow of speeds Ω and U, respectively. A quasi-cylindrical approximation is made in order to solve the steady Euler equation, mainly the Bragg–Hawthorne equation. As in our previous work on the cylindrical region downstream [R González et al., Phys. Fluids **20**, 24106 (2008); R. González et al., Phys. Fluids **22**, 74102 (2010), R González et al., J. Phys.: Conf. Ser. **296**, 012024 (2011)], the steady flow in the transition region shows a Beltrami flow structure. The Beltrami flow is defined as a field $\mathbf{v}_B$ that satisfies $\boldsymbol{\omega}_B = \nabla \times \mathbf{v}_B = \gamma \mathbf{v}_B$, with $\gamma = constant$. We say that the flow has a *Beltrami flow structure* when it can be put in the form $\mathbf{v} = U\mathbf{e}_z + \Omega r \mathbf{e}_\theta + \mathbf{v}_B$, being U and Ω constants, i.e it is the superposition of a solid body rotation and translation with a Beltrami one. Therefore, those findings about flow stability hold. The quasi-cylindrical solutions do not branch off and the results do not depend on the chosen transition profile in view of the boundary conditions considered. By comparing this with our earliest work, we relate the critical Rossby number ϑ_{cs} (stagnation) to the corresponding one at the fold ϑ_{cf} [J. D. Buntine et al., Proc. R. Soc. Lond. A **449**, 139 (1995)].

I. Introduction

We have recently conducted studies on the formation of Kelvin waves and some of their features when an axisymmetric Rankine flow experiences a soft expansion between two cylindrical pipes [1, 2]. One of the significant characteristics of this phenomenon is that the downstream flow

*E-mail: rgonzale@ungs.edu.ar

[1] Instituto de Desarrollo Humano, Universidad Nacional de General Sarmiento, Gutierrez 1150, 1613 Los Polvorines, Pcia de Buenos Aires, Argentina.

[2] Departamento de Física FCEyN, Universidad de Buenos Aires, Pabellón I, Ciudad Universitaria, 1428 Buenos Aires, Argentina .

[3] Instituto de Ciencias, Universidad Nacional de General Sarmiento, Gutierrez 1150, 1613 Los Polvorines, Pcia de Buenos Aires, Argentina.

shows a Rankine flow superposing a Beltrami flow (Beltrami flow structure [4])). Yet, upstream and downstream cylindrical geometries were considered without taking into account the flow in the expansion. This work considered that the base upstream flow, formed by a vortex core surrounded by a potential flow, would have the same Beltrami structure at the expansion and downstream. Nevertheless, the flow at the expansion was not determined. However, it has been seen that this flow is only possible when no reversed flow is present and if its parameters do not take the values where a vortex breakdown appears [6–8]. The starting point in the study of the expansion flow is an axysimmetric steady state resulting from the Bragg–Hawthorne equation [7, 9–11] for both the vortex breakdown and the formation of waves. Therefore, the solution behavior, whether it branches off or shows a possible stagnation point on the axis, will be determinant to delimit both phenomena.

Our previous research focused on the formation of Kelvin waves with a Beltrami flow structure downstream [1–3], when the upstream flow was a Rankine one. This present investigation considers only a solid body rotation flow with uniform axial flow at the inlet. As a first step in the study of the flow at the expansion, we only study the rotational flow. However, comparisons with our previous work [1] will be drawn.

The aim of this present work is to obtain the steady flow structure at the expansion, considering a quasi-cylindrical approximation when the inlet flow is a solid body rotation with uniform axial flow of speeds Ω and U, respectively. If a is the radius of the cylindrical region upstream, a relevant parameter is the Rossby number $\vartheta = \frac{U}{\Omega a}$. Thus, we would like to determine how this flow depends on the Rossby number, on the geometrical parameters of the expansion and on the critical values of the parameters. We focus on finding the parameter values for which a stagnation point emerges on the axis, or for which the solution of the Bragg–Hawthorne equation branches off. We take them as the conditions for the vortex breakdown to develop.

First, this paper presents the inlet flow and the corresponding Bragg–Hawthorne equation written for the transition together with the boundary conditions in section II. Second, it works on the quasi-cylindrical approximation for the Bragg–Hawthorne equation and its solution is developed in section III. Third, results and discussions are offered in section IV together with a comparison with our previous work [1]. Finally, conclusions are presented in section V.

II. The Bragg–Hawthorne equation

We assume an upstream flow in a pipe of radius a as an inlet flow in an axisymmetric expansion of length L connecting to another pipe with radius b, $b > a$. The inlet flow filling the pipe consists of a solid body rotation of speed Ω with a uniform axial flow of speed U:

$$\mathbf{v} = U\mathbf{e}_z + \Omega r\mathbf{e}_\theta, \qquad (1)$$

U and Ω being constants. The equilibrium flow in the whole region is determined by the steady Euler equation which can be written as the Bragg–Hawthorne equation [10]

$$\frac{\partial^2 \psi}{\partial z^2} + r\frac{\partial}{\partial r}\left(\frac{1}{r}\frac{\partial \psi}{\partial r}\right) + r^2\frac{\partial H}{\partial \psi} + C\frac{\partial C}{\partial \psi} = 0, \quad (2)$$

where ψ is the defined stream function

$$v_r = -\frac{1}{r}\frac{\partial \psi}{\partial z}, \quad v_z = \frac{1}{r}\frac{\partial \psi}{\partial r}, \qquad (3)$$

and $H(\psi), C(\psi)$ are the total head and the circulation, respectively

$$H(\psi) = \frac{1}{2}(v_r^2 + v_\theta^2 + v_z^2) + \frac{p}{\rho}, \ C(\psi) = rv_\theta. \quad (4)$$

To solve Eq. (2), the boundary conditions must be established. These consist of giving the inlet flow, of being both the centerline and the boundary wall, streamlines, and of being the axial velocity positive ($v_z > 0$). For the upstream flow, the stream function is $\psi = \frac{1}{2}Ur^2$, and $H(\psi), C(\psi)$ are given by

$$H(\psi) = \frac{1}{2}U^2 + \Omega\gamma\psi, \ C(\psi) = \gamma\psi, \qquad (5)$$

$\gamma = \frac{2U}{\Omega}$ being the eigenvalue of the flow with Beltrami structure [3]. Thus, by considering the inlet flow, Eqs. (5) are valid for the whole region. The second condition regarding the streamlines implies the following relations

$$\psi(r = 0, z) = 0,$$
$$\psi(r = \sigma(z), z) = \frac{1}{2}Ua^2, \ 0 \le z \le L \qquad (6)$$

where $r = \sigma(z)$ gives the axisymmetric profile of the pipe expansion. Deducing from Eq. (6), the boundary conditions are determined by the inlet flow. Additionally, curved profiles are considered, so

$$\frac{\partial \psi}{\partial z}(r, z = L) = 0, \ 0 \le r \le b. \qquad (7)$$

III.   Quasi-cylindrical approximation

If we consider that $\frac{\partial^2 \psi}{\partial z^2} = 0$, the solutions to Eqs. (2) and (5) for the cylindrical regions are given by [10]

$$\psi = \frac{1}{2}Ur^2 + ArJ_1[\gamma r], \qquad (8)$$

where A is a constant. The quasi-cylindrical approximation consists of taking the dependence of $A(z)$ on z but with the condition $\frac{\partial^2 \psi}{\partial z^2} \approx 0$ compared with the remaining terms of (2). The amplitude $A(z)$ is then obtained by imposing the boundary conditions (6) which depend on the wall profile $r = \sigma(z)$, giving

$$A(z) = \frac{1}{2}\frac{U\left(a^2 - \sigma^2(z)\right)}{\sigma(z)J_1[\gamma\sigma(z)]}. \qquad (9)$$

By using the dimensionless quantities $\tilde{r} = \frac{r}{a}$, $\tilde{z} = \frac{z}{a}$, $\tilde{v} = \frac{v}{U}$ the stream function in the quasi-cylindrical approximation can be written as

$$\tilde{\psi} = \frac{1}{2}\tilde{r}^2 + \tilde{A}(\tilde{z})\tilde{r}J_1[\frac{2}{\vartheta}\tilde{r}],$$

$$\tilde{A}(\tilde{z}) = \frac{1}{2}\frac{\left(1 - \tilde{\sigma}^2(\tilde{z})\right)}{\tilde{\sigma}(\tilde{z})J_1[\frac{2}{\vartheta}\tilde{\sigma}(\tilde{z})]}, \qquad (10)$$

where $\vartheta = \frac{U}{\Omega a}$ is the Rossby number. Hence the velocity field becomes

$$\tilde{v}_r(\tilde{r}, \tilde{z}) = -\tilde{A}'(\tilde{z})J_1[\frac{2}{\vartheta}\tilde{r}] \qquad (11)$$

$$\tilde{v}_\theta(\tilde{r}, \tilde{z}) = \frac{1}{\vartheta}\tilde{r} + \frac{2}{\vartheta}\tilde{A}(\tilde{z})J_1[\frac{2}{\vartheta}\tilde{r}] \qquad (12)$$

$$\tilde{v}_z(\tilde{r}, \tilde{z}) = 1 + \frac{2}{\vartheta}\tilde{A}(\tilde{z})J_0[\frac{2}{\vartheta}\tilde{r}], \qquad (13)$$

where $\tilde{A}'(\tilde{z}) = d\tilde{A}(\tilde{z})/d\tilde{z}$.

Finally, it is necessary to give the wall profile $\tilde{\sigma}(z)$ to completely determine the flow. Two kinds of profiles were seen:

i- conical profile

$$\tilde{\sigma}(\tilde{z}) = 1 + \left(\frac{\eta - 1}{\tilde{L}}\right)\tilde{z},$$

$$0 \leq \tilde{z} \leq \tilde{L} \text{ and } \eta = \frac{b}{a}. \qquad (14)$$

ii- curved profile

$$\tilde{\sigma}(\tilde{z}) = \frac{1+\eta}{2} - \left(\frac{\eta - 1}{2}\right)\cos\left(\frac{\pi\tilde{z}}{\tilde{L}}\right),$$

$$0 \leq \tilde{z} \leq \tilde{L}. \qquad (15)$$

The latter meets the boundary condition (7) as well. Therefore, Eqs. (11-15) together with the boundary conditions (6,7) allow to determine the flow structure for both the conical and curved wall profile.

IV.   Results and discussion

We note that the flow keeps a Beltrami flow structure in the quasi-cylindrical approximation. Effectively, giving (11-13)

$$\tilde{v}_r(\tilde{r}, \tilde{z}) = \tilde{v}_{Br}(\tilde{r}, \tilde{z}) \qquad (16)$$

$$\tilde{v}_\theta(\tilde{r}, \tilde{z}) = \frac{1}{\vartheta}\tilde{r} + \tilde{v}_{B\theta}(\tilde{r}, \tilde{z}) \qquad (17)$$

$$\tilde{v}_z(\tilde{r}, \tilde{z}) = 1 + \tilde{v}_{Bz}(\tilde{r}, \tilde{z}), \qquad (18)$$

it is easy to see that under this approximation $\nabla \times \mathbf{v}_B(\tilde{r}, \tilde{z}) = \frac{2}{\vartheta}\mathbf{v}_B(\tilde{r}, \tilde{z})$ and so, the whole flow is the sum of a solid body rotation flow with a uniform axial flow plus a Beltrami flow, given the latter in a system with uniform translation velocity $\mathbf{U} = 1.\hat{\mathbf{z}}$ and uniform rigid rotation velocity $\mathbf{V} = \frac{1}{\vartheta}\tilde{r}\hat{\boldsymbol{\theta}}$.

Given the flow field and its structure, the parameters are considered by evaluating the behavior of $\tilde{v}_z(\tilde{r}, \tilde{z}_0)$ with $\tilde{z}_0 = \tilde{L}$ i.e., taken at outlet, and with $\tilde{L} = 1$. In order to do so, a wall profile is selected (14 or 15) and three different values of the expansion parameter are taken, mainly $\eta_1 = 1.1$, $\eta_2 = 1.2$ and $\eta_3 = 1.3$.

The first step is to analyze the flow dependence on the Rossby number. In Fig. 1, the contour flows corresponding to the conical and curved profiles for $\eta_1 = 1.1$, $\vartheta_1 = 0.695$ are shown. Graphics in Fig. 2 represent the same configuration but for $\vartheta = 0.68 < \vartheta_1$. The broken lines represent points for which $\tilde{v}_z = 0$. Inflow and recirculation are present but it is not a real flow because the model fails when considering inflow. It can be seen

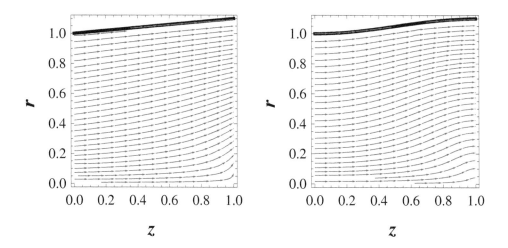

Figure 1: Contour flow in the transition region for conical and curved profiles for $\eta_1 = 1.1$, $\vartheta_1 = 0.695$.

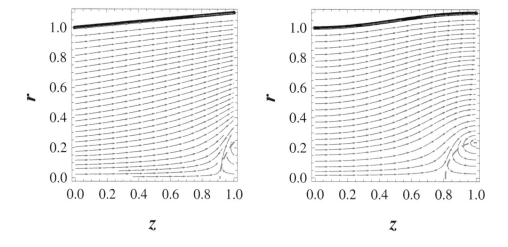

Figure 2: Contour flow in the transition region for conical and curved profiles for $\eta_1 = 1.1$, $\vartheta_1 = 0.68$. The broken lines represent points with $\tilde{v}_z = 0$.

that for $\vartheta_1 = 0.695$, $\tilde{v}_z = 0$ at the outlet, on the axis. For the Rossby numbers with $\vartheta \geq \vartheta_c$, the azimuthal flow vorticity is negative ($\omega_\phi < 0$), resulting in an increase in the axial velocity with the radius, and so having a minimum on the axis where the stagnation point appears [6]. Therefore, the critical Rossby number can be defined ϑ_c as the value where $\tilde{v}_z$ is zero at the outlet on the axis i.e., where the flow shows a stagnation point. This is the necessary condition to produce a vortex breakdown [6]. We find the same critical Rossby number for both wall profiles and so we will not treat them separately from now on. The critical Rosssby values for $\eta_2 = 1.2$ and $\eta_3 = 1.3$ are $\vartheta_2 = 0.869$ and $\vartheta_3 = 1.052$, respectively.

Given the previous analysis, the second step is to show the behavior of $\tilde{v}_z$ on the axis at the outlet as a function of ϑ for each η in order to study the existence of folds in the Rossby number-continuation parameter (equivalent to the swirl parameter in [5,7,11]); indeed, we have seen that $\tilde{v}_z$ has the minimum on the axis. Besides, when using Eq. (13) when $r = 0$, it is easy to see that $\tilde{v}_z$ decreases with z and so it reaches the minimum at the outlet being $\tilde{v}_z \geq 0$. In Fig. 3, the radial dependence of $\tilde{v}_z$ is plotted at the outlet for η_1, η_2, η_3 and its variation with ϑ when it is slightly shifted from ϑ_1. In Fig. 4, it can be seen that the minimum of $\tilde{v}_z$ on the axis increases with ϑ so there is no fold of $\tilde{v}_{zmin}$ as defined by Buntine and Saffman in a similar approximation [5].

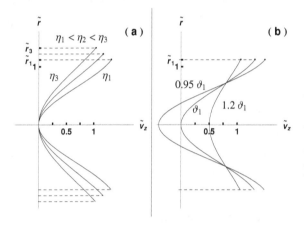

Figure 3: (a) $\tilde{v}_z$ at the outlet as a function of r for η_1, η_2, η_3 and the corresponding critical Rossby numbers $\vartheta_1, \vartheta_2, \vartheta_3$. (b) $\tilde{v}_z$ at the outlet as a function of r for ϑ_1 and for values of ϑ slightly shifted from ϑ_1 . In each case, the minimum of $\tilde{v}_z$ is reached on the axis.

The dependence of the results on L is analyzed. It can be seen that when $z = L$ in Eqs. (14) and (15), $\tilde{\sigma}(\tilde{L}) = \eta$ is obtained. By replacing this in Eq. (13) for $z = L$ and $r = 0$ it gives

$$\tilde{v}_z min = 1 + \frac{(1-\eta^2)}{\vartheta \eta J_1 \left[\frac{2}{\vartheta}\eta\right]}, \qquad (19)$$

and so ϑ_c is obtained as a function of η by solving the last equation when $\tilde{v}_z min = 0$, as shown in Fig. 5. This result seems to be surprising, but it is not so if it is considered as derived from the quasi-cylindrical approximation: the dependence of

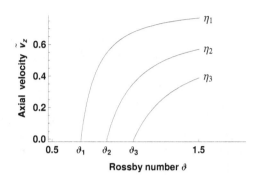

Figure 4: $\tilde{v}_z$ at the outlet on the axis as a function of the Rossby number ϑ for $\eta_1 = 1.1, \eta_2 = 1.2, \eta_3 = 1.3$. Here $\vartheta_1 = 0.695, \vartheta_2 = 0.869$ and $\vartheta_3 = 1.052$ correspond to stagnation points.

the flow on z is obtained through the boundary conditions expressed by Eq. (6). At the same time, these boundary conditions depend on the inlet flow and on the parameter η. This explains the fact that the same results, for both conical and curved profiles, have been obtained and that the condition given by Eq. (7) at the outlet has not influenced them.

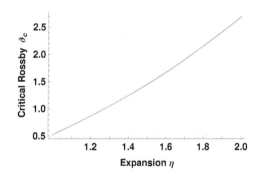

Figure 5: Critical Rossby number ϑ_c as a function of η.

Differences with Batchelor's seminal work should be marked [10]. Mainly, he works in cylindrical geometry and does not consider the dependence of the flow on z . We introduce this z dependence through the quasi-cylindrical approximation. This, therefore, allows us to find the structure of the flow in the transition together with the Rossby critical number defined by considering this structure and by showing that the minimum of $\tilde{v}_z$ is reached at the outlet on the axis. Nevertheless, once the flow reaches the pipe downstream, the analysis coincides because, as shown, the problem depends on the inlet flow and on the parameter expansion η. This allows us to consider the issue of the vortex core that we have not considered at the inlet flow. As we know the structure of the flow in the downstream cylindrical region [1] and by assuming a quasi-cylindrical approximation for the vortex core in the transition region, the minimum of v_{core_z} at the outlet on the axis is given by

$$v_{core_z min} = 1 + \frac{(1 - \hat{\eta}^2)}{\hat{\vartheta}\hat{\eta}J_1[\frac{2}{\hat{\vartheta}}\hat{\eta}]}, \qquad (20)$$

where $\hat{\vartheta} = \frac{\vartheta}{\iota}$, $\hat{\eta} = \frac{\xi}{\iota}$ and ξ and ι are the dimensionless radius of the core downstream and upstream, respectively. We note that $\hat{\eta}$ is the expansion parameter of the core. Hence Eqs. (19) and (20) have the same structure. In the present work, we have not found any fold in the Rossby number-continuation parameter of $\tilde{v}_z$, as found in our previous work [1] where the fold was associated with a critical Rossby number called ϑ_{cf} by Buntine and Saffman [5]. As we have already done, we define the Rossby critical number for which $v_{core_z min} = 0$ where there is a stagnation point, and we will call it ϑ_{cs}. In [1], for $\iota = 0.272$ and pipe expansion parameters η_1, η_2, η_3, we have found that ϑ_{cf} were 0.35, 0.44 and 0.53, respectively, while the core expansion parameters $\hat{\eta}$ were 1.25, 1.47 and 1.65, respectively.

By replacing these values in Eq. (20) when $v_{core_z min}$ is zero, we get the corresponding $\hat{\vartheta}_{cs}$ and then ϑ_{cs} for the vortex core. These are respectively 0.26, 0.38 and 0.49. That is to say that in all the cases we have $\vartheta_{cs} < \vartheta_{cf}$. Therefore, at the fold $\tilde{v}_z > 0$. This coincides with the results found by Buntine and Saffman [5] in their analysis using a three-parameter family inlet flow.

V. Conclusions

The main conclusions drawn from the previous sections are:

1. In the quasi-cylindrical approximation, the steady flow in the transition expansion region corresponding to a solid body rotation with

uniform axial flow as inlet flow has the same Beltrami flow structure as in the pipe downstream, which is compatible with the boundary conditions. Therefore, findings from our previous work on stability [1–3] can hold.

2. For fixed values of η and $\vartheta \geq \vartheta_c$, $\omega_\phi < 0$ and then $\tilde{v}_z$ in the transition region is an increasing function of r and a decreasing function of z reaching its the minimum on the axis at the outlet.

3. For fixed values of η, the minimum of $\tilde{v}_z$ on the axis is an increasing function of ϑ (Fig. 4), where the stagnation point corresponds to ϑ_c.

4. As a consequence, no branching off takes place for the solutions of Bragg–Hawthorne equation.

5. The critical Rossby number ϑ_c corresponding to stagnation is an increasing function of η (Fig. 5).

6. The whole picture can be reached by putting together these results with those obtained in [1], where there is a branching owing to the boundary conditions at the frontier between the vortex and the irrotational flow. Moreover, since the results in [1] for the rotational flow depend on the inlet flow as well as on the rotational expansion parameter $\hat{\eta}$ defined in Eq. (20), given a quasi-cylindrical approximation, it can be concluded that this expression is the minimum of v_z in the core. Therefore, we can get the critical Rossby number ϑ_{cs} and compare it with that corresponding to the fold ϑ_{cf}. This present work verifies that $\vartheta_{cs} < \vartheta_{cf}$, in accordance with Buntine and Saffman's results [5].

7. In the quasi-cylindrical approximation, previous results do not depend on the chosen profile. This can be explained by the boundary conditions chosen depending on the inlet flow and on the parameter expansion.

Acknowledgements - We would like to thank Unversidad Nacional de General Sarmiento for its support for this work, and our colleague Gabriela Di

Gesú for her advice on the English version of this paper.

———————

[1] R González, G Sarasúa, A Costa, *Kelvin waves with helical Beltrami flow structure*, Phys. Fluids **20**, 24106 (2008).

[2] R González, A Costa, E S Santini, *On a variational principle for Beltrami flows*, Phys. Fluids **22**, 74102 (2010).

[3] R González, E S Santini, *The dynamics of beltramized flows and its relation with the Kelvin waves*, J. Phys.: Conf. Ser. **296**, 012024 (2011).

[4] The Beltrami flow is defined as a field $\mathbf{v}_B$ that satisfies $\boldsymbol{\omega}_B = \nabla \times \mathbf{v}_B = \gamma \mathbf{v}_B$, with $\gamma = constant$. We say that the flow has a *beltrami flow structure* when it can be put in the form $\mathbf{v} = U\mathbf{e}_z + \Omega r\mathbf{e}_\theta + \mathbf{v}_B$, being U and Ω constants, i.e it is the superposition of a solid body rotation and translation with a Beltrami one. For a potential flow $\gamma = 0$.

[5] J D Buntine, P G Saffman, *Inviscid swirling flows and vortex breakdown*, Proc. R. Soc. Lond. A **449**, 139 (1995).

[6] G L Brown, J M Lopez, *Axisymmetric vortex breakdown Part 2. Physical mechanisms*, J. Fluid Mech. **221**, 573 (1990).

[7] B Benjamin, *Theory of the vortex breakdown phenomenon*, J. Fluid Mech. **14**, 593 (1962).

[8] R Guarga, J Cataldo, *A theoretical analysis of symmetry loss in high Reynolds swirling flows*, J. Hydraulic Res. **31**, 35 (1993).

[9] S L Bragg, W R Hawthorne, *Some exact solutions of the flow through annular cascade actuator discs*, J. Aero. Sci. **17**, 243 (1950).

[10] G K Batchelor, *An introduction to fluids dynamics*, Cambridge University Press, Cambridge (1967).

[11] S V Alekseenko, P A Kuibin, V L Okulov, *Theory of concentrated vortices. An introduction*, Springer-Verlag, Berlin Heidelberg (2007).

24

A criterion to identify the equilibration time in lipid bilayer simulations

Rodolfo D. Porasso,[1*] J.J. López Cascales,[2]

With the aim of establishing a criterion for identifying when a lipid bilayer has reached steady state using the molecular dynamics simulation technique, lipid bilayers of different composition in their liquid crystalline phase were simulated in aqueous solution in presence of $CaCl_2$ as electrolyte, at different concentration levels. In this regard, we used two different lipid bilayer systems: one composed by 288 DPPC (DiPalmitoylPhosphatidyl-Choline) and another constituted by 288 DPPS (DiPalmitoylPhosphatidylSerine). In this sense, for both type of lipid bilayers, we have studied the temporal evolution of some lipids properties, such as the surface area per lipid, the deuterium order parameter, the lipid hydration and the lipid-calcium coordination. From their analysis, it became evident how each property has a different time to achieve equilibrium. The following order was found, from faster property to slower property: coordination of ions $\approx$ deuterium order parameter > area per lipid $\approx$ hydration. Consequently, when the hydration of lipids or the mean area per lipid are stable, we can ensure that the lipid membrane has reached the steady state.

I. Introduction

Over the last few decades, different computational techniques have emerged in different fields of science, some of them being extensively implemented and used by a great number of scientists around the globe. Among others, the Molecular Dynamics (MD) simulation is a very popular computational technique, which is widely used to obtain insight with atomic detail of steady and dynamic properties in the fields of biology, physics and chemistry. In this regard, a critical aspect that must be identified in all the MD simulations is related to the required equilibration time to achieve a steady state.

This point is crucial in order to avoid simulation artifacts that could lead to wrong conclusions. Currently, with the increment of the computing power accessible to different investigation groups, much longer simulation trajectories are being carried out to obtain reliable information about the systems, with the purpose of approaching the time scale of the experimental phenomena. However, even when this fact is objectively desirable without further objections, nowadays, much longer equilibration times are arbitrarily being required by certain reviewers during the revision process. From our viewpoint, this should be thoroughly revised due to the following two main reasons: first, because it results in a limiting factor in the use of this technique by other research groups which cannot access to very expensive computing centers (assuming that authors provide enough evidence of the equilibration of the system). Second, to avoid wasting expensive computing time in the study of certain properties which do not require such long equilibration times, once the steady state of the system has been properly

*E-mail: rporasso@unsl.edu.ar

[1] Instituto de Matemática Aplicada San Luis (IMASL) - Departamento de Física, Universidad Nacional de San Luis/CONICET, D5700HHW, San Luis, Argentina.

[2] Universidad Politécnica de Cartagena, Grupo de Bioinformática y Macromoléculas (BioMac) Aulario II, Campus de Alfonso XIII, 30203 Cartagena, Murcia, Spain.

identified.

Phospholipid bilayers are of a high biological relevance, due to the fact that they play a crucial role in the control of the diffusion of small molecules, cell recognition, and signal transduction, among others. In our case, we have chosen the Phosphatidyl-Choline (PC) bilayer because it has been very well studied by MD simulations [1–7] and experimentally as well [8–14]. Furthermore, studies of the effects of different types of electrolytes on a PC bilayer have also been studied, experimentally [15–24] and by simulation [25–33].

As mentioned above, the Molecular Dynamics (MD) simulations have emerged during the last decades as a powerful tool to obtain insight with atomic detail of the structure and dynamics of lipid bilayers [34–36]. Several MD simulations of membranes under the influence of different salt concentrations have been carried out. One of the main obstacles related to these studies has been the time scale associated to the binding process of ions to the lipid bilayer. Considering the literature, a vast dispersion of equilibration times associated to the binding of ions to the membrane has been reported, where values ranging from 5 to 100 ns have been suggested for monovalent and divalent cations [25, 27–29, 32, 37, 38]. In this regard, we carried out four independent simulations of a lipid bilayer formed by 288 DPPC in aqueous solutions, for different concentrations of $CaCl_2$ to provide an overview of their equilibration times. Among other properties, the surface area per lipid, the deuterium order parameters, lipid hydration and lipid-calcium coordination were studied.

Finally, in order to generalize our results, a bilayer formed by 288 DPPS in its liquid crystalline phase, in presence of $CaCl_2$ at 0.25N, was simulated as well .

II.   Methodology

Different molecular Dynamics (MD) simulations of lipid bilayer formed by 288 DPPC were carried out in aqueous solutions for different concentrations of $CaCl_2$, from 0, up to 0.50 N. Furthermore, with the aim of generalizing our results, a bilayer of 288 DPPS in presence of $CaCl_2$ at 0.25 N was simulated as well. Note that the concentration of $CaCl_2$ in terms of normality is defined as:

Type of Lipid	[CaCl$_2$] N	Ca^{2+}	Cl$^-$	Water
DPPC	0	0	0	10068
DPPC	0.06	5	10	10053
DPPC	0.13	12	24	10032
DPPC	0.25	23	46	9999
DPPC	0.50	46	92	9930
DPPS	0.25	204	120	26932

Table 1: The simulated bilayer systems. Note that the salt concentration is given in normal units. The numerals describe the number of molecules contained in the simulation box.

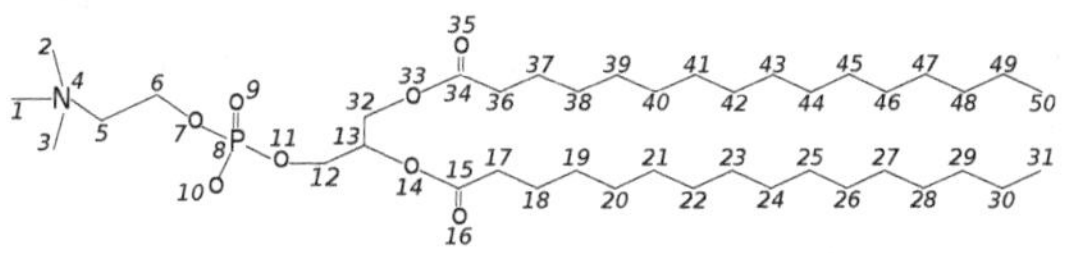

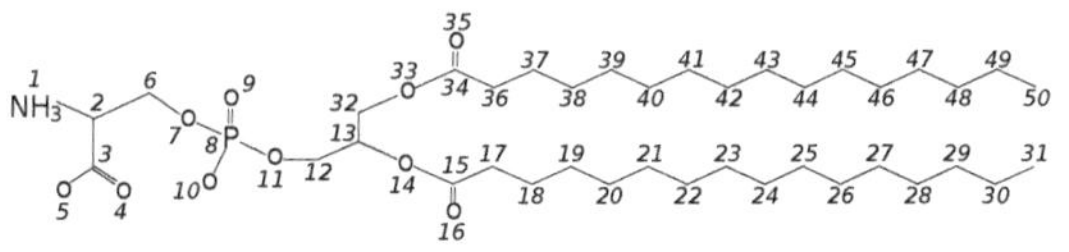

Figure 1: Structure and atom numbers for DPPC and DPPS used in this work.

$$normality = \frac{n_{equivalent\ grams}}{l_{solution}} \quad (1)$$

where $n_{equivalent\ grams} = \frac{gr(solute)}{equivalent\ weight}$ and equivalent weight $= \frac{Molecular\ weight}{n}$, being n the charge of the ions in the solution.

In Table 1, the number of molecules that constitute each system, applying Eq. (1), is summarized.

To build up the original system, a single DPPC lipid molecule, or DPPS lipid (Fig. 1), was placed with its molecular axis perpendicular to the membrane surface (xy plane). Next, each DPPC, or DPPS, was randomly rotated and copied 144 times on each leaflet of the bilayer. Finally, the gaps existing in the computational box (above and below the phospholipid bilayer) were filled using an equilibrated box containing 216 water molecules of the extended simple point charge (SPC/E) [39] water model.

Thus, the starting point of the first system of Table 1 was formed by 288 DPPC in absence of

CaCl$_2$. Once this first system was generated, the whole system was subjected to the steepest descent minimization process to remove any excess of strain associated with overlaps between neighboring atoms of the system. Thereby, the following DPPC systems in presence of CaCl$_2$ were generated as follows: to obtain a [CaCl$_2$]=0.06 N, 15 water molecules were randomly substituted by 5 Ca^{2+} and 10 Cl$^-$. An analogous procedure was applied to the rest of the systems, where 36, 69 and 138 water molecules were substituted by 12, 23 and 46 Ca^{2+} and 24, 46 and 92 Cl$^-$, to obtain a [CaCl$_2$] concentration of 0.13 N, 0.25 N and 0.50 N, respectively. Finally, the DPPS bilayer was generated following the same procedure described above for the DPPC, starting from a single DPPS molecule and once the lipid bilayer in presence of water passed the minimization process, 324 water molecules were substituted by 204 Ca^{2+} and 120 Cl$^-$ (note that 144 of the 204 calcium ions were added to balance the negative charge associated with the DPPS).

GROMACS 3.3.3 package [40,41] was used in the simulations, and the properties showed in this work were obtained using our own code. The force field proposed by Egberts et al. [2] was used for the lipids, and a time step of 2 fs was used as integration time in all the simulations. A cut-off of 1.0 nm was used for calculating the Lennard-Jones interactions. The electrostatic interaction was evaluated using the particle mesh Ewald method [42,43]. The real space interaction was evaluated using a 0.9 nm cut-off, and the reciprocal space interaction using a 0.12 nm grid with a fourth-order spline interpolation. A semi-isotropic coupling pressure was used for the coupling pressure bath, with a reference pressure of 1 atm which allowed the fluctuation of each axis of the computer box independently. For the DPPC bilayer, each component of the system (i.e., lipids, ions and water) was coupled to an external temperature coupling bath at 330 K, which is well above the transition temperature of 314 K [44,45]. For DPPS bilayer, each component of the system was coupled to an external temperature coupling bath at 350 K, which is above the transition temperature [46,47]. All the MD simulations were carried out using periodic boundary conditions. The total trajectory length of each simulated system was of 80 ns of MD simulation, where the coordinates of the system were recorded every 5 ps for their appropriate analysis.

Finally, in order to study the effect of the temperature, only the case corresponding to 0.25 N CaCl$_2$ was investigated at two additional temperatures, 340 K and 350 K.

III. Results and discussion

i. Effect of the CaCl$_2$ concentration

a. Surface area per lipid

Surface area per lipid $\langle A \rangle$ is a property of lipid bilayers which has been accurately measured from experiments [48]. The calculation of mean area per lipid can be determined from the MD simulation as:

$$\langle A \rangle = \frac{x \cdot y}{N} \qquad (2)$$

where x and y represent the box sizes in the direction x and y (perpendicular to the membrane surface) over the simulation, and N is the number of lipids contained in one leaflet, in our case $N = 144$.

Focusing on the study of the time evolution of the area per lipid, Figure 2 depicts the running

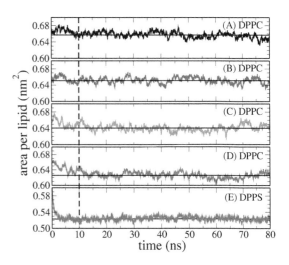

Figure 2: Running area per lipid at T = 330 K in presence of [CaCl$_2$] at (A) 0.06 N, (B) 0.13 N, (C) 0.25 N, (D) 0.50 N and (E) 0.25 N (in this case, T = 350 K). Solid lines represent the mean area obtained from the last 70 ns of the simulated trajectories (see text for further explanation). The type of lipid is indicated in the legends.

surface area per lipid for different concentrations of CaCl$_2$ and type of lipid. In general, for the 5 bilayers formed by DPPC or DPPS, the area per lipid achieved a steady state after 10 ns of simulation, being this equilibration time almost independent of the concentration of CaCl$_2$ and type of lipid which composed the membrane.

In absence of salt, an average area per lipid of $\langle A \rangle = 0.663 \pm 0.008$ nm^2 was calculated from the last 70 ns of the simulated trajectory, discarding the first 10 ns corresponding to the equilibration time. This value agrees with experimental data, where values in a range from 0.55 to 0.72 nm^2 have been measured [10, 11, 48–51]. Table 2 shows the mean surface area per lipid (again, after discarding the equilibration time of 10 ns) with their corresponding error bar.

From the simulation results, a shrinking in the surface area per lipid with the increment of the ionic strength of the solution is observed. This shrinking is expected and attributed to the complexation of lipid molecules by calcium, such as it has been pointed out in previous studies [28,29,52].

b. *Deuterium order parameter*

The deuterium order parameter, S_{CD}, is measured from ^{2}H-NMR experiments. This parameter provides relevant information related to the disorder of the hydrocarbon region in the interior of the lipid bilayers by measuring the orientation of the hydrogen dipole of the methylene groups with respect to the perpendicular axis to the lipid bilayer. Due to the fact that hydrogens of the lipid methylene groups (CH$_2$) have not been taken into account (in an explicit way) in our simulations, the order parameter $-S_{CD}$ on the $i+1$ methylene group was defined as the normal unitary vector to the vector defined from the i to the $i+2$ CH$_2$ group and contained in the plane formed by the methylene groups i, $i+1$ and $i+2$. Thus, the deuterium order parameter $-S_{CD}$ on the $i-th$ of the CH$_2$ group can be estimated by Molecular Dynamics simulations as follows:

$$-S_{CD} = \frac{1}{2}\langle 3\cos^2(\theta) - 1 \rangle \qquad (3)$$

where θ is the angle formed between the unitary vector defined above and the z axis. The expression in brackets $\langle \ldots \rangle$ denotes an average over all

Type of Lipid	[CaCl$_2$] N	$\langle A \rangle$ (nm^2)	Hydration Number
DPPC	0	0.663 ±0.008	1.758 ±0.009
DPPC	0.06	0.658 ±0.008	1.740 ±0.009
DPPC	0.13	0.651 ±0.007	1.719 ±0.010
DPPC	0.25	0.641 ±0.009	1.680 ±0.015
DPPC	0.50	0.628 ±0.010	1.610 ±0.015
DPPS	0.25	0.522 ±0.007	2.552 ±0.010

Table 2: Area per lipid and lipid hydration number as a function of salt concentration (see text for further explanation). Note that the salt concentration is given in normal units. Error bars were calculated for each system separately from subtrajectories of 10 ns length. Simulation temperature = 330 K.

the lipids and time. Hence, note that the $-S_{CD}$ can adopt any value between -0.5 (corresponding to a parallel orientation to the lipid/water interface) and 1 (oriented along the normal axis to the lipid bilayer).

Figure 3 shows the running $-S_{CD}$ for different carbons of the DPPC and DPPS tails and salt concentrations. Only the carbons which correspond to the initial (hydrocarbons 2 and 6), the middle (hydrocarbon 10) and final (hydrocarbons 13 and 15) methylene groups of the lipid tails were depicted in this figure. Each point of the figure represents the average values of $-S_{CD}$ over 5 ns of subtrajectory length, and the lines represent the mean values calculated from the last 70 ns of the trajectories simulated. From this figure, it is observed how in all the cases, the required equilibration time is less than 10 ns of simulation time, independently of the salt concentration and the type of lipid. Finally, it is noted that Figure 3 exhibits an increase in the deuterium order parameters with the salt concentration, consistent with the shrinking of the area per lipid described above.

c. *Lipid hydration*

To analyze the lipid hydration, the radial distribution function $g(r)$ of water around one of the oxygens of the phosphate group (atom number 10 in Fig. 1 for DPPC and DPPS) was calculated. The radial distribution function $g(r)$ is defined as follows:

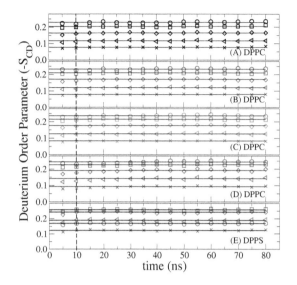

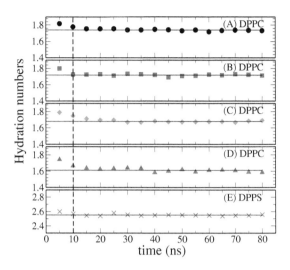

Figure 3: Running deuterium order parameter, $-S_{CD}$, in presence of [CaCl$_2$] at (A) 0.06 N, (B) 0.13 N, (C) 0.25 N, (D) 0.50 N and (E) 0.25 N. DPPC simulations were performed at 330 K and DPPS simulation temperature was 350 K. Solid lines represent mean values $-S_{CD}$ obtained from the last 70 ns of the simulated trajectories. The type of lipid is indicated in the legends. Symbols: ∘ hydrocarbon 2; ⋄ hydrocarbon 6; ◁ hydrocarbon 10; + hydrocarbon 13 and × hydrocarbon 15. Note that the error bars have the same size as symbol.

$$g(r) = \frac{N(r)}{4\pi r^2 \rho \delta r} \qquad (4)$$

where $N(r)$ is the number of atoms in a spherical shell at distance r and thickness δr from a reference atom. ρ is the density number taken as the ratio of atoms to the volume of the total computing box.

From numerical integration of the first peak of the radial distribution function, the hydration numbers can be estimated for different atoms of the DPPC or DPPS. Figure 4 depicts the hydration number of phosphate oxygen (atom 10 in Fig. 1 for DPPC and DPPS) in presence of CaCl$_2$, where each point represents the average of 5 ns subtrajectory length. These results show how this property reached a steady state for the cases (A), (B) and (E), after 10 ns of simulation. However, for the cases (C) and (D), 5 ns of extra simulation trajectory were required to reach a steady state. Table 2

Figure 4: Hydration number of the phosphate oxygen (atom 10 in Fig. 1) along the simulated trajectories in presence of [CaCl$_2$] at (A) 0.06 N, (B) 0.13 N, (C) 0.25 N, (D) 0.50 N (for DPPC T = 330 K) and (E) 0.25 N. In this case, T = 350 K. Solid lines represent the mean value of the hydration number calculated from the last 70 ns of the simulated trajectories. The type of lipid is indicated in the legends. Note that the error bars have the same size as symbol.

shows the hydration numbers for the last 70 ns of the trajectory length, corresponding to four concentrations of CaCl$_2$ and both types of lipids, DPPC and DPPS. In this regard, from Fig. 4, the significant lipid dehydration with the increment of the ionic strength of the solution is evident, in good accordance with previous results [52].

d. Phospholipid-calcium coordination

Some authors have reported how the lipid coordination by divalent cations widely varies . Thus, on the one hand, some authors [25] have reported that this is a very slow process, which requires about 85 ns of simulation time, but, on the other hand, other authors [26] have suggested that this process results much more rapid, taking less than 1 ns. In this sense, the coordination of DPPC-Ca^{2+} was studied by monitoring the oxygen-calcium coordination of the carbonyl oxygens (atoms 16 and 35 in Fig. 1) and phosphate oxygens (atoms 9 and 10 of DPPC

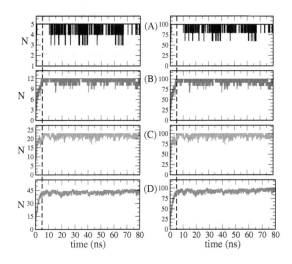

T (K)	$\langle A \rangle$ (nm^2)	Hydration Number
330	0.642 $\pm$0.009	1.680 $\pm$ 0.010
340	0.650 $\pm$0.007	1.683 $\pm$ 0.020
350	0.666 $\pm$0.008	1.689 $\pm$ 0.015

Table 3: Area per lipid and the lipid hydration number as a function of temperature. Error bars were calculated from subtrajectories of 10 ns length. DPPC bilayer in presence of [CaCl$_2$] = 0.25 N.

Figure 5: Left column represents the number of Ca^{2+} coordinated to lipids in presence of [CaCl$_2$] at (A) 0.06 N, (B) 0.13 N, (C) 0.25 N and (D) 0.50 N, for T = 330 K. Right column shows the quantity of calcium ions coordinated to lipids expressed in percentage, along the simulated trajectory.

in Fig. 1), as a function of time. The left column in Fig. 5 represents the oxygen coordination number, while the right one depicts the percentage of calcium ions involved in the coordination process with respect to the total number of calcium ions present in the aqueous solution. Figure 5 shows how the DPPC coordination by calcium is a quick process, taking less than 5 ns of simulation time to achieve a steady state. The kinetic of this process appears to be related to the ratio between calcium/lipid. Thus, after the first 5 ns of simulation time, the Ca–lipid coordination presents some fluctuation along the rest of the simulated trajectory. However, in Fig. 5 (A) and (B) (for the cases of lower concentration), it is observed how the percentage of coordination fluctuates between a 60% and a 100%. We consider that this broad fluctuation is related to the limited sample size of our simulations that introduces a certain noise in our results.

ii. Effect of temperature

This section focuses on the study of the role played by temperature on the equilibration process. In this regard, only the system corresponding to a concentration of 0.25 N in CaCl$_2$ was studied, for a range of temperatures from 330 K to 350 K (all of them above the transition temperature of 314 K [44, 45] for the DPPC).

Figure 6 shows the running area along the trajectory. In this case, it was noticed how the systems achieve a steady state after a trajectory length of roughly 10 ns, where Table 3 shows the mean area per lipid calculated from the last 70 ns of simulation time. Figure 7 shows the deuterium order parameter of the methylene groups along the lipid tails, calculated from Eq. (3). Figure 7, on the one hand, clearly shows that for the three temperatures the systems have reached the steady state before the first 10 ns of simulation. On the other hand, it shows an increase in the disorder of the lipid tails with temperature, which is closely related with the increase of the area per lipid, such as it was pointed out above. Figure 8 depicts the results of the hydration numbers of DPPC for the three temperatures studied, where the equilibrated state was achieved after 10 ns of simulation time. Table 3 provides the calculated hydration numbers in the equilibrium, showing how the lipid hydration remained invariable with the rising of the temperature. Concerning the lipid-calcium coordination, Fig. 9 represents the lipid-calcium coordination number, and the right column represents the calcium that participates in the coordination expressed in percentage respect the total of calcium ions in solution. From simulation, it becomes evident how calcium ions required less than 5 ns to achieve an equilibrated state for the three temperatures studied. In summary, for all the properties studied in this section, a slight decrease in the equilibration time with the increasing temperature was appreciated.

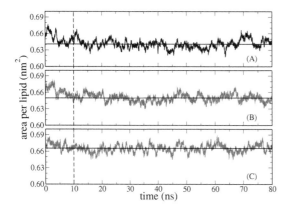

Figure 6: Running area per lipid for $[CaCl_2] = 0.25$ N at different temperatures, (A) T = 330 K, (B) T = 340 K and (C) T = 350 K. Solid lines represent the mean values obtained from the last 70 ns of simulation.

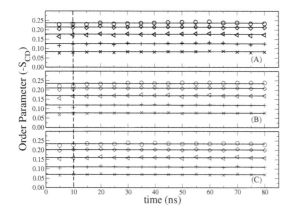

Figure 7: Deuterium Order Parameter, $-S_{CD}$, along the simulated trajectory for a concentration of $[CaCl_2] = 0.25$ N, for the following temperatures: (A) T = 330 K, (B) T = 340 K and (C) T = 350 K. Solid lines represent the average order parameter for the last 70 ns of simulation. Note that the error bars have the same size as symbol.

IV.   Conclusions

The present work deals with the simulation time required to achieve the steady state for a lipid bilayer system in presence of $CaCl_2$. In this regard, we studied two different systems: one with DPPC and another one with DPPS bilayer; both systems in

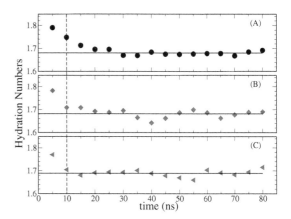

Figure 8: Hydration number of phosphate oxygen (atom 10 in Fig. 1) along the simulated trajectories for a $[CaCl_2] = 0.25$ N at different temperatures: (A) T = 330 K, (B) T = 340 K and (C) T = 350 K. Solid lines represent the average hydration number for the last 70 ns of simulation. Note that the error bars have the same size as symbol.

presence of $CaCl_2$ (at different level concentration). The salt free case was also studied, as control. The analysis of various lipid properties studied here indicates that some properties reach the steady state more quickly than others. In this sense, we found that the area per lipid and the hydration number are slower than the deuterium order parameter and the coordination of cations. Consequently, to ensure that a system composed by a lipid bilayer has reached a steady state, the criterion that we propose is to show that the area per lipid or the hydration number have reached the equilibrium.

From our results, two important aspects should be remarked:

1. The equilibration time is strongly dependent on the starting conformation of the system. Wrong starting conformations will require much longer equilibration times, even of one order of magnitude higher than the requested from a more refined starting conformation.

2. Temperature is a critical parameter for reducing the equilibration time in our simulations, due to the fact that higher temperatures increase the kinetic processes, i.e., the sampling of the configurational space of the system.

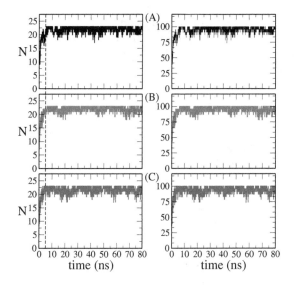

Figure 9: Left column represents the number of calcium ions involved in the lipid coordination along time for a concentration of $[CaCl_2]=0.25$ N at different temperatures: (A) T = 330 K, (B) T = 340 K and (C) T = 350 K. The right column shows the same information expressed as a percentage of the total number of calcium ions in solution.

Acknowledgements - Authors wish to thank the assistance of the Computing Center of the Universidad Politécnica de Cartagena (SAIT), Spain. RDP is member of 'Carrera del Investigador', CONICET, Argentine.

[1] O Berger, O Edholm, F Jahnig, *Molecular dynamics simulations of a fluid bilayer of dipalmitoylphosphatidylcholine at full hydration, constant pressure, and constant temperature*, Biophys. J. **72**, 2002 (1997).

[2] E Egberts, S J Marrink, H J C Berendsen, *Molecular dynamics simulation of a phospholipid membrane*, Eur. Biophys. J. **22**, 423 (1994).

[3] U Essmann, L Perera, M L Berkowitz, *The origin of the hydration interaction of lipid bilayers from MD simulation of dipalmitoylphosphatidylcholine membranes in gel and crystalline phases*, Langmuir **11**, 4519 (1995).

[4] S E Feller, Y Zhang, R W Pastor, R B Brooks, *Constant pressure molecular dynamics simulations: The Langevin piston method*, J. Chem. Phys. **103**, 4613 (1995).

[5] W Shinoda, T Fukada, S Okazaki, I Okada, *Molecular dynamics simulation of the dipalmitoylphosphatidylcholine (DPPC) lipid bilayer in the fluid phase using the Nosr-Parrinello-Rahman NPT ensemble*, Chem. Phys. Lett. **232**, 308 (1995).

[6] D P Tieleman, H J C Berendsen, *Molecular dynamics simulations of a fully hydrated dipalmitoylphosphatidylcholine bilayer with different macroscopic boundary conditions and parameters*, J. Chem. Phys. **105**, 4871 (1996).

[7] K Tu, D J Tobias, M L Klein, *Constant pressure and temperature molecular dynamics simulation of a fully hydrated liquid crystal phase dipalmitoylphosphatidylcholine bilayer*, Biophys. J. **69**, 2558 (1995).

[8] M F Brown, *Theory of spin-lattice relaxation in lipid bilayers and biological membranes. Dipolar relaxation*, J. Chem. Phys. **80**, 2808 (1984).

[9] M F Brown, *Theory of spin-lattice relaxation in lipid bilayers and biological membranes. ^2H and ^{14}N quadrupolar relaxation*, J. Phys. Chem. **77**, 1576 (1982).

[10] J F Nagle, R Zang, S Tristam-Nagle, W S Sun, H I Petrache, R M Suter, *X-ray structure determination of fully hydrated L. phase dipalmitoylphosphatidylcholine bilayers*, Biophys. J. **70**, 1419 (1996).

[11] R P Rand, V A Parsegian, *Hydration forces between phospholipid bilayers*, Biochim. Biophys. Acta **988**, 351 (1989).

[12] J Seelig, *Deuterium magnetic resonance: Theory and application to lipid membranes*, Q. Rev. Biophys. **10**, 353 (1977).

[13] J Seelig, A Seelig, *Lipid conformation in model membranes and biological systems*, Q. Rev. Biophys. **13**, 19 (1980).

[14] W J Sun, R M Suter, M A Knewtson, C R Worthington, S Tristram-Nagle, R Zhang, J F Nagle, *Order and disorder in fully hydrated unoriented bilayers of gel phase dipalmitoylphosphatidylcholine*, Phys. Rev. E. **49**, 4665 (1994).

[15] H Akutsu, J Seelig, *Interaction of metal ions with phosphatidylcholine bilayer membranes*, Biochemistry **20**, 7366 (1981).

[16] M G Ganesan, D L Schwinke, N Weiner, *Effect of Ca^{2+} on thermotropic properties of saturated phosphatidylcholine liposomes*, Biochim. Biophys. Acta **686**, 245 (1982).

[17] L Herbette, C A Napolitano, R V McDaniel, *Direct determination of the calcium profile structure for dipalmitoyllecithin multilayers using neutron diffraction*, Biophys. J. **46**, 677 (1984).

[18] D Huster, K Arnold, K Gawrisch, *Strength of Ca^{2+} binding to retinal lipid membrane: Consequences for lipid organization*, Biophys. J. **78**, 3011 (2000).

[19] Y Inoko, T Yamaguchi, K Furuya, T Mitsui, *Effects of cations on dipalmitoyl phosphatidylcholine/cholesterol/water systems*, Biochim. Biophys. Acta **413**, 24 (1975).

[20] R Lehrmann, J J Seelig, *Adsorption of Ca^{2+} and La^{3+} to bilayer membranes: Measurement of the adsorption enthalpy and binding constant with titration calorimetry*, Biochim. Biophys. Acta **1189**, 89 (1994).

[21] L J Lis, W T Lis, V A Parsegian, R P Rand, *Adsorption of divalent cations to a variety of phosphatidylcholine bilayers*, Biochemistry **20**, 1771 (1981).

[22] L J Lis, V A Parsegian, R P Rand, *Binding of divalent cations to dipalmitoylphosphatidylcholine bilayers and its effect on bilayer interaction*, Biochemistry **20**, 1761 (1981).

[23] T Shibata, *Pulse NMR study of the interaction of calcium ion with dipalmitoylphosphatidylcholine lamellae*, Chem. Phys. Lipids. **53**, 47 (1990).

[24] S A Tatulian, V I Gordeliy, A E Sokolova, A G Syrykh, *A neutron diffraction study of the influence of ions on phospholipid membrane interactions*, Biochim. Biophys. Acta **1070**, 143 (1991).

[25] R A Böckmann, H Grubmüller, *Multistep binding of divalent cations to phospholipid bilayers: A molecular dynamics study*, Angewandte Chemie **43**, 1021 (2004).

[26] J Faraudo, A Travesset, *Phosphatidic acid domains in membranes: Effect of divalent counterions*, Biophys. J. **92**, 2806 (2007).

[27] A A Gurtovenko, *Asymmetry of lipid bilayers induced by monovalent salt: Atomistic molecular-dynamics study*, J. Chem. Phys. **122**, 244902 (2005).

[28] P Mukhopadhyay, L Monticelli, D P Tieleman, *Molecular dynamics simulation of a palmitoyl-oleoyl phosphatidylserine bilayer with Na^+ counterions and NaCl*, Biophys. J. **86**, 1601 (2004).

[29] S A Pandit, D Bostick, M L Berkowitz, *Molecular dynamics simulation of a dipalmitoylphosphatidylcholine bilayer with NaCl*, Biophys. J. **84**, 3743 (2003).

[30] U R Pedersen, C Laidy, P Westh, G H Peters, *The effect of calcium on the properties of charged phospholipid bilayers*, Biochim. Biophys. Acta **1758**, 573 (2006).

[31] J N Sachs, H Nanda, H I Petrache, T B Woolf, *Changes in phosphatidylcholine headgroup tilt and water order induced by monovalent salts: Molecular dynamics simulations*, Biophys. J. **86**, 3772 (2004).

[32] K Shinoda, W Shinoda, M Mikami, *Molecular dynamics simulation of an archeal lipid bilayer whit sodium chloride*, Phys. Chem. Chem. Phys. **9**, 643 (2007).

[33] N L Yamada, H Seto, T Takeda, M Nagao, Y Kawabata, K Inoue, *SAXS, SANS and NSE studies on unbound state in $DPPC/water/CaCl_2$ system*, J. Phys. Soc. Jpn. **74**, 2853 (2005).

[34] D Frenkel, B Smit, *Understanding molecular simulations*, Academic Press, New York (2002).

[35] J J López Cascales, J García de la Torre, S J Marrink, H J C Berendsen, *Molecular dynamics simulation of a charged biological membrane*, J. Chem. Phys. **104**, 2713 (1996).

[36] W F van Gunsteren, H J C Berendsen, *Computer simulations of molecular dynamics: Methodology, applications and perspectives in chemistry*, Angew. Chem Int. Ed. Engl. **29**, 992 (1990).

[37] R A Böckmann, A Hac, T Heimburg, H Grubmüller, *Effect of sodium chloride on a lipid bilayer*, Biophys. J. **85**, 1647 (2003).

[38] A A Gurtovenko, I Vattulainen, *Effect of NaCl and KCl on phosphatidylcholine and phosphatidylethanolamine lipid membranes: Insight from atomic-scale simulations for understanding salt-induced effects in the plasma membrane*, J. Phys. Chem. B. **112**, 1953 (2008).

[39] H J C Berendsen, J R Grigera, T P Straatsma, *The missing term in effective pair potentials*, J. Phys. Chem. **91**, 6269 (1987).

[40] H J C Berendsen, D van der Spoel, R van Drunen, *A message-passing parallel molecular dynamics implementation*, Comp. Phys. Comm. **91**, 43 (1995).

[41] E Lindahl, B Hess, D van der Spoel, *GROMACS 3.0: A package for molecular simulation and trajectory analysis*, J. Mol. Mod. **7**, 306 (2001).

[42] T Darden, D York, L Pedersen, *Particle mesh Ewald: An N.log(N) method for Ewald sums in large systems*, J. Chem. Phys. **98**, 10089 (1993).

[43] U Essmann, L Perea, M L Berkowitz, T Darden, H Lee, L G Pedersen, *A smooth particle mesh Ewald method*, J. Chem. Phys. **103**, 8577 (1995).

[44] L R De Young, K A Dill, *Solute partitioning into lipid bilayer-membranes*, Biochemistry **27**, 5281 (1988).

[45] A Seelig, J Seelig, *The dynamic structure of fatty acyl chains in a phospholipid bilayer measured by deuterium magnetic resonance*, Biochemistry **13**, 4839 (1974).

[46] G Cevc, A Watts, D Marsh, *Titration of the phase transition of phosphatidilserine bilayer membranes. Effect of pH, surface electrostatics, ion binding and head-group hydration*, Biochemistry **20**, 4955 (1981).

[47] H Hauser, F Paltauf, G G Shipley, *Structure and thermotropic behavior of phosphatidylserine bilayer membranes*, Biochemistry **21**, 1061 (1982).

[48] J F Nagle, S Tristam-Nagle, *Structure of lipid bilayers*, Biochim. Biophy. Acta **1469**, 159 (2000).

[49] B A Lewis, D M Engelman, *Lipid bilayer thickness varies linearly with acyl chain length in fluid phosphatidylcholine vesicles*, J. Mol. Biol. **166**, 211 (1983).

[50] R J Pace, S I Cham, *Molecular motions in lipid bilayer. I. Statistical mechanical model of acyl chains motion*, J. Chem. Phys. **76**, 4217 (1982).

[51] R L Thurmond, S W Dodd, M F Brown, *Molecular areas of phospholipids as determined by 2H NMR spectroscopy*, Biphys. J. **59**, 108 (1991).

[52] R D Porasso, J J López Cascales, *Study of the effect of Na^+ and Ca^{2+} ion concentration on the structure of an asymmetric DPPC/DPPS + DPPS lipid bilayer by molecular dynamics simulation*, Coll. and Surf. B. Bioint. **73**, 42 (2009).

25

Natural and laser-induced cavitation in corn stems: On the mechanisms of acoustic emissions

E. Fernández,[1] R. J. Fernández,[1] G. M. Bilmes[2]*

Water in plant xylem is often superheated, and therefore in a meta-stable state. Under certain conditions, it may suddenly turn from the liquid to the vapor state. This cavitation process produces acoustic emissions. We report the measurement of ultrasonic acoustic emissions (UAE) produced by natural and induced cavitation in corn stems. We induced cavitation and UAE *in vivo,* in well controlled and reproducible experiments, by irradiating the bare stem of the plants with a continuous-wave laser beam. By tracing the source of UAE, we were able to detect absorption and frequency filtering of the UAE propagating through the stem. This technique allows the unique possibility of studying localized embolism of plant conduits, and thus to test hypotheses on the hydraulic architecture of plants. Based on our results, we postulate that the source of UAE is a transient "cavity oscillation" triggered by the disruptive effect of cavitation inception.

I. Introduction

The cohesion-tension theory suggests that water in the xylem of transpiring plants is under tension with a hydrostatic pressure below atmospheric and, thus, most of the time at "negative" values [1]. Negative pressures means that water in the xylem has a reduced density compared to equilibrium [2]. According to its phase diagram, water under these conditions is overheated (i.e., in a meta-stable state). Therefore, it should not be in the liquid but in the vapor phase [3]. The molecules in the liquid phase are further away from each other, but their mutual attraction allows the system to remain unchanged.

Under sufficiently high tension (i.e., low pressures caused by water deficit), xylem may fail to maintain this state, causing liquid water to turn into vapor in a violent way. This phenomenon, usually known as cavitation, causes the embolism of the conduits, reducing tissue hydraulic conductivity and exacerbating plant physiological stress [4,5]. Some herbaceous species are known to sustain cavitation almost every day, repairing embolism during the night, while most woody species preclude cavitation occurrence by a combination of stomatal behavior and anatomical and morphological adjustment [6].

Cavitation events in xylem produce sound [7,8]. In 1966, Milburn and Johnson developed a technique to detect sound by registering 'clicks' in a record player pick-up head attached to stressed plants and connected to an amplifier [9]. They associated this sound emission with the rupture of the water column in xylem vessels. Since then, several authors have used this audible acoustic emis-

*E-mail: gabrielb@ciop.unlp.edu.ar

[1] IFEVA, Facultad de Agronomía, Universidad de Buenos Aires y CONICET, Av. San Martín 4453, C1417DSE Buenos Aires, Argentina.

[2] Centro de Investigaciones Opticas (CONICET-CIC) and Facultad de Ingeniería, Universidad Nacional de La Plata, Casilla de Correo 124, 1900 La Plata, Argentina.

sion technique to measure xylem cavitation [10,11]. Later on, some authors have improved the technique by detecting ultrasonic acoustic emissions (UAE) [12–17]. These authors have demonstrated a good correlation between UAE and cavitation. However, the connection between audible or ultrasonic acoustic emissions and cavitation phenomena on xylem vessels remains unexplained. Tyree and Dixon [12] proposed four possible sources of acoustic emissions that we will consider in the Discussion section. Others authors have developed explanations based on alternatives to the cohesion-tension theory [3,18]

One of the problems of studying cavitation in plants is the spontaneous character of the phenomenon, so far precluding our ability to produce it in a controlled way. Most cavitation experiments use transpiration to raise xylem water tension, the trigger for cavitation events. In some cases, xylem tension was increased by centrifugation [19,20], but even there, cavitation events took place rather randomly along the water column. On the other hand, in order to study bubble behavior in isolated physical systems, several authors explored the generation of cavitation phenomena using lasers [21–24]. In these experiments, cavitation bubbles are generated in a very well defined location taking advantage of the accuracy of the laser beam. Even though this technique was developed to generate cavitation in transparent environments, we wondered whether it could be used in biological systems to generate cavitation at specific locations along the stem.

In this article, we report spontaneous UAE produced by natural cavitation in xylem vessels of corn (*Zea mays* L.) stems and we characterized and classified the signals. We also developed a method to produce laser-induced cavitation and UAE events in a controlled way by irradiating plants with a continuous-wave laser. We performed experiments with this method to study the generation and propagation of UAE. Our results allowed us to explain the connection between cavitation and UAE, as well as the relationship between signal frequency and the localization of the source in the stem.

II. Materials and methods

Corn plants were grown in a greenhouse in 3l pots containing sand. They were watered at field capac-

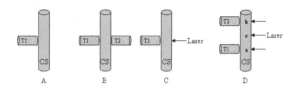

Figure 1: Set-up for the different experiments. (A) Experiment 1. (B) Experiments 2 and 4. (C) Experiment 3. (D) Experiment 5. CS: corn stem; T1 and T2: PZT transducers for UAE detection.

ity every 1-2 days with nutritive solution (3 g l^{-1} of KSC II – Roulier). After four months, tasseling plants (around 1 m high and 12 mm stem width) were used to perform experiments under different conditions: total darkness (D); room diffuse light (RL; PAR ca. 100 mE m^{-2} s^{-1}); leaf illumination with a 150 W incandescent lamp (IL; placed ca. 0.5 m away), and laser irradiation (L). In the latter case, experiments were carried out by directing the beam of a 50 mW He–Ne red laser (630 nm), or a continuous-wave (CW) Ar-ion laser (Spectra Physics Model 165/09) directly on to the stems. Most experiments were conducted sequentially in 3-5 plants, and we report the full range of observed results.

Ultrasonic acoustic emissions generated in the stems of plants were recorded by home-made PZT-based piezoelectric transducers (4 × 4 mm, 230 kHz) [25] coated with glycerin and clamped to the bare stem by a three-prong thumbtack. Signals, of the order of 1 mV, were amplified (gain 10^3) and recorded in a storage digital oscilloscope. Different transducer positions in the stems were explored, as well as simultaneous measurement of UAE with two detectors attached to different points of the stems, providing a method to trace the origin of the signals (Fig. 1).

III. Results

i. Measurements of spontaneous UAE

In the first experiment (*Experiment 1*), a transducer was attached to the bare stem on an internode with 5–7 developed leaves above it. UAE were monitored in the dark (D), under room light (RL) and under incandescent lamp (IL) illumination. The experiment was performed with several

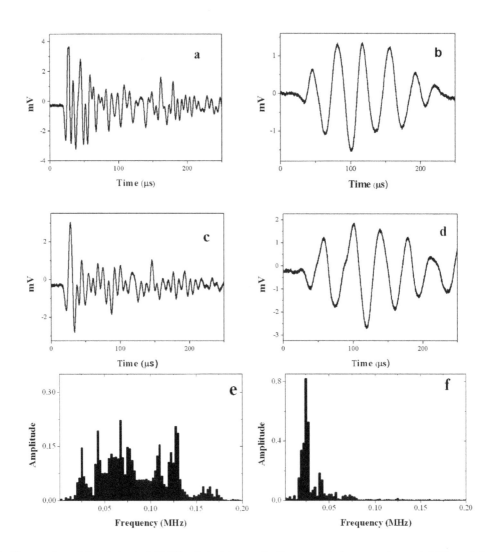

Figure 2: Examples of the detected UAE related to cavitation events in corn stem. (a) Type 1 broadband frequency emission signals detected at room light. (b) Type 2 low frequency signals detected at room light. (c) Type 1 signals detected with the laser impinging near the detector. (d) Type 2 signals detected with the laser impinging far from the detector. (e) and (f) Frequency spectra of type 1 and type 2 signals detected with laser. See the similarity of the signals produced with laser and those detected at room light.

plants changing the sequence conditions of light (D–RL–IL; RL–D–IL, etc.). We registered no emissions in the dark. In experiments 2–3 h long, a rate of 1.15 ± 0.09 emissions $\min^{-1}$ were detected when the plant was transpiring under ambient light. This rate was increased to 1.45 ± 0.15 emissions $\min^{-1}$ when transpiration was stimulated with an incandescent lamp. The change in the rate of emissions between RL and IL took place less than a minute after turning the lamp on or off. When two transducers were attached to the bare stem at the same height but in different radial positions [Fig. 1(B)], the rate of emission and the type of signals (see below) were the same for both detectors.

In a second set of experiments performed under room light, and each lasting ca. 2 h (*Experiment 2*), two transducers were attached to the bare stem at different heights. Both transducers registered UAE, but the rate of emissions was dependent on the transducer position: near the leaves was higher than closer to the plant base. For instance, when T1 was located at 14 cm from the plant base and T2 at 35–40 cm, no signals were detected by T1, while 0.3–3.5 emissions min^{-1} were detected by T2. The amplitude of the signals detected by each transducer was registered as a function of time, and the UAE were classified by their form and main frequency. Two types of signals were identified: those who have a broad band of frequencies up to 0.2 MHz, named type 1, [Fig. 2(a)], and low frequency signals, with values below 0.075 MHz, named type 2, [Fig. 2(b)].

ii. Laser induced UAE

With the aim of developing a method to induce UAE in a controlled way, in the next series of experiments (*Experiment 3*) we impinged a laser beam at a point on a corn stem with a transducer attached on the opposite side [Fig. 1(C)]. We started measuring UAE in the dark, and without laser irradiation. Under these conditions, no UAE were detected. Then, again in the dark, we irradiated the stem with the He–Ne red laser, but even at its maximum power, no UAE were detected. After that, the CW Ar-ion laser was tested at different wavelengths and powers. We found that with powers up to 600 mW, only the blue line at 488 nm produced results. Under these conditions, when the laser was turned on, acoustic signals were registered and when it was turned off, the rate of emission decayed and disappeared after a few seconds (Fig. 3).

This sequence (switching the laser on and off, always impinging on the same point of the stem) was repeated with the same qualitative results, although the rate of UAE decreased with every cycle (in Fig. 3 compare the slope of the sequence starting at minute 40 with the one starting at minute 55). Even when the rate of emissions in different plants encompassed a wide range (ca. 2–17 emissions min^{-1} with the laser on), the same pattern always held (i.e., emissions when laser is on, and no emissions a few seconds after the laser is off). The same behavior was observed when the laser

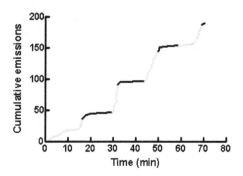

Figure 3: Laser induced UAE in corn stem. The beam of a CW Ar ion laser (600 mW) at 488 nm impinges on the stem opposite to the transducer. Grey line: the laser is on. Black line: the laser is off.

impinged at a right angle from the transducer axis.

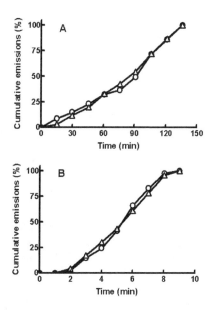

Figure 4: (A) Spontaneous and (B) laser induced UAE in corn stems measured simultaneously with two transducers attached at the same height of the stem. In (B) the laser impinged between both transducers. Open triangles: transducer T1. Open circles: transducer T2.

The signals were classified according to their form and frequencies. Figure 2(c) shows a typical signal generated by the laser in this experiment.

As it can be seen, these signals are similar to the broadband frequency signals [the type 1 shown in Fig. 2(a)] measured in experiment 2 with room light.

With the aim of comparing spontaneous and laser-induced UAE, we attached two transducers to the bare stem on opposite sides, at the same height [*Experiment 4*, Fig. 1(B)]. We first registered acoustic emissions detected simultaneously by both transducers at room light, without laser irradiation [Fig. 4(A)]. After that, in the dark, we measured the UAE generated after impinging the CW laser between both detectors, in a direction perpendicular to their axes [Fig. 4(B)]. The characteristic signals observed in both cases were type 1 signals (broadband frequency signals).

Then, we proceeded to study how the distance between detector and source modified the rate and shape of the UAE (*Experiment 5*). Two transducers were attached to the bare stem: one at 8.5 cm (T1) and the other at 12.5 cm (T2) from the base. The CW laser beam impinged on different points of the stem. Points **a** and **b** were at the same height of T1 and T2, respectively, but at the opposite side; point **c** was between T1 and T2 [Fig. 1(D)].

When the laser beam impinged on **a**, both transducers detected UAE. Broadband frequency signals (type 1) were observed with T1 and low frequency signals (type 2) were observed with T2. Figure 2(d) shows an example of type 2 signals generated with laser. As it can be seen, these signals are similar to those detected at room light [Fig. 2(b)].

Besides, the number of emissions detected by T1 was higher than that detected by T2 [Fig. 5(A)]. When the laser beam impinged on **b**, once again, both transducers detected UAE. In this case, T2 detected type 1 signals while T1 detected type 2 signals, and the number of emissions detected by T2 was higher than that detected by T1 [Fig. 5(B)]. When the laser beam impinged on **c**, both transducers simultaneously detected UAE of low frequency similar to those described as type 2 [Fig. 5(C)].

IV.   Discussion and conclusions

Experiments 1 and 2 show that the spontaneous UAE can be attributed to natural cavitation events occurring in the xylem vessels of the corn stem: no

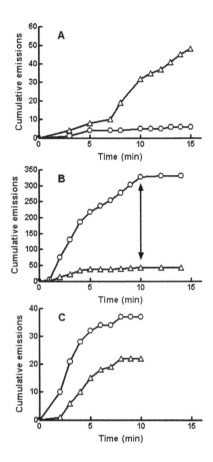

Figure 5: Laser induced UAE as a function of the transducer position. Two PZT transducers were attached to the stem at two heights as shown in Fig. 1(D). Open triangles: transducer T1. Open circles: transducer T2. (A) The Laser beam impinged near T1. (B) The Laser beam impinged near T2. (C) The Laser beam impinged between T1 and T2. The arrow in (B) indicates the laser was off.

emissions were observed when the plant was in the dark. Around 1 emission min^{-1} was detected under room light, and a rate ca. 25% higher under the lamp. This behavior is in agreement with the cohesion-tension theory and current plant cavitation models. As transpiration rate increases, xylem tension rises and cavitation events are expected to increase, as it happens in our experiments. Besides, the UAE signals registered (Fig. 2) were very similar to those described [12]. These authors demonstrated that these kinds of emissions are strongly related to cavitation events [12, 13, 26].

In the transpiring plant, the tension developed in the water stream generates a metastable equilibrium. When liquid water is subjected to a sufficiently low pressure, this equilibrium can be broken, and form a cavity. This initial stage of the cavitation phenomenon is termed cavitation inception. When the plant is in the dark, water in the xylem is slightly under tension at a pressure value close to atmospheric. Under these conditions, the local pressure does not fall enough, compared to the saturated vapor pressure, to produce cavitation inception.

As the CW laser impinges on the stem, this absorbs light and release energy to the xylem, heating it. This extra energy allows the phase change to gas in the water column, triggering cavitation inception. In this sense, the physical process of cavitation inception is similar to boiling, the major difference being the thermodynamic path which precedes the formation of the vapor. We found that UAE generated using a CW laser are of the same kind of those registered on transpiring plants. We can conclude that this method allows, for the first time, to induce cavitation events in xylem in a controlled and reproducible way.

Regarding the mechanisms of UAE generation, either natural or laser-induced, previous work has clearly shown that once cavitation inception is produced, embolism of the xylem immediately takes place. This means that the formed cavity remains, and there is no collapse of the void in the water column (as would occur in the so called inertial cavitation). Then, UAE generation can be produced by an oscillating source activated by the rupture of the water column. As mentioned in the introduction, Tyree et al. [12] proposed four possible UAE sources. The first one, oscillation of hydrogen bonds in water after tension release, seems unlikely because of its very low magnitude, undetectable by the kind of transducers we used. The second one, oscillations caused by a "snap back" of vessel walls, is also unlikely because of the rigidity of the xylem, and especially hard to explain under laser-induced cavitation inception in the dark, when xylem tension was nil or very small. The third one, torus aspiration, is impossible in our case because of the absence of these structures in corn. Finally, the fourth one, structural failure in the sapwood, was elegantly rejected by Tyree himself [13], who ex-

posed xylem to pressure and detected a different kind of emission.

We postulate that another possible source of the UAE must be taken into account. It is the local oscillation of the liquid–gas interface of the water column produced by the expansion and compression of the formed cavity, i.e., the stress wave generated by rapid bonding energy release. During cavitation inception, after the cavity expands, it is expected to be compressed almost immediately by the water column. This "cavity oscillation" starts as a high frequency burst produced by the disruptive effect of the cavitation inception. As a consequence, ultrasonic acoustic signals are produced.

In order for cavitation inception to occur, the cavitation "bubbles" generally need a surface on which they can nucleate. This could be provided by impurities in the liquid or the xylem walls, or by small undissolved micro-bubbles within the water, but most likely by air seeding through pit membranes [4]. These act as capillary valves that allow or prevent air seeding by adjusting local curvatures and interface positions [27]. Air seeding induced by the heating at pit membranes under CW laser irradiation should also be taken into account as an initial stage in laser induced cavitation inception.

The CW laser-induced cavitation opens the opportunity to study embolism in plants in a controlled manner. It also has the advantage of tracing the source, allowing the characterization of the signals and studying their propagation. By directing the laser beam to one point in the stem and recording acoustic emissions at different distances, we found that when cavitation was produced near the transducer, broadband frequency emissions were registered. But, if the transducer was installed further away, the rate and frequency of the emissions decreased with the distance to the cavitation source. This means that during signal propagation, absorption by the tissue takes place (rate decay) as well as frequency filtering. Figures 2(e) and 2(f) show the frequency spectra of type 1 and type 2 signals. When comparing these figures, the frequency filtering effect is evident.

Our results confirm the hypothesis by Ritman and Milburn [28], who proposed that cavitation of xylem sap generally results in the production of a broadband acoustic emission with lower cutoff frequency determined by the dimensions of the resonating element. The larger a conduit dimen-

sion, the lower the frequency of its major resonance. Thus, small cavitating elements, such as
corn stem xylem, are expected to produce acoustic signals with a broadband frequency spectrum.
Our results can also explain the observations by
Tyree and Dixon [12] who found and classified UAE
of different frequencies (between 0.1 and 1 MHz).
According to our experiments, the different signals
would be generated by cavitation events produced
in different regions of the stem. Broad band frequency signals would come from near the transducer while low frequency signals would come from
regions far from to the transducer.

According to these results, one might use the
waveform of the emissions to determine the location of each cavitation event. In that case, a whole
new field would be opened in the study of hydraulic
architecture of plants.

Acknowledgements - The authors are indebted to
Dr. H. F. Ranea Sandoval of FCE-UNCBA-Tandil-
Argentina, Professor Silvia E. Braslavsky from
Max-Planck-Institut für Bioanorganische Chemie
Mülheim an der Ruhr, Germany and Dr. J.
Alvarado-Gil from CINVESTAV-Unidad, Merida,
Merida, Mexico for fruitful comments and sug-
gestions. This work was partially supported by
ANPCyT, UNCPBA, UBA and UNLP. G.M.B. is
member of the Carrera del Investigador Científico
CIC-BA, and R.J.F. of CONICET.

[1] H Lambers, F S Chapin, T L Pons, *Plant physiological ecology*, Springer Verlag, New York
(1998).

[2] F Caupin, E Herbert, *Cavitation in water: a
review*, C. R. Phys. **7**, 1000 (2006).

[3] U Zimmermann, H Schneider, L H Wegner, A
Haase, *Water ascent in tall trees: Does evolution of land plants rely on a highly metastable
state?* New Phytol. **162**, 575 (2004).

[4] M T Tyree, *The cohesion-tension theory of sap
ascent: Current controversies*, J. Exp. Bot. **48**,
1753 (1997).

[5] J S Sperry, F R Adler, G S Campbell, J P
Comstock, *Limitation of plant water use by*
rhizosphere and xylem conductance: results
from a model*, Plant Cell Environ. **21**, 347
(1998).

[6] P H Maseda, R J Fernández, *Stay wet or
else: Three ways in which plants can adjust hy-
draulically to their environment*, J. Exp. Bot.
57, 3963 (2006).

[7] H H Dixon, *Transpiration and the ascent of
sap in plants*, McMillan & Co., New York
(1914).

[8] H N V Temperley, *The behaviour of water under hydrostatic tension: III.*, P. Phys. Soc. **59**,
199 (1947).

[9] J A Milburn, R P C Johnson, *The conduction
of sap. II. Detection of vibrations produced by
sap cavitation in Ricinus xylem*, Planta **69**, 43
(1966).

[10] D S Crombie, J A Milburn, M F Hipkins,
*Maximum sustainable xylem sap tensions in
Rhododendron and other species*, Planta **163**,
27 (1985).

[11] V G Williamson, J A Milburn, *Cavitation
events in cut stems kept in water: Implications
for cut flower senescence*, Sci. Hortic. (Amsterdam) **64**, 219 (1995).

[12] M T Tyree, M A Dixon, *Cavitation events
in Thuja occidentalis L.? Ultrasonic acoustic
emissions from the sapwood can be measured*,
Plant Physiol. **72**, 1094 (1983).

[13] M T Tyree, M A Dixon, R G Thompson, *Ultrasonic acoustic emissions from the sapwood
of Thuja occidentalis measured inside a pressure bomb*, Plant Physiol. **74**, 1046 (1984).

[14] M T Tyree, E L Fiscus, S D Wullschleger, M
A Dixon, *Detection of xylem cavitation in corn
under field conditions*, Plant Physiol. **82**, 597
(1986).

[15] G M A Lo, S Salleo, *Three different methods
for measuring xylem cavitation and embolism:
A comparison*, Ann. Bot. (London) **67**, 417
(1991).

[16] G E Jackson, J Grace, *Field measurements of xylem cavitation: Are acoustic emissions useful?* J. Exp. Bot. **47**, 1643 (1996).

[17] S B Kikuta, P Hietz, H Richter, *Vulnerability curves from conifer sapwood sections exposed over solutions with known water potentials*, J. Exp. Bot. **54**, 2149 (2003).

[18] R Laschimke, M Burger, H Vallen, *Acoustic emission analysis and experiments with physical model systems reveal a peculiar nature of the xylem tension*, J. Plant Physiol. **163**, 996 (2006).

[19] W T Pockman, J S Sperry, J W O'Leary, *Sustained and significant negative water pressure in xylem*, Nature **378**, 715 (1995).

[20] H Cochard, G Damour, C Bodet, I Tharwat, M Poirier, T Ameglio, *Evaluation of a new centrifuge technique for rapid generation of xylem vulnerability curves*, Physiol. Plantarum **124**, 410 (2005).

[21] P Kafalas, A P Ferdinand Jr., *Fog droplet vaporization and fragmentation by a 10.6 mm laser pulse*, Appl. Optics **12**, 29 (1973).

[22] W Hentschel, W Lauterborn, *Acoustic emission of single laser-produced cavitation bubbles and their dynamic*, Appl. Sci. Res. **38**, 225 (1982).

[23] S I Kudryashov, K Lyon, S D Allen, *Photoacoustic study of relaxation dynamics in multibubble systems in laser-superheated water*, Phys. Rev. E **73**, 055301 (2006).

[24] R Zhao, R Q Xu, Z H Shen, J Lu, X W Ni, *Experimental investigation of the collapse of laser-generated cavitation bubbles near a solid boundary*, Opt. Laser Technol. **39**, 968 (2007).

[25] A C Tam, *Applications of photoacoustic sensing techniques*, Rev. Mod. Phys. **58**, 381 (1986).

[26] M T Tyree, M A Dixon, E L Tyree, R Johnson, *Ultrasonic acoustic emissions from the sapwood of Cedar and Hemlock: An examination of three hypotheses regarding cavitations* Plant Physiol. **75**, 988 (1984).

[27] A G Meyra, V A Kuz, G J Zarragoicoechea, *Geometrical and physicochemical considerations of the pit membrane in relation to air seeding: The pit membrane as a capillary valve,* Tree Physiol. **27**, 1401 (2007).

[28] K T Ritman, J A Milburn, *Acoustic emissions from plants. Ultrasonic and audible compared,* J. Exp. Bot. **39**, 1237 (1988).

Permissions

All chapters in this book were first published in PIP, by Papers in Physics; hereby published with permission under the Creative Commons Attribution License or equivalent. Every chapter published in this book has been scrutinized by our experts. Their significance has been extensively debated. The topics covered herein carry significant findings which will fuel the growth of the discipline. They may even be implemented as practical applications or may be referred to as a beginning point for another development.

The contributors of this book come from diverse backgrounds, making this book a truly international effort. This book will bring forth new frontiers with its revolutionizing research information and detailed analysis of the nascent developments around the world.

We would like to thank all the contributing authors for lending their expertise to make the book truly unique. They have played a crucial role in the development of this book. Without their invaluable contributions this book wouldn't have been possible. They have made vital efforts to compile up to date information on the varied aspects of this subject to make this book a valuable addition to the collection of many professionals and students.

This book was conceptualized with the vision of imparting up-to-date information and advanced data in this field. To ensure the same, a matchless editorial board was set up. Every individual on the board went through rigorous rounds of assessment to prove their worth. After which they invested a large part of their time researching and compiling the most relevant data for our readers.

The editorial board has been involved in producing this book since its inception. They have spent rigorous hours researching and exploring the diverse topics which have resulted in the successful publishing of this book. They have passed on their knowledge of decades through this book. To expedite this challenging task, the publisher supported the team at every step. A small team of assistant editors was also appointed to further simplify the editing procedure and attain best results for the readers.

Apart from the editorial board, the designing team has also invested a significant amount of their time in understanding the subject and creating the most relevant covers. They scrutinized every image to scout for the most suitable representation of the subject and create an appropriate cover for the book.

The publishing team has been an ardent support to the editorial, designing and production team. Their endless efforts to recruit the best for this project, has resulted in the accomplishment of this book. They are a veteran in the field of academics and their pool of knowledge is as vast as their experience in printing. Their expertise and guidance has proved useful at every step. Their uncompromising quality standards have made this book an exceptional effort. Their encouragement from time to time has been an inspiration for everyone.

The publisher and the editorial board hope that this book will prove to be a valuable piece of knowledge for researchers, students, practitioners and scholars across the globe.

List of Contributors

Damián H. Zanette
Consejo Nacional de Investigaciones Científicas y T´ecnicas, Centro Atómico Bariloche e Instituto Balseiro, 8400 Bariloche, Río Negro, Argentina

M. F. Assaneo and M. A. Trevisan
Laboratorio de Sistemas Dinámicos, Depto. de Física, FCEN, Universidad de Buenos Aires. Pabellón I, Ciudad Universitaria, 1428EGA Buenos Aires, Argentina

Horacio S. Wio
IFCA (UC-CSIC), Avda. de los Castros s/n, E-39005 Santander, Spain

Roberto R. Deza
IFIMAR (UNMdP-CONICET), Funes 3350, 7600 Mar del Plata, Argentina

Carlos Escudero
Depto. Matemáticas & ICMAT (CSIC-UAM-UC3M-UCM), Cantoblanco, E-28049 Madrid, Spain

Jorge A. Revelli
FaMAF-IFEG (CONICET-UNC), 5000 Córdoba, Argentina

E. V. Bonzi
Facultad de Matemática, Astronomía y Física, Universidad Nacional de Córdoba, Ciudad Universitaria, 5000 Córdoba, Argentina
Instituto de Física Enrique Gaviola (CONICET), Ciudad Universitaria, 5000 Córdoba, Argentina

G. B. Grad
Facultad de Matemática, Astronomía y Física, Universidad Nacional de Córdoba, Ciudad Universitaria, 5000 Córdoba, Argentina

A. M. Maggi
Escuela de Fonoaudiología, Facultad de Ciencias Médicas, Universidad Nacional de Córdoba, Ciudad Universitaria, 5000 Córdoba, Argentina

M. R. Muñóz
Escuela de Fonoaudiología, Facultad de Ciencias Médicas, Universidad Nacional de Córdoba, Ciudad Universitaria, 5000 Córdoba, Argentina

A. Chacoma
Instituto Balseiro and Centro Atómico Bariloche, 8400 San Carlos de Bariloche, Río Negro, Argentina

D. H. Zanette
Instituto Balseiro and Centro Atómico Bariloche, 8400 San Carlos de Bariloche, Río Negro, Argentina
Consejo Nacional de Investigaciones Cientícas y Técnicas, Argentina

C. F. M. Magalhães
Universidade Federal de Itajubá - Campus de Itabira, Rua Irmã Ivone Drumond, 200, 35900-000 Itabira, Brazil

A. P. F. Atman
Departamento de Física e Matemática, Centro Federal de Educacão Tecnológica de Minas Gerais, Av. Amazonas, 7675, 30510-000 Belo Horizonte, Brazil Instituto Nacional de Ciência e Tecnologia Sistemas Complexos, 30510-000 Belo Horizonte, Brasil

G. Combe
UJF-Grenoble 1, Grenoble-INP, CNRS UMR 5521, 3SR Lab. Grenoble F-38041, France

J. G. Moreira
Universidade Federal de Minas Gerais, Caixa Postal 702,30161-970 Belo Horizonte, Brasil

D. R. Parisi
Instituto Tecnológico de Buenos Aires, 25 de Mayo 444, 1002 Ciudad Autónoma de Buenos Aires, Argentina
Consejo Nacional de Investigaciones Científcas y Téecnicas, Av. Rivadavia 1917, 1033 Ciudad Autónoma de Buenos Aires, Argentina

P. A. Negri
Consejo Nacional de Investigaciones Científcas y Téecnicas, Av. Rivadavia 1917, 1033 Ciudad Autónoma de Buenos Aires, Argentina
Universidad Argentina de la Empresa, Lima 754, 1073 Ciudad Autónoma de Buenos Aires, Argentina

I. C. Ramos and C. B. Briozzo
Facultad de Matemática, Astronomía y Física, Universidad Nacional de Córdoba, X5000HUA Córdoba, Argentina

Dante R. Chialvo
Consejo Nacional de Investigaciones Cientícas y Tecnológicas (CONICET), Rivadavia 1917, Buenos Aires, Argentina

Ana María Gonzalez Torrado
Institut Universitari d'Investigacions en Ciéencies de la Salut (IUNICS) & Universitat de les Illes Balears (UIB), Palma de Mallorca, Spain

Ewa Gudowska-Nowak
M. Kac Complex Systems Research Center and M. Smoluchowski Institute of Physics, Jagiellonian University, Krakóow, Poland

Jeremi K. Ochab
Biocomplexity Department, Ma lopolska Center of Biotechnology, Jagiellonian University, Krakóow, Poland

Pedro Montoya
Institut Universitari d'Investigacions en Ciéencies de la Salut (IUNICS) & Universitat de les Illes Balears (UIB), Palma de Mallorca, Spain

Maciej A. Nowak
M. Kac Complex Systems Research Center and M. Smoluchowski Institute of Physics, Jagiellonian University, Krakóow, Poland
Biocomplexity Department, Ma lopolska Center of Biotechnology, Jagiellonian University, Krakóow, Poland

Enzo Tagliazucchi
Institute for Medical Psychology, Christian Albrechts University, Kiel, Germany

G. P. Suárez and M. Hoyuelos
Instituto de Investigaciones Físicas de Mar del Plata (IFIMAR - CONICET) and Departamento de Físicas, Facultad de Ciencias Exactas y Naturales, Universidad Nacional de Mar del Plata, Deán Funes 3350, 7600 Mardel Plata, Argentina

D. R. Chialvo
Consejo Nacional de Investigaciones Científcas y Técnicas (CONICET), Godoy Cruz 2290, Buenos Aires, Argentina

Ignacio H. López Grande and Miguel A. Larotonda
DEILAP-UNIDEF (CITEDEF-CONICET), J. B. de La Salle 4397, B1603ALO Villa Martelli, Buenos Aires, Argentina

Christian T. Schmiegelow
Laboratorio de Iones y Átomos Fríos, Departamento de Física, Facultad de Ciencias Exactas y Naturales, Universidad de Buenos Aires & IFIBA-CONICET, Pabellón 1, Ciudad Universitaria, 1428 C.A.B.A., Argentina

Alan Murray
SAE Creative Media Institute, Sydney, Australia

Fernando Alonso-Marroquin
School of Civil Engineering, The University of Sydney, NSW 2006, Australia

E. Freyssingeas, D. Frelat, Y. Dossmann and J.-C. Géminard
Univ Lyon, Ens de Lyon, Univ Claude Bernard, CNRS, Laboratoire de Physique, F-69342 Lyon, France

E. A. Jagla
Comisión Nacional de Energía Atómica, Instituto Balseiro (UNCu), and CONICET. Centro Atómico Bariloche, Avda. E. Bustillo 9500, 8400 Bariloche, Argentina

Y. Núñez Fernández and K. Hallberg
Centro Atómico Bariloche and Instituto Balseiro, CNEA, CONICET, Avda. E. Bustillo 9500, 8400 San Carlos de Bariloche, Río Negro, Argentina

S. Ghosh and A. Roy
Discipline of Physics, School of Basic Sciences, Indian Institute of Technology Indore, Khandwa Road, Simrol, MP-453552, India

Leopoldo R. Gómez, Nicolás A. García and Daniel A. Vega
Instituto de Física del Sur (IFISUR), Consejo Nacional de Investigaciones Científicas y Técnicas (CONICET), Universidad Nacional del Sur, 8000 Bahía Blanca, Argentina

Richard A. Register
Department of Chemical and Biological Engineering, Princeton University, Princeton, New Jersey, 08544, USA

A. Allerdt and A. E. Feiguin
Department of Physics, Northeastern University, Boston, Massachusetts MA 02115, USA

Manuel Á. Gonzáalez, Alfonso Gómez and Miguel Á. González
Universidad de Valladolid, 47011 Valladolid, Spain

Kyle Forinash and Raymond Wisman
Indiana University Southeast, 4201 Grantline Road, New Albany, Indiana 47150, USA

A. Hilberer, G. Laurent, A. Lorin, A. Partier, J. J. M. Fischbach and B. Pilette
Magistére de Physique Fondamentale, Département de Physique, Univ. Paris-Sud, Université Paris-Saclay,91405 Orsay Campus, France

J. Bobroff, F. Bouquet, C. Even, M. Monteverde and Q. Quay
Laboratoire de Physique des Solides, CNRS, Univ. Paris-Sud, Université Paris-Saclay, 91405 Orsay Campus, France

C. A. Marrache-Kikuchi
CSNSM, Univ. Paris-Sud, CNRS/IN2P3, Université Paris-Saclay, 91405 Orsay, France

Mario Everaldo de Souza
Universidade Federal de Sergipe, Departamento de Física, Av. Marechal Rondon, s/n, Campus Universitãrio, Jardim Rosa Elze 49100-000, São Cristovão, Brazil

Rafael González
Instituto de Desarrollo Humano, Universidad Nacional de General Sarmiento, Gutierrez 1150, 1613 Los Polvorines, Pcia de Buenos Aires, Argentina Departamento de Física FCEyN, Universidad de Buenos Aires, Pabellón I, Ciudad Universitaria, 1428 Buenos Aires, Argentina

Ricardo Page
Instituto de Ciencias, Universidad Nacional de General Sarmiento, Gutierrez 1150, 1613 Los Polvorines, Pcia de Buenos Aires, Argentina

Andrés S. Sartarelli
Instituto de Desarrollo Humano, Universidad Nacional de General Sarmiento, Gutierrez 1150, 1613 Los Polvorines, Pcia de Buenos Aires, Argentina

Rodolfo D. Porasso
Instituto de Matemática Aplicada San Luis (IMASL) - Departamento de Física, Universidad Nacional de San Luis/CONICET, D5700HHW, San Luis, Argentina

J. J. López Cascales
Universidad Politécnica de Cartagena, Grupo de Bioin formática y Macromoléculas (BioMac) Aulario II, Campus de Alfonso XIII, 30203 Cartagena, Murcia, Spain

E. Fernández and R. J. Fernández
IFEVA, Facultad de Agronomía, Universidad de Buenos Aires y CONICET, Av. San Martín 4453, C1417DSE Buenos Aires, Argentina

G. M. Bilmes
Centro de Investigaciones Opticas (CONICET-CIC) and Facultad de Ingeniería, Universidad Nacional de La Plata, Casilla de Correo 124, 1900 La Plata, Argentina

Index